CALIFORNIA

CO CESS

· E ·

C A L I F O R N I A
COASTAL ACCESS
· G U I D E ·

STATE OF CALIFORNIA
Gray Davis, Governor

CALIFORNIA COASTAL COMMISSION

Peter Douglas, *Executive Director*   Mike Reilly, *Chair*   Susan Hansch, *Chief Deputy Director*

Steve Scholl
Project Director, Sixth Edition

Editors

Erin Caughman   Jo Ginsberg
design, cartography   *research writing*

Staff

cartography, sixth edition	*research writing*	*revision, sixth edition*
Gregory M. Benoit	Stephen J. Furney-Howe	Sylvie Bloch
Andrew McIntyre	Sabrina S. Simpson	Joy Chase
Jonathan Van Coops	Jeffrey D. Zimmerman	

Jane Heaphy
illustrator

Linda Locklin, *Coastal Access Program Manager*

UNIVERSITY OF CALIFORNIA PRESS
Berkeley   Los Angeles   London

First Edition, © 1981

Revised Commemorative Edition, © 1982

Third Edition, © 1983

Fourth Edition, Revised, © 1991

Fifth Edition, Revised, © 1997

Printed in Canada

ISBN 978-0-520-24098-8

2   3   4   5   6   7   8   9

The paper used in this publication meets the minimum requirements of ANSI/NISO Z39.48-1992 (R 1997)   *(Permanence of Paper).*

Contents

Foreword

Introduction

Feature Articles

ac·cess *n.*, liberty to approach

Foreword

The 1,100-mile California coast, from the majestic redwoods and rocky shores in the north to the palm trees and wide, sandy beaches in the south, is an area of unsurpassed beauty and diversity. In 1972, the voters of California approved Proposition 20, the Coastal Initiative, which led to the creation of the California Coastal Commission. The California Coastal Act of 1976 made the Commission permanent and gave it a mandate to conserve, protect, and restore coastal environmental resources and values. A fundamental Commission goal is to enhance public access and recreational opportunities along the coast.

The first edition of the *California Coastal Access Guide* was published in 1981 with the dual purposes of identifying areas along the coast that are open to the public and explaining the public's rights and responsibilities regarding the use of coastal resources. Later revisions of the guide have recognized the creation of new public access sites and points of interest. The *California Coastal Resource Guide*, also prepared by the California Coastal Commission and published in 1987, includes extensive information on natural, cultural and economic resources of the California coast and serves as a companion to this volume.

Preparing this comprehensive guide to California's magnificent coast has been a challenging and worthwhile task for the Coastal Commission. The Commission would like to express its appreciation to the University of California Press and to the state, federal, and local agencies and individuals who assisted in creating this guide.

This edition of the *California Coastal Access Guide* is dedicated to the memory of Deborah Bové, Tom Crandall, and Pat Stebbins, whose decades of tireless and dedicated public service promoting shoreline access are a remarkable legacy to future generations. We acknowledge also the work of those countless individuals who have contributed in a variety of ways to improve and expand opportunities for coastal public access and recreation. As California's population and development continue to grow, the efforts of citizen activists will continue to be crucial in ensuring that the coast is shared with all the people.

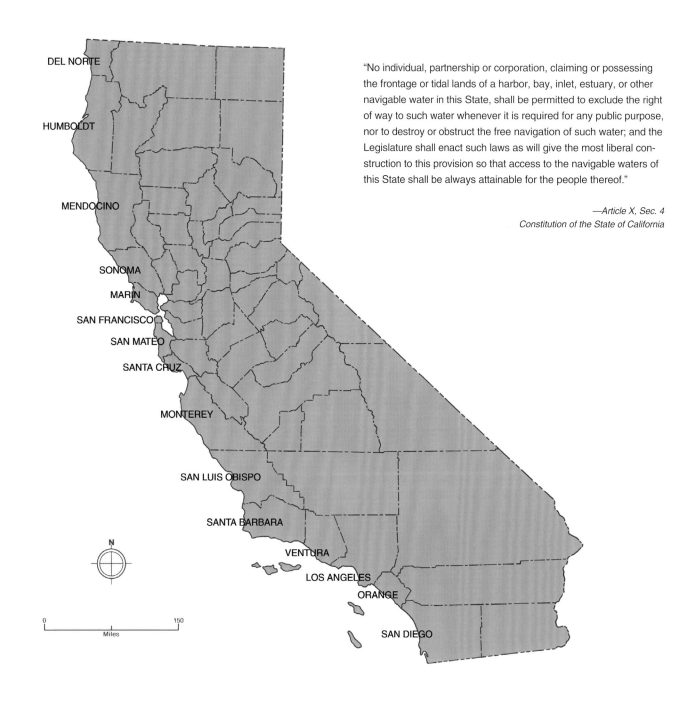

DEL NORTE

HUMBOLDT

MENDOCINO

SONOMA

MARIN

SAN FRANCISCO

SAN MATEO

SANTA CRUZ

MONTEREY

SAN LUIS OBISPO

SANTA BARBARA

VENTURA

LOS ANGELES

ORANGE

SAN DIEGO

"No individual, partnership or corporation, claiming or possessing the frontage or tidal lands of a harbor, bay, inlet, estuary, or other navigable water in this State, shall be permitted to exclude the right of way to such water whenever it is required for any public purpose, nor to destroy or obstruct the free navigation of such water; and the Legislature shall enact such laws as will give the most liberal construction to this provision so that access to the navigable waters of this State shall be always attainable for the people thereof."

—Article X, Sec. 4
Constitution of the State of California

N

0   150
Miles

Introduction

Coastal access means getting to the coast; this Guide tells you where to go on the coast, how to get there, and what facilities and type of environment you will find at each location. It is meant for all beachgoers—hikers, campers, swimmers, surfers, divers, wheelchair-users, joggers, boaters, anglers—and is intended to introduce the richness and diversity of the California coast.

The Guide is a thorough and easy-to-use handbook that lists and describes all open public accessways, beaches, parks, and recreational areas along the coast; it provides addresses, phone numbers, hours of use, and descriptions of facilities and type of environment for each site. The Guide is divided into sections for each county and local area, with website and transit information and with accompanying maps that clearly locate each accessway and major streets. Photos and illustrated feature articles on history, coastal resources, and recreational activities are found throughout the book.

Natural conditions along the California coast are always changing. The width of sandy beaches fluctuates with the seasons, and the topography of coastal bluffs is altered by erosion. The heavy rain and high winds and waves that often accompany ocean storms can change the physical features of the coastline drastically in a matter of days, or even hours.

Coastal access and recreation facilities are inevitably subjected to the same physical forces as the shoreline itself. Unfortunately, these facilities often are not as resilient as the shoreline, and consequently can suffer debilitating damage. From time to time, some of the facilities described in this book, including piers, trails, stairways, parking areas, campgrounds, and restrooms, may be closed temporarily while undergoing repairs. In some cases where storm damage has been particularly severe, an entire park may be closed.

When planning any trip to the coast, but especially right after a storm, it is advisable to check ahead whenever possible to make sure that the coastal area you choose to visit is currently accessible and usable.

The Commission intends to prepare regional access publications, such as the *Orange County Access Map* that was published in 2002, and to periodically update the *California Coastal Access Guide.*

Coastal Access Program

Installation of the sign announcing the grand opening of a public accessway in Malibu, a 1981 achievement of the California Coastal Commission and the State Coastal Conservancy. The accessway was named after the cartoon character, Zonker Harris, created by Gary Trudeau as a part of the Doonesbury comic strip. Zonker was known for his love of sun tanning and his free, sun-seeking spirit.

Proposition 20 of 1972, and the California Coastal Act and State Coastal Conservancy Act, both enacted in 1976, contain strong access policies and programs. The Coastal Commission implements these policies through its requirement of providing public shoreline access as a condition of certain coastal development permits, and through local coastal programs (LCPs), the coastal area plans prepared by local governments, which contain provisions for acquiring, improving, and managing access areas. The Coastal Conservancy provides grants and technical assistance to local governments and citizens' groups to acquire, develop, operate, and manage new accessways.

In 1979, legislation was enacted directing the California Coastal Commission and State Coastal Conservancy to establish a comprehensive program to maximize public coastal access. These two agencies are responsible for coordinating all local, state, and federal efforts to implement the access program. For example, in cooperation with the California Department of Transportation and the Conservation Corps, coastal access signs have been installed along the coast, indicating where accessways are located. Legislation also mandated that the Commission prepare this guide to all coastal accessways.

This logo has been adopted by the California Coastal Commission and State Coastal Conservancy as the official symbol to appear on coastal access signs.

Public and Private Rights

Article 10, Section 4 of the California Constitution guarantees the public's right to access to the state's navigable waters; each year, over 50 million people visit the California coast. Approximately 42% of this shoreline is publicly owned and accessible, while the remaining 58% is owned privately or is held by federal, state, or local government and is not open to the public.

The State of California owns the tide and submerged lands seaward of what is called the "mean high tide line." The precise boundary between these public tidelands and private lands is determined by using the mean, or average, of high tides over approximately 19 years, and an analysis of where this mean high tide intersects the land, taking into account both long-term and seasonal changes in the shoreline profile. State law gives primary responsibility for making this determination to the State Lands Commission, which has administered the state's tidelands since 1938. Although it is difficult to ascertain the boundary between public and private lands, a general rule to follow is that visitors have the right to walk on the wet beach.

The public's right of access to or along the state's tidelands can be obtained by three methods. The simplest way is through the purchase of shoreline lands for public use by federal, state, and local governments, or private organizations.

Public access to the tidelands can also be obtained through deed restrictions or dedications by the landowner, which grant the public the right to cross private property. "Dedicated" lands, however, become open to the public only after an agency or private party has accepted responsibility for liability and maintenance. The Coastal Commission and many local governments require access as a condition of approving development permits to compensate the people of California for the adverse impacts of new development on public access. Access along the ocean, parallel to the shore, is called "lateral" access, and is usually provided through deed restrictions or dedications of an easement between the mean high tide line and the seaward edge of the development. Access from the nearest public road or area to the shoreline is called "vertical" access, and such access through private property is usually a ten-foot-wide path to the water. In almost every case, a buffer area of approximately five to ten feet exists between private structures and public areas.

Most of these "dedicated" strips of land remain undeveloped and unmanaged by government agencies and do not yet form a continuous system of access to and along the shore. Visitors are asked not to trespass onto private property adjoining public accessways and to respect the rights of property owners. If any controversy results from the use of shoreline land, please obey the requests of the property owner. You can determine the extent of public rights in the area by obtaining the street address of the parcel in question and consulting the Coastal Commission.

The public can also acquire the right to gain access to and use shoreline property through the legal doctrines of "implied dedication" and "prescriptive rights." The existence of these rights was confirmed by the California Supreme Court in 1970 (*Gion* v. *City of Santa Cruz*, 2Cal3d129) and reconfirmed in 1980 (*County of Los Angeles* v. *Berk*, 26Cal3d201). Because these rights are acquired through the use of property without the owner's permission, and because of problems associated with proving historic use, the determination of their existence is difficult and controversial.

According to court decisions, in order for the public to obtain an easement by way of implied dedication, the essential elements that must be established are that the public has used the land:

- for a *continuous* period of five years as if it were public land,

- without asking or receiving permission from the owner,

- with the actual or presumed knowledge of the owner, and

- without significant objection or significant attempts by the owner to prevent or halt such use.

The ultimate determination of prescriptive rights, if they are challenged, takes place in court. However, Section 30211 of the Coastal Act requires the Coastal Commission to take evidence of historic use of a given area into account in making its regulatory determinations. Other responsible public agencies, including affected local governments, the California Department of Parks and Recreation, the Wildlife Conservation Board, the State Lands Commission, and the Attorney General, are attempting to protect public prescriptive rights through a variety of programs. There are many areas along the California coast that may be subject to prescriptive rights. However, only those that are managed for public use have been included in the *California Coastal Access Guide*.

Hazards, Safety, and Liability

Fritz Eichenberg, from *The Long Loneliness* by Dorothy Day © 1952, Harper & Row Publishers, Inc. Reprinted by permission of the publisher.

Many portions of the California coast are hazardous; steep bluffs and cliffs, agricultural and port activities, and treacherous coastal waters can all be dangerous. California's ocean waters, particularly in the north, are extremely cold, and many beaches have rip currents. A number of beaches do not have lifeguards, and numerous drownings occur annually. Even on the calmest days, huge waves can unexpectedly crash ashore; therefore, avoid turning your back on the sea, and never leave children alone on beaches or bluffs. For your own safety, consult tide tables before exploring rocky beaches and tidepools, check marine weather forecasts, and stay on trails and paths.

Before 1963, public agencies responsible for improved and unimproved property open to the public were liable for injuries resulting from dangerous conditions not visible on the land. However, in that year the Legislature added section 831.4 to the California Tort Claims Act to grant the responsible agency immunity in the case of such injuries when the unimproved trails and accessways were being used by the public for recreational purposes. At the same time, the Legislature enacted Civil Code Section 846, which granted similar immunity to private landowners.

In 1979, the Legislature further amended section 831.4 to eliminate the distinction between paved and unimproved accessways. This law creates absolute immunity from liability for injuries occurring on paved as well as unimproved accessways as long as the responsible agency makes reasonable attempts, such as posting signs, to provide adequate warning of existing health or safety hazards.

In 1980, the Legislature enacted Government Code Section 831.5, which created immunities under most circumstances for coastal public land trusts that maintain coastal accessways and have entered into a specific agreement with the State Coastal Conservancy. This legislation also extended the immunity that protects against liability for non-paid users of land so that it applies to owners of accessways. The trend in accessway legislation clearly indicates that the favored public policy is to increase coastal access even if that requires changes to traditional rules of tort liability.

Children and the Coast

The coast is perhaps enjoyed most by young children. The natural wonders of tidepools and marine life and the pleasures of swimming and playing in the surf and building sand castles are our children's birthright to enjoy. Learning opportunities abound at the water's edge, making excursions to the beach delightful events for the whole family.

The best beach trips with children include some planning ahead. Suitable toys, beach shoes, a change of clothes, snacks, and sunscreen help make the trip enjoyable and comfortable. Trips with younger children should be planned for a short visit, to lessen the amount of damage the sun can do to young skin and to prevent overtiredness of children. The continued activity can be exhausting for children as well as adults, since children at the beach must be accompanied continuously by an adult. Many hazards exist at the beach, and constant supervision is the best way to avoid those hazards. In addition to those safety pointers mentioned throughout this guide, parents and guardians should keep in mind the following considerations:

- Keep small children in sight at all times. Dress your children in bright clothes and remember what they are wearing. Identify landmarks to help children remember where their group is located.

- When playing near the surf, always keep an eye on the waves. Along the California coast, particularly in the north, sleeper or sneaker waves occur irregularly and without warning. These waves are much larger and more powerful than other waves, and can reach onto shoreline rocks or low bluffs.

- Children are especially sensitive to sunburn. Sunscreens and protective creams and oils are rated by degree of protection. Apply higher numbered water-resistant products frequently and liberally during exposure. Hats also help prevent burns and sunstroke or heat exhaustion.

- Always accompany children into the water, even if it is shallow. Swim only near lifeguards and do not depend on flotation devices for safety.

- Watch out for rip currents, which are strong but narrow seaward flows. If you get caught in one, don't panic; swim parallel to shore until you get out of the current, then return to shore. If you can't escape the current, call or wave for help.

- Logs and other debris in the surf zone should be avoided because the waves can hurl them with great force.

- Don't let family members dive unless you know how deep the water is and whether there are any underwater hazards.

- When exploring tidepools, be careful on slippery rocks, avoid stepping on sea urchins, and watch for the incoming tide.

- Jellyfish should be avoided on land as well as in water, as they can still sting after several hours on a dry beach. Children's feet are best protected from jellyfish and other debris by rubber thongs or sneakers.

- Stingrays often bury themselves in shallow water. Usually splashing feet will scare them away. However, if a stingray's tail stinger strikes, get emergency medical help as soon as possible.

- Milk products, infant formula, and other foods spoil quickly in the sun and summer heat. Keep perishable foods in a cooler.

- Watch for broken glass and other debris on the beach. Please pick up and safely dispose of any dangerous fragments.

- Direct children away from surf-casting fishermen. Do not attempt to pull out fishhooks from the skin, but seek medical help immediately.

- Sea cliffs and caves are potentially dangerous. Keep young children in hand on steep trails and away from the base of cliffs where rocks may fall.

- Playing in firepits can be dangerous. Coals can remain hot overnight when fires have been extinguished only with sand. Douse all fires with water.

- Children playing in deep holes in the sand can be injured by a cave-in. Refill all deep pits.

- Keep children away from storm water and sewage outfalls. Unsanitary and toxic wastes are health hazards.

- Eating sand is not harmful unless it is contaminated by waste. If there is any question, see a doctor.

- To remove tar and oil, use mineral or food oil rather than harsh turpentine or other spirits.

Many educational institutions and scientific organizations provide opportunities to enrich children's visits to the coast. Several guidebooks specifically describe trips that are especially enjoyable for children, and more detailed county and regional guides are available at most bookshops and libraries. These books list local museums, historical exhibits, natural resource displays, and commercial endeavors that children will enjoy and find beneficial. For example, a visit to the artificial tidepool at Steinhart Aquarium in San Francisco's Golden Gate Park, to Pacific Grove's or Santa Barbara's Museum of Natural History, or to the Cabrillo Marine Aquarium in San Pedro can make a later visit to the coast more rewarding or further enhance and explain a previous trip.

State historic parks, such as the ones at Fort Humboldt, at Fort Ross, and in Monterey, attractively present the cultural history of the coast. Several forest product companies have demonstration forests and logging museums, and offer tours of their plants. These displays show the current and historical commercial development of an important coastal industry. Marinas, ports, and commercial fishing harbors are exciting places where children can learn the importance of the coast to the economy of the state. Historical museums and displays are often provided at coastal military facilities and lighthouses.

Although Playland-at-the-Beach in San Francisco, Pacific Ocean Park in Los Angeles, and the Pike in Long Beach are gone, Monterey Bay Aquarium, Santa Cruz Beach and Boardwalk, Santa Monica Pier, Aquarium of the Pacific in Long Beach, and San Diego's Sea World remain popular coastal amusements for children of all ages.

Children, along with their families, often desire to help protect and improve the condition of a favorite beach. Those who are interested in helping the coast may be interested in California's Adopt-A-Beach program, organized by the California Coastal Commission and cooperating organizations throughout the state, which gives people of all ages the opportunity to learn about and actively participate in conserving coastal resources. A key element of the program is promoting the involvement of young people. By becoming Adopt-A-Beach volunteers, young people demonstrate their commitment to keeping California's coast beautiful and safe. For more information on Adopt-A-Beach, call: 1-800-262-7848 or 1-800-COAST-4U.

Protecting Marine Wildlife

Visitors to California's coast take on the important responsibility of helping to keep the beach safe and beautiful. Litter at the beach and in the ocean is not just ugly, it's dangerous. No one wants to come across a piece of broken glass or an aluminum pop-top while barefoot in the sand. Furthermore, people are not the only victims of litter. Marine wildlife suffers too, and in ways that are not always obvious to the coastal visitor. Whales and other marine mammals have died as the result of plastic waste lodged in their stomachs or intestines. Scientists have found that sea turtles, fish, and birds also ingest plastic bags and floating balloons, probably mistaking them for jellyfish. Birds and other sea animals can become entangled in discarded fishing line and netting.

Plastic strapping and six-pack rings pose a particularly serious threat to marine mammals, and seals are among the major victims. A "collar" of plastic strapping can, at the very least, cause lacerations that are prone to serious infection. If the entangled seal is a pup, its growth causes more constriction, eventually resulting in strangulation. A plastic band stuck around the snout of a seal or sea lion can cause starvation.

Some entangled or injured animals are lucky enough to be rescued, and a statewide network of organizations is in place to help save stranded marine creatures. If you come across a stranded or entangled animal on the beach or in the ocean, follow the instructions below, suggested by The Marine Mammal Center in Marin County.

1. Do not touch, pick up, or feed the animal. Do not return the animal to the water. Seals and sea lions temporarily "haul out" on land to rest and mothers briefly leave their pups while at sea. However, a beached whale or dolphin should be reported immediately.

2. Observe the animal from a distance of at least 50 feet. Keep people and dogs away.

3. Before calling the nearest stranding agency, note the physical characteristics such as size, absence or presence of external ears, and fur color. This will help the stranding agency determine the species and what rescue equipment and volunteers are needed. Note the animal's condition. Is it weak and gaunt? Are there any open wounds? Does it have any identification tags or markings?

4. Determine the exact location of the animal for accurate directions to stranding agency personnel.

5. Call the nearest stranding network agency:

In Del Norte and Humboldt counties: Crescent City Marine Mammal Rehabilitation Center, (707) 465-6265 or 465-MAML.

From Mendocino to San Luis Obispo counties: The Marine Mammal Center in Marin County, (415) 289-7325 or 289-SEAL.

In Santa Barbara and Ventura counties: Marine Mammal Center of Santa Barbara, (805) 687-3255.

In the Los Angeles area: Fort MacArthur Marine Mammal Care Center, (310) 548-5677.

In the Orange County area: Friends of the Sea Lion Marine Mammal Center, (949) 494-3050.

In the San Diego area: Sea World Department of Animal Care, 1-800-541-7325.

Harbor seals, Carpinteria, Santa Barbara County

Caring for the Coast

California is famous for its beaches and the bounty of recreational activities they afford. Less well known are the many ways the public can participate in activities that give back to the coast and ocean. The Coastal Commission and other organizations are working to raise public awareness about the coast and how the public can get involved in its conservation.

On the annual Coastal Cleanup Day every 3rd Saturday in September, tens of thousands of volunteers fan out across California's beaches to collect hundreds of thousands of pounds of trash, much of which is then recycled. Through its Adopt-A-Beach Program, the California Coastal Commission also provides groups with the resources to organize their own beach cleanups. Many beaches in California, from the Oregon border to San Diego, await adoption. By participating in a beach cleanup, Californians can help maintain a clean and save environment for the marine creatures that inhabit the beach and coastal waters, and at the same time improve the beach experience for human visitors.

In addition to the Coastal Commission, a multitude of organizations—government agencies, non-profit organizations and private entities—are working to connect the public with California's coastal resources. The Commission's on-line Marine, Coastal and Watershed Resource Directory is a way to learn about these organizations statewide, and the resources, programs, internships, and volunteer opportunities they offer.

California drivers can make a contribution to the coast by purchasing a Whale Tail License Plate. Designed by the environmental artist Wyland, the plate features an image of a whale diving into the ocean. Proceeds from plate sales support California Coastal Cleanup Day, the Adopt-A-Beach Program, and grants for marine education and the purchase of beach wheelchairs. Gray whale motorcycle plates are also available for purchase.

See www.coastforyou.org or call 1-800-262-7848 (1-800-COAST-4U) to learn more about how you can help care for the coast.

Children conducting beach cleanup

Whale Tail License Plate

Decal for motorcycle plates

Boating and Boating Safety

California's 1,100 miles of shoreline provide numerous opportunities for recreational boating, including pleasure motor boating and water-skiing. Extended fair weather periods and a number of sheltered coastal areas contribute to favorable sailing conditions, and at various times of the year fishing boats can be seen on coastal rivers and ocean waters.

Today there are numerous locations along the coast with berthing and launching facilities for recreational boating. Over the years, California's boating community has continued to grow as boating has become increasingly popular. Between 1960 and 2001, the number of boats per 1,000 Californians steadily increased from 11 to 28; over 1 million boats are now registered to California residents. As the number of boaters has increased, so too has the variety of boating activities that are popular in California, as shown by the increasing numbers of kayakers and jet skiers out on the water today.

With the continuing increase of newcomers enjoying various boating activities, the importance of practicing boating safety cannot be over-emphasized. Before departing on a trip, boaters should familiarize themselves with existing and predicted weather conditions and the tide stages, along with the safety devices of the vessel.

The U.S. Coast Guard Auxiliary offers free vessel safety inspections that help mariners determine whether they have properly equipped their vessels and have maintained them in safe and seaworthy condition. Guidelines for safety and the "rules of the road" applicable to all navigable waterways in California are available from the California Department of Boating and Waterways. These rules consist of standard operating procedures designed to prevent collisions, signal danger, or indicate a vessel in distress. Boaters can also enroll in one of the boating safety courses offered annually by the U.S. Coast Guard Auxiliary or the U.S. Power Squadrons.

In addition to protecting passengers, boaters are encouraged to help protect the waterways in which they recreate. It is important to preserve these natural and scenic resources, both for the sake of boating and for the ecological health of the coast.

While boating, mariners are advised not to discharge untreated sewage illegally within the 3 mile territorial limit, and to instead use restrooms (as much as possible), dump stations for port-o-potties, or sewage pumpouts to empty holding tanks. Never discharge sewage (treated or untreated) in a designated no-discharge area or in marina waters.

Boaters can prevent fuel spills by filling tanks slowly and carefully at the fuel dock. Oil absorbents should be on hand to catch drips and remove an oil spill or sheen. Major repairs and boat maintenance should be conducted in a boat repair facility with proper waste collection and treatment services. Boaters should limit in-water repairs and maintenance to small tasks and take precautions to minimize the discharge of pollutants and debris to the air and water.

For more information about safe and environmentally sound boating in California, call:

California Department of Boating and Waterways
(safety and environmental information)
1-888-326-2822

U.S Coast Guard Auxiliary
(boating safety classes and vessel safety inspections)
1-877-875-6296

U.S. Power Squadrons
(boating safety classes and boater education)
1-888-367-8777, or 1-888-FOR-USPS

California Integrated Waste Management Board
(information and locations for recycling and disposal of wastes)
1-800-253-2687, or 1-800-CLEANUP

See also the California Coastal Commission's website: www.coastal.ca.gov.

Beach Water Quality

You can go to the beach, but can you go into the water? Concerns about beach water quality led the California legislature in 1997 to require more frequent monitoring of heavily used beaches and to develop statewide beach water quality standards. As a result, land managers are deciding to close beaches and to post warning signs on a more consistent basis, when required by poor water quality conditions. More frequent beach monitoring has demonstrated that many beaches do not consistently meet water quality standards, and furthermore that it can be very costly to determine the cause of the poor water quality.

Information on beach water quality is available by telephone hotline in many counties: San Diego (619) 338-2073, Orange (714) 667-3752, Los Angeles 1-800-525-5662, Ventura (805) 662-6555, Santa Barbara (805) 681-4949, San Luis Obispo (805) 788-3411, Monterey 1-800-347-6363, Santa Cruz (831) 454-3188, San Mateo (650) 599-1266, San Francisco (415) 242-2214, Sonoma, (707) 565-6552, and Humboldt (707) 445-6215. For other communities, the best contact is typically the county health department. Websites with California beach water quality information include www.healthebay.org/brc and www.earth911.org.

There are several guidelines for minimizing the health risk of swimming or wading in the ocean. Avoid swimming in the ocean for two to three days after rainstorms, and always avoid swimming near flowing storm drains. Also, stay out of the ponds and lagoons that form on beaches at the end of creeks or storm drains. These ponds often attract children as they are usually shallow, waveless, and warm, but they are the ultimate recipients of contaminants from our streets, yards, and pet wastes.

Beach water quality is affected by activities elsewhere in the watershed, including overuse of fertilizers and pesticides and dumping of waste into storm drains. Waste from sinks and toilets flows to sewers and is treated before discharge to surface waters, but in most areas storm drains flow directly to creeks and the ocean without treatment. Hence the slogan, "Only Rain Down the Storm Drain!"

Most people visit beaches during the dry season, the time of year when pollutants flowing into coastal streams and the ocean are often at their highest levels of concentration. During the dry months, beaches would benefit from efforts to minimize or eliminate runoff from property elsewhere in the watershed. Landscape irrigation should be managed so that no excess water flows to storm drains or surface waters; sidewalks and driveways should be swept clean instead of being hosed off; and cars should be washed on the lawn or at a carwash. Both fish and people may end up swimming in whatever enters storm drains. Everyone's efforts to reduce runoff from yards and other places are important for improving and protecting California's beach water quality.

Santa Barbara

Access for Persons with Disabilities

The California State Department of Rehabilitation estimates that the total number of disabled persons in California exceeds two million; since 80% of California's population lives within 30 miles of the coast, wheelchair-accessible coastal facilities should be an important part of accessway development.

Disabilities, as defined by the Department of Rehabilitation, range from visual and hearing impairments, to ambulatory and other orthopedic disabilities that limit a person's mobility, to certain mental and developmental disabilities. In addition to the two million officially "disabled," countless others among the very young, elderly, and temporarily injured may require special improvements to facilities in order to be afforded access to coastal resources.

Federal, state, and local agencies and organizations throughout California are making efforts to meet the needs of all of these people. For example, the State Department of Parks and Recreation has retrofitted facilities in a number of parks; more improvements are planned to insure accessibility to disabled persons. Numerous piers, overlooks, and highway pull-outs allow people with all ranges of disabilities to enjoy the sight, smell, and sounds of the ocean. Wheelchair ramps and modified restrooms for the disabled are provided at many coastal accessways; these facilities are noted in the descriptive grid for each area under "Facilities for the Disabled."

Some facilities that are accessible to wheelchair users are not specifically listed; it is advisable to contact the facility directly for the most up-to-date and accurate information.

Prairie Creek Redwoods State Park in Humboldt County has a nature walk for the blind, Revelation Trail, with guidebooks available in Braille. At Año Nuevo State Reserve in San Mateo County, a portable trail of rubber belting can be set up across the sand to provide access to a platform from which the elephant seals can be seen in season at their rookery. The public beach in Marina del Rey is equipped with a ramp for access to the water. The city of Santa Cruz has installed a beach access ramp adjacent to the wharf at Cowell Beach that allows access by persons with limited mobility to the packed sand portion of the beach. Huntington State Beach has six such ramps.

Wheelchairs with large balloon tires that are designed to be pushed across the sand are offered at many beaches, including Half Moon Bay State Beach in San Mateo County; Natural Bridges and Sunset State Beaches in Santa Cruz County; Morro Bay, Pismo, and Cayucos State Beaches in San Luis Obispo County; El Capitan, Refugio, and Carpinteria State Beaches in Santa Barbara County; San Buenaventura and McGrath State Beaches in Ventura County; Topanga, Will Rogers, Venice Beach, Marina Beach (Marina del Rey), Dockweiler, Manhattan Beach, Torrance Beach, and Cabrillo

Beach (Cabrillo Marine Aquarium), all in Los Angeles County; and Huntington and San Clemente State Beaches in Orange County. Motorized beach wheelchairs are available at Mission and Imperial Beaches in San Diego County, as well as at Avila Beach in San Luis Obispo County. Where available, the chairs can be borrowed from lifeguards or concession operators; call ahead for reservations and availability.

Wheelchair-accessible campsites exist at MacKerricher State Park, Pfeiffer Big Sur State Park, Morro Bay State Park, Pismo State Beach, Carpinteria State Beach, South Carlsbad State Beach, and many other sites along the coast. Reservations may be made by calling: 1-800-444-7275, or TDD 1-800-274-7275. For information, contact:

California Department of Parks and Recreation
P.O. Box 942896
Sacramento, CA 94296-0001   (916) 653-6995

Many organizations offer trips, transportation programs, and education programs specifically for disabled people, and are aware of facility improvements. Information about these programs and facilities can be obtained from:

California Foundation for Independent Living Centers
660 J St., Suite 270
Sacramento, CA 95814   (916) 325-1690
www.cfilc.org   (916) 325-1695 TTY

Whole Access
517 A Lincoln Avenue
Redwood City, CA 94061-1732   (650) 363-2647
www.wholeaccess.org   (Voice/Text Telephone/TDD)

A Wheelchair Rider's Guide: Los Angeles and Orange County Coast by Coastwalk and the Coastal Conservancy is available on the Conservancy's website: www.scc.ca.gov, or for a paper copy at no charge write or call:

California State Coastal Conservancy
1330 Broadway Suite 1100
Oakland, CA 94612   (510) 286-1015

Transit agencies may have Dial-A-Ride and other wheelchair-accessible service available in the coastal areas; check the county entry for more information.

Beach Wheelchair, Hermosa Beach

National Parks and Recreation Areas

National parks and recreation areas along the California coast include Redwood National Park in Del Norte and Humboldt counties, Point Reyes National Seashore in Marin County, Golden Gate National Recreation Area (GGNRA) in Marin, San Francisco, and San Mateo counties, Channel Islands National Park in Santa Barbara and Ventura counties, and Santa Monica Mountains National Recreation Area in Ventura and Los Angeles counties.

Since each area has a variety of features, such as forests, mountains, headlands, lakes, rivers, cliffs, beaches, and tidepools, it is difficult to apply general rules to all of the parks. Normally those rules that apply to state parks apply to national parks as well. For example, animal and plant protection rules apply to all parks, while rules regarding pets are specific to each park, and often to particular areas within a park. In the Golden Gate National Recreation Area, pets must be on a leash or are not allowed at all. Always check at park offices or information centers to be sure of rules for that park. See www.nps.gov for more information.

National Park Headquarters:

Redwood National Park
1111 2nd Street
Crescent City, CA 95531   (707) 464-6101

Point Reyes National Seashore
Point Reyes, CA 94956   (415) 464-5100

Golden Gate National Recreation Area
Fort Mason, Bldg. 201
San Francisco, CA 94123   (415) 561-4700

Channel Islands National Park
1901 Spinnaker Drive
Ventura, CA 93001   (805) 658-5730

Santa Monica Mountains National Recreation Area
401 W. Hillcrest Drive
Thousand Oaks, CA 91360   (805) 370-2301

There are two other federal land areas along the coast that are open for public use. The Bureau of Land Management (BLM) operates the King Range National Conservation Area in Humboldt and Mendocino counties; for information, write or call:

District Manager of BLM   www.blm.gov
1695 Heindon Road
Arcata, CA 95521   (707) 825-2300

Los Padres National Forest, operated by the U.S. Forest Service, is located along the coast only in Monterey County, although it extends into five other Southern California counties. For information, write or call:

Monterey Ranger District Office   www.fs.fed.us
406 South Mildred Avenue
King City, CA 93930   (831) 385-5434

Woodblock print by Mallette Dean, *Monterey Peninsula*, Federal Writers' Project © 1941.

State Parks and Beaches

There are many beautiful state parks and beaches near or along the California coast. In order to keep them beautiful and safe for visitors, please observe the following rules and regulations:

- Dogs with proof of rabies inoculation are permitted in state park campgrounds and day-use areas, but are not allowed on most state beaches; check with the ranger. Pets must be leashed unless otherwise indicated.

- Open fires are not allowed. Most state park camping and picnic areas have stoves or barbecues, and many beaches have fire rings.

- Dune buggies, motorcycles, and other vehicles may be driven on the beach at Oceano Dunes State Vehicle Recreation Area. In other state park units all vehicles must stay on designated roads.

- Many state beaches do not have lifeguards; where provided, lifeguards are usually on duty only during the summer.

- Surfing is allowed only in designated areas at most state beaches.

- Plants and animals in state park units are protected. Researchers may apply for collecting permits at the Department of Parks and Recreation headquarters in Sacramento at 1416 Ninth Street, 14th Floor, 95814.

- A valid California fishing license is required in order to fish in state park units; consult the California Sport Fishing Regulations for season, size, and bag limits.

- Individuals may collect driftwood from state beaches; commercial collection is not allowed.

- Please stay on designated trails to prevent erosion and for your own safety.

- Camping is allowed only in designated campsites.

- Please don't litter–if you bring it in, take it back out.

The *Official State Parks Map,* including information on facilities at each park unit, is available for a small fee from the California State Parks e-store at www.parks.ca.gov, by telephone at 1-800-777-0369, or by mail from:

California State Parks e-store
1416 9th Street
Sacramento, CA 95814

Most campsites at state park campgrounds can be reserved in advance for a service fee through ReserveAmerica™. Reservations, which are recommended for popular areas, can be made 7 months in advance on the first day of the month beginning at 8:00 a.m. Pacific Standard Time in one of three ways: online at www.reserveamerica.com; by mail to ReserveAmerica™, P.O. Box 1510, Rancho Cordova, CA 95741-1510; or by phone: 1-800-444-7275 (TDD 1-800-274-7275). See www.parks.ca.gov for more information on different types of camping, how reservations work, and for a list of parks with first-come, first-served campsites.

Enroute camping in self-contained recreational vehicles is available for one-night-only stays at certain coastal parks, including: Van Damme State Park in Mendocino County; Half Moon Bay State Beach in San Mateo County; New Brighton, Seacliff, and Sunset State Beaches in Santa Cruz County; Moss Landing State Beach in Monterey County; Morro Bay and Montaña de Oro State Parks in San Luis Obispo County; El Capitan and Carpinteria State Beaches in Santa Barbara County; Emma Wood State Beach in Ventura County; Dockweiler State Beach in Los Angeles County; Bolsa Chica State Beach in Orange County; and San Onofre and Silver Strand State Beaches in San Diego County. Camping fees apply; call ahead to make sure that your vehicle can be accommodated.

State park day-use and camping facilities listed in the Guide are subject to change. For general information on the state park system, call: (916) 653-6995 or 1-800-777-0369.

Andrew Molera State Park

Pescadero State Beach

National Marine Sanctuaries

Four national marine sanctuaries have been designated off the coast of California; designation by the Secretary of Commerce as a marine sanctuary allows coordinated management of the area's ecological, research, recreational, aesthetic, and historical resources. No drilling for oil or gas is permitted in marine sanctuaries off California, and discharge of materials such as harbor dredge spoils is prohibited. Fishing in the marine sanctuaries requires a California fishing license, although in a few areas, removal of any marine organism is forbidden.

The Channel Islands National Marine Sanctuary was the first sanctuary designated off the California coast. Some 1,252 square nautical miles are encompassed by the sanctuary, which lies about 20 nautical miles southwest of Santa Barbara. (A nautical mile is 6076.1 feet, or 1.15 statute miles.) The main purpose of the Channel Islands Sanctuary is to protect the resources of the waters surrounding Anacapa, Santa Cruz, Santa Rosa, San Miguel, and Santa Barbara islands,

including the area from high tide to six nautical miles offshore. Many species of fish and invertebrates inhabit the waters surrounding the islands, along with many species of seabirds. Some 27 species of whales and dolphins spend part of the year here, including endangered gray, blue, humpback, and sei whales. Cultural resources in the sanctuary include the remains of over 100 shipwrecks.

The Gulf of the Farallones National Marine Sanctuary surrounds the Farallon Islands, extending up to 40 miles off the mainland, but it also encompasses nearshore waters and bays on the mainland coast, such as Tomales Bay and the Esteros Americano and San Antonio in Marin County. The waters of the sanctuary are home to harbor seals, California sea lions, and elephant seals. Commercial fisheries in the sanctuary include rockfish, Pacific herring, and Dungeness crab. The largest concentration of breeding seabirds in the continental United States is found on the Farallon Islands, which are a national wildlife refuge.

To the northwest of the Farallon Islands is the Cordell Bank National Marine Sanctuary, which was designated in 1989. Cordell Bank is a seamount, or submerged mountain top, where the ocean floor is as little as 114 feet below the water's surface, although the bank lies some 50 miles northwest of San Francisco. The bank lies at the edge of the Continental Shelf; water depths plunge to 6,000 feet only a few miles away. The wide range of water depths and the upwelling of deep-ocean nutrient-laden waters have encouraged a wide range of algae, invertebrates, fishes, marine mammals, and seabirds to inhabit the area. Divers and fishermen also visit the waters of the sanctuary.

The Monterey Bay National Marine Sanctuary encompasses 4,024 square nautical miles of marine waters along the Central California Coast. It extends from the south boundary of the Gulf of the Farallones National Marine Sanctuary, near San Francisco, south past San Simeon to Cambria Rock and includes the area from the high tide mark to approximately 50 miles offshore. Designated in September 1992, it is the Nation's largest marine sanctuary.

The Monterey Bay Sanctuary protects expansive kelp forests near shore. The Sanctuary also protects the Monterey Submarine Canyon farther offshore. At more than 10,000 feet

Gulf of the Farallones National Marine Sanctuary

deep at its seaward end, it is one of North America's largest submarine canyons, and is the closest-to-shore deep ocean environment in the continental United States. The area is characterized by a narrow continental shelf surrounded by a variety of ocean bottom formations, and by nutrient-rich currents providing for a diversity of highly productive marine habitats. It supports sea otters, seals, shorebirds, seabirds, the California gray whale, finback, humpback, and sperm whales, and thousands of other species. The most diverse algal community in the U.S. is found in the rich ocean waters of Monterey Bay.

Scientists study resources of the sanctuary for the 13 research educational facilities located adjacent to it. Fishing, tidepool viewing, diving, and whale watching are popular recreational activities. Commercial fishing for species including anchovies, albacore, squid, and flatfish is also done within the Monterey Bay National Marine Sanctuary.

A fundamental challenge of managing the national marine sanctuaries is to accommodate within them both conservation and consumption of resources. The National Oceanic and Atmospheric Administration of the U.S. Department of Commerce administers the marine sanctuaries under the 1972 Marine Protection, Research, and Sanctuaries Act. Designated or proposed sanctuaries outside California include coral reefs in Florida; the site of the sunken Civil War naval vessel, the U.S.S. *Monitor*, off Cape Hatteras, North Carolina; and Fagatele Bay, a tropical reef in American Samoa.

All of the sanctuaries have regulations to help protect their particular resources. The guidelines for use of each area are available from the local sanctuary office. Many of the habitats and species in sanctuaries are endangered or threatened, and visitors can help protect the area by observing the local rules. It is against the law to disturb or remove marine mammals, or to remove tidepool life in any marine sanctuary. Aircraft are restricted from flying at altitudes of less than 1,000 feet in order to avoid disturbance to wildlife.

For information on the nationwide sanctuary program:

National Marine Sanctuaries and Reserves Division
National Oceanic and Atmospheric Administration
1305 East–West Highway
Silver Spring, MD 20910   (301) 713-3125
www.noaa.gov/ocean.html

For information on the sanctuaries offshore California:

Channel Islands National Marine Sanctuary
113 Harbor Way
Santa Barbara, CA 93109   (805) 966-7107

Cordell Bank National Marine Sanctuary
Gulf of the Farallones National Marine Sanctuary
Fort Mason, Building 201
San Francisco, CA 94123   (415) 556-3509

Monterey Bay National Marine Sanctuary
299 Foam Street, Suite D
Monterey, CA 93940   (831) 647-4201

For information on the Farallon Islands National Wildlife Refuge:

U. S. Fish and Wildlife Service
P. O. Box 524
Newark, CA 94560   (510) 792-0222

The California Coastal National Monument, created in January 2000, includes all undesignated rocks, islands, exposed reefs, and pinnacles above mean high tide extending 12 nautical miles off the California coastline. Protection of geologic features, scenic values, and habitat for sea birds, mammals, and other forms of sea life are among the goals cited by the presidential proclamation that created the monument. The Bureau of Land Management is preparing a management plan for the rocks and islands, in cooperation with other agencies such as the California Department of Fish and Game. For more information, contact:

Rick Hanks, Manager
California Coastal National Monument
299 Foam Street
Monterey, CA 93940

Santa Barbara Island, Channel Islands National Marine Sanctuary

Natural Reserves, Preserves, and Refuges

Historically, the California coast has been an area of diverse landforms, vegetation, animal life, and cultural resources. Today, many coastal ecosystems and resources have been altered or destroyed by human activity. In an effort to protect and preserve remaining undisturbed areas of significant natural or cultural value, various public agencies have established a number of reserves, preserves, and refuges along the California coast. These protected areas provide excellent education and research opportunities and allow the general public to observe and develop an understanding of the state's natural environment and cultural background.

The University of California operates natural land and water reserves within the coastal zone ranging from the Pygmy Forest Reserve along the Mendocino County coast to the Kendall-Frost Mission Bay Marsh Reserve in San Diego County; the purpose of these reserves is to protect for teaching and research purposes a series of undisturbed natural areas representing California's ecological diversity. Primary users of these reserves are public and private institutions of higher learning; other interested groups may use the reserves only if written permission has been obtained from the reserve manager, if there is no conflict with primary users, and if adequate supervision is provided.

To obtain more specific information on the U.C. Reserve System, write:

University of California
Natural Reserve System
1111 Franklin Street, 6th Floor
Oakland, CA 94607-5200   (510) 987-0150

The California Department of Fish and Game is responsible for the protection and management of California's terrestrial and aquatic wildlife and their habitats. Fish and Game operates more than 40 marine life refuges and reserves within the California coastal zone, some of which are located in or adjacent to publicly accessible areas such as state parks, while others are restricted to access for education or scientific research.

The reserves and refuges have been established to protect and preserve endangered, threatened, or ecologically significant species of marine life or their habitats. As a general rule, marine reserves restrict the taking of most marine invertebrates (e.g., clams, abalone, lobster) and marine plant life. With respect to marine life refuges, it is generally unlawful to take or possess the marine life for which the refuge is named or to have in one's possession an implement designed to catch such animals.

For example, in the Pacific Grove Marine Gardens Fish Refuge, it is unlawful to catch fish or to possess any fish or fishing gear. Specific rules and regulations may vary from one reserve or refuge to another; for more specific information concerning marine life reserves and refuges, consult the Department of Fish and Game bulletins on marine resources or call or write the marine resources branch of the Department of Fish and Game office nearest you.

Entire state park units or distinct areas within them may be classified by the California Department of Parks and Recreation as reserves or preserves. The purpose of a state reserve is to provide day-use areas for public enjoyment and education while preserving the reserve's unique natural features. It is illegal to disturb or take any living or non-living resource within a state park reserve except by authorized

personnel for scientific or management purposes. Año Nuevo State Reserve in San Mateo County, which protects elephant seals, is an example of a natural reserve within the State Park System.

Preserves, which are distinct portions within a state park unit, are established for the purpose of maintaining an outstanding natural, scenic, scientific, or cultural feature in its natural condition. For example, a 19th and early 20th century dairy farm within Wilder Ranch State Park has been classified as a cultural preserve; the farm has been restored and is operated as a museum displaying early California dairying methods.

The California Nature Conservancy manages or owns several protected areas along the coast, including the Pygmy Forest Preserve in Mendocino County, Elkhorn Slough in Monterey County, Santa Cruz Island in Santa Barbara County, and the Irvine Ranch Land Reserve in Orange County. To obtain more information, write or call:

The Nature Conservancy of California
California Regional Office
201 Mission Street, 4th floor
San Francisco, CA 94105   (415) 777-0487
www.tnccalifornia.org

In addition to the reserves, preserves, and refuges discussed above, other protected areas have been established by public agencies ranging from the federal government to local park districts. Visitors to any reserve, preserve, or refuge are asked to observe all rules and regulations; look for signs containing information on the proper use of the facility. Any questions concerning the use of a reserve, preserve, or refuge should be addressed to the local manager or the agency responsible for the operation of the area.

Environmental Camping

The California Department of Parks and Recreation offers environmental campsites at certain locations, in order to provide visitors with the opportunity to enjoy the scenic and natural features of California's state parks without the distractions often associated with conventional camping, such as automobiles and noise from nearby camps.

An environmental campsite consists of a tent space, a table, and an enclosed pit toilet. Some campsites also include stoves or fire rings and a water source. Stoves are not provided in areas of high fire danger, but campers are allowed to use their portable camp stoves. The campsites are separated from each other and the regular campgrounds by natural features such as vegetation and landforms. Campers are required to carry in supplies from a designated parking area; automobiles are not permitted at the campsite.

Each environmental campsite has been located in an area containing significant scenic and natural qualities. Along the coast, environmental campsites currently exist at Lake Earl and Nickel Creek in Del Norte County; Prairie Creek Redwoods State Park, Humboldt Lagoons State Park, and Flint Ridge at Redwood National Park in Humboldt County; Van Damme State Park and Manchester State Beach in Mendocino County; Sonoma Coast State Beaches (at Willow Creek and Pomo Canyon) in Sonoma County; Mount Tamalpais State Park (both tent sites and simple cabins at Steep Ravine Beach) in Marin County; Andrew Molera and Julia Pfeiffer Burns State Parks in Monterey County; Montaña de Oro State Park in San Luis Obispo County; Crystal Cove State Park in Orange County; and San Onofre State Beach in San Diego County. The Department of Parks and Recreation plans to install additional environmental campsites in other coastal park units.

The environmental camping experience includes assuming responsibility for the maintenance of your campsite. Campers are expected to keep and leave the campsite and facilities clean; cleaning materials are located in the toilet structure.

Montaña de Oro State Park

Garbage should be placed in trash receptacles (where provided) or carried out to the nearest trash container; do not dump garbage in the toilets. Unless otherwise posted, wood gathering in state parks is prohibited; therefore, campers should bring their own fuel. Except for guide dogs, pets are not allowed in environmental campsites. Do not drink untreated water.

Conditions at the environmental camps may vary with the season and from site to site. Stays are limited to seven nights. For reservations, contact ReserveAmerica™ by phone: 1-800-444-7275 (TDD 1-800-274-7275); on-line at www.reserveamerica.com; or by mail to ReserveAmerica™, P.O. Box 1510, Rancho Cordova, CA 95741-1510.

Coastal Hostels

Golden Gate Hostel

Coastal hostels offer travelers a distinctive type of low-cost overnight lodging. Hostels vary in type and size; all are based on the international principle of community accommodations and on the premise that a person should leave a place in better condition than he or she finds it, whether in the natural or human-made environment.

Features generally found in hostels include a community living room; bathing, washing, and laundry facilities; a fully equipped kitchen and dining room where guests can cook and eat their own meals; dormitory rooms with bedding; and a tradition of participation in the work required to maintain the facility. Hostels in the U. S. are open to people of all ages, not only to "youth." In fact, many U.S. hostels include family sleeping quarters that may be reserved in advance.

Hostels encourage non-motorized forms of tourism such as hiking, bicycling, skiing, and canoeing; however, most hostels are also accessible by public transportation or private automobile, and have limited parking available.

Since the first American hostel opened in Massachusetts in December of 1934, hundreds of additional facilities have been established nationwide. Hostels established by Hostelling International – USA and by other entities provide a low-cost lodging alternative for travelers along the California coast. Many of these hostels are located only 20 to 30 miles apart, providing non-motorized travelers with the ability to "hostel-hop" the California coast.

For instance, more than 10 hostels are located along the coast in California's three most southerly counties, including facilities in San Diego in the lively Gaslamp District and at Pacific Beach. In the San Francisco Bay Area, five hostels operated by Hostelling International – USA cover some 120 miles of magnificent shoreline from the remote beaches of Point Reyes National Seashore to the urban waterfront in San Francisco. Facilities include two working lighthouse stations, and each Bay Area hostel offers distinctive accommodations in converted or renovated structures. The Marin Headlands Hostel, for example, now accommodates 60 overnight guests in a former 1902 officers' quarters building, and sits amid eucalyptus trees, rolling hills, and meadows, only a short walk from Rodeo Beach and Lagoon and only four miles from the north end of the Golden Gate Bridge.

For further information on California coastal hostels or to make overnight reservations, contact Hostelling International - USA in the coastal area where you will be traveling, see www.hiayh.org, or call the hostels listed below.

Golden Gate Council
425 Divisadero Street #307
San Francisco, CA 94117   (415) 701-1320

Central California Council
P.O. Box 2538
Monterey, CA 93942   (831) 899-1252

Los Angeles Council
1434 Second Street
Santa Monica, CA 90401   (310) 393-3413

San Diego Council
437 J Street, Suite #315
San Diego, CA 92101   (619) 338-9981

Del Norte County

Redwood-DeMartin House
14480 Highway 101
Klamath, CA 95548   (707) 482-8265

Humboldt County

Arcata Hostel
1390 "I" Street
Arcata, CA 95521   (707) 822-9995

Marin County

Point Reyes Hostel
P.O. Box 247
Point Reyes Station, CA 94956   (415) 663-8811

Marin Headlands Hostel
Building 941
Fort Barry
Sausalito, CA 94965   (415) 331-2777

San Francisco County

San Francisco International Hostel
Building 240
Fort Mason
San Francisco, CA 94123   (415) 771-7277

San Mateo County

Montara Lighthouse Hostel
P.O. Box 737
16th Street at Cabrillo Highway (Hwy. 1)
Montara, CA 94037   (650) 728-7177

Pigeon Point Lighthouse Hostel
Pigeon Point Road
Pescadero CA 94060   (650) 879-0633

Santa Cruz County

Santa Cruz Hostel
P.O. Box 1241 (321 Main Street)
Santa Cruz, CA 95061   (831) 423-8304

Monterey County

HI Monterey, Carpenters Hall
778 Hawthorne Street
Monterey, CA 93940   (831) 649-0375

San Luis Obispo County

Cambria Hostel/Bridge Street Inn
4314 Bridge Street
Cambria, CA 93428   (805) 927-7653

Hostel Obispo
1617 Santa Rosa Street
San Luis Obispo, CA 93401   (805) 544-4678

Point Reyes Hostel

Montara Lighthouse Hostel

Santa Barbara County

Santa Barbara International Tourist Hostel
134 Chapala Street
Santa Barbara, CA 93109   (805) 963-3586

Los Angeles County

Hostel Los Angeles / Santa Monica
1436 Second Street
Santa Monica, CA 90401   (310) 393-9913

Venice Beach Cotel
25 Windward Ave.
Venice, CA 90291   (310) 399-7649

Venice Beach Hostel
1515 Pacific Ave.
Venice, CA 90291   (310)452-3052

Hostel California
2221 Lincoln Boulevard
Venice, CA 90291   (310) 305-0250

Los Angeles Surf City Hostel
26 Pier Ave.
Hermosa Beach, CA 90254   (310) 798-2323

Hostel Los Angeles South Bay
3601 South Gaffey Street
Building 613
San Pedro, CA 90731   (310) 831-8109

Orange County

Colonial Inn Hostel
421 8th Street
Huntington Beach, CA 92648   (714) 536-3315

San Diego County

Mission Beach Hostel
3204 Mission Boulevard
Mission Beach, CA 92109   (858) 539-0043

Point Loma Hostel
3790 Udall Street
San Diego, CA 92107   (619) 223-4778

San Diego Hostel International
521 Market Street
San Diego, CA 92101   (619) 525-1531

Geology of the Coast

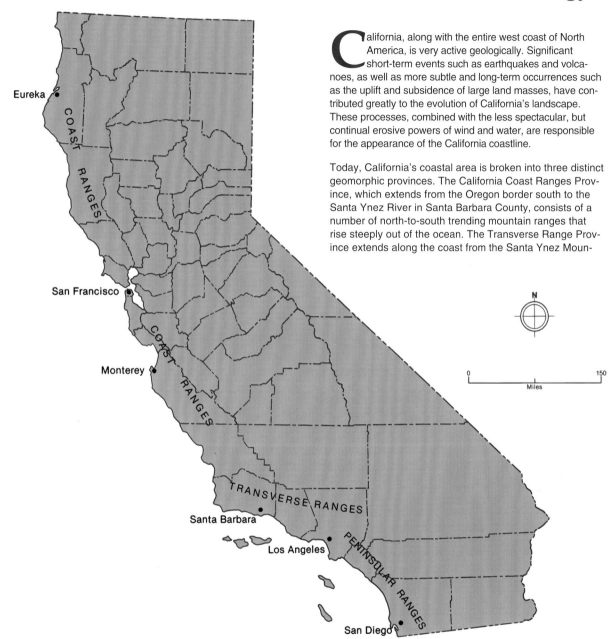

Eureka

San Francisco

Monterey

Santa Barbara

Los Angeles

San Diego

COAST RANGES

COAST RANGES

TRANSVERSE RANGES

PENINSULAR RANGES

N

0   150
Miles

California, along with the entire west coast of North America, is very active geologically. Significant short-term events such as earthquakes and volcanoes, as well as more subtle and long-term occurrences such as the uplift and subsidence of large land masses, have contributed greatly to the evolution of California's landscape. These processes, combined with the less spectacular, but continual erosive powers of wind and water, are responsible for the appearance of the California coastline.

Today, California's coastal area is broken into three distinct geomorphic provinces. The California Coast Ranges Province, which extends from the Oregon border south to the Santa Ynez River in Santa Barbara County, consists of a number of north-to-south trending mountain ranges that rise steeply out of the ocean. The Transverse Range Province extends along the coast from the Santa Ynez Moun-

tains in Santa Barbara County south to the Los Angeles Basin. Unlike the general north-to-south alignment of all the other mountain ranges in California, the mountains of the Transverse Range Province extend from east to west. Some of the oldest rocks in North America are found in the Transverse Range Province.

The third geomorphic province along the coast, the Peninsular Ranges Province, is one of the largest geomorphic areas in western North America. This province extends from the Los Angeles Basin south into Mexico to the tip of Baja California. Like the Coast Ranges, the Peninsular Ranges are aligned generally in a north-to-south direction.

The development of these geomorphic provinces is explained by the theory of plate tectonics. According to this theory, the earth's outermost layer, or crust, is made up of a number of rigid segments or plates that are able to move over the layer beneath it, called the mantle. Each plate may consist of a continent, an adjacent sea floor, or both, and is driven by internal earth forces.

Geologists believe that about 200 million years ago the world's seven continents existed as one land mass. The continents have since pulled apart and slowly moved to their present locations. About 150 million years ago, the California coast was located in the area now occupied by the Sierra Nevada. As the North American plate moved westward, it overrode the Pacific Plate. Subsequently, the California coast was extended west more than 100 miles and the Coast Ranges were built from sedimentary material scraped from the Pacific Plate as it moved beneath the western edge of the North American continent.

The San Andreas fault zone, one of the most widely known earthquake-producing regions in the world, marks the current boundary in California where the North American Plate meets the Pacific Plate. Earthquakes along the San Andreas fault, such as the devastating 1906 earthquake centered near San Francisco, are caused by the friction and stress created as the two plates grind past one another. The fault runs through the interior of California from the Gulf of California until it meets the coast at Mussel Rock near Daly City. Continuing north, the fault passes through Stinson Beach, Tomales Bay, Bodega Bay, and Point Arena before leaving the coast at Shelter Cove, near Point Delgada in Humboldt County.

About 25 million years ago, the Pacific Plate began to move northward along the San Andreas fault. The central California coast was formed from a part of Baja California that attached to the Pacific Plate and moved north along the fault. Continued northward movement compressed the earth's crust and created the Transverse Range Province about five million years ago. The city of Los Angeles is located on the Pacific Plate, west of the San Andreas fault, while San Francisco is on the North American Plate, east of the San Andreas fault. If the two plates continue to move in the same direction and at the same rate as they have in the geologic past, in about 10 million years Los Angeles will be due west of San Francisco.

One of the worst natural disasters in recent California history was the Loma Prieta earthquake, centered near Watsonville, which struck on October 17, 1989. This quake was caused by a 30-mile-long rupture on the San Andreas fault, starting about 11 miles below the earth's surface and ending 3.7 miles below the surface. The strong shaking lasted 6 to 10 seconds with a surface wave of magnitude 7.1 on the Richter scale. Much of the damage from this quake occurred in areas of soft soils, such as San Francisco's Marina district, where ground movement was magnified, and in low-lying sand and mud, where the soils became saturated and liquefied.

The geologic units created by plate tectonics are gradually but continually being altered by other physical processes such as changes in sea level and erosion. The formation of the many lagoons in San Diego County, for example, illustrates how these processes affect the coastal landscape.

Tectonic activity resulted in the development of the Peninsular Ranges. Subsequently, ocean waves eroded the seaward edge of the ranges, resulting in the development of marine terraces. During the past ice age, sea level was lowered, thereby exposing the marine terraces. Terrestrial water runoff then cut deep channels in the newly exposed terraces. As the glaciers melted, sea level rose, and the seaward ends of the channels were flooded, creating estuaries. While streams carried water and sediment to the estuaries, the ocean currents distributed the sediments across the estuary mouths and created the bay mouth bars and lagoon systems visible today along the San Diego coast.

San Andreas Fault, view northwest from Bolinas Lagoon to Tomales Bay.

Erosion

Waves, nearshore currents, and coastal sand supply affect beaches in three basic ways: first, if the amount of sand supplied to a beach is greater than the amount removed by waves and currents, the beach accretes, or extends seaward; second, if sediment supply is generally equivalent to the amount removed by the ocean, the beach is considered to be in equilibrium, although the beach's width may fluctuate in response to storm and calm periods; finally, if more sand is removed from a beach than is supplied, the beach erodes, or retreats landward.

The reduction in the amount of sand supplied to a beach may be caused by the damming of streams or the installation of structures that interrupt or inhibit the flow of sediment along the coast, such as breakwaters or jetties; the resultant erosion can have significant adverse impacts on the affected beach area. As the beach narrows from erosion, attacking waves threaten shoreline structures such as roads, houses, and other buildings. It then becomes necessary to protect these structures by installing shoreline protection devices such as seawalls, or structures designed to trap sand such as groins, or by widening the beach with sand from other sources. These protective measures are not only costly, but often increase erosion at adjacent beaches.

Coastal bluffs, which form the inland extent of many California beaches, are also subject to erosion. Advancing waves undercut the base of the bluffs, resulting in the collapse of bluff material and the subsequent landward retreat of the bluff itself. Bluff erosion is also caused by water runoff flowing over and through the bluffs; the water removes bluff material and increases the potential for landslides, slumping, or other forms of terrestrial erosion. As coastal bluffs erode and retreat landward, bluff stability, and the stability of structures located on the bluffs, becomes threatened. Subsequently, expensive stabilization, protection, and repair efforts are often required to slow or stop bluff retreat.

Many coastal access facilities include blufftop areas and trails down the bluffs to the beach. Although beachgoers can do little to stop the erosive power of waves, they can assist in reducing terrestrial erosion. Visitors should not park or walk near bluff edges except in designated locations and should use only formal trails and stairs. In addition to being unsafe, activity near bluff edges and the creation of informal trails removes vegetation and loosens bluff material, which increases erosion. Visitors who follow these simple rules will aid in reducing facility repair and maintenance costs and will prolong the availability of safe, operable access facilities.

Pacific Palisades

Climate

California's coastal climate generally can be classified as Mediterranean—that is, characterized by temperate wet winters and warm dry summers. Only one percent of the world has this climate, and it is found nowhere else in the United States. The two chief influences on the coastal climate are the persistent high-pressure system located west of California over the Pacific Ocean, called the Pacific High, and the effect of the ocean itself.

The Pacific High is stronger and in a more northerly position in the summer. It prevents storms from approaching the West Coast and is responsible for the dry weather along the shore from May through September. In the winter this high-pressure system moves south and occasionally weakens, allowing storms to reach the state.

As storms most frequently approach from the north, that area receives more rainfall than the south, which remains partially shielded by the high, even in the winter. As a result, Crescent City receives an average of 70 inches of rain annually, while San Francisco receives an average of 20 inches, and San Diego averages only 12 inches. There are also more days with rain in the north than in the south, with San Francisco having an average of 63 days of rain per year while Los Angeles has 34. As the high migrates south, Northern California receives precipitation that is entirely blocked from the south; thus, its rainy season is longer.

As might be expected, the temperatures at the beach also vary from north to south. In Eureka in Humboldt County the average temperature in January is 47 degrees and in August 57 degrees, while in San Diego the average is 55 degrees in January and 70 degrees in August. The temperatures along the coast are limited to a relatively narrow seasonal and daily range by the moderating influence of the ocean; thus temperatures of below freezing or above 100 degrees are unusual. Long Beach, because of its south-facing shore alignment, has an average of 15 days a year of daily high temperatures above 90 degrees; however, the rest of the Southern California coast gets that warm less than five times a year.

The marine air also keeps the humidity high year round, with 65 percent relative humidity being about normal. Contributing to this is the frequent occurrence of fog or low cloudiness, which is common in the spring and summer months; however, the fog and low clouds usually "burn off" by noon.

The Pacific High results in an overall wind pattern from the northwest. In most areas, onshore sea breezes increase in the afternoon. Before and during most storms, the wind shifts so that it is blowing from the south.

While the summer climate in California is fairly stable, wind, temperature, and rainfall averages can be deceptive when compared to what one actually encounters on a yearly or daily basis. California regularly experiences cycles of both drought and flood conditions, of varying lengths and intensity, resulting in rainfall amounts that vary greatly from year to year. Local topography has a pronounced effect on the climate as well. For example, where the shoreline is backed by mountains, precipitation is usually heavier. Another example is that south-facing shores such as those at Santa Cruz, Santa Barbara, Malibu, and Long Beach usually have warmer temperatures than the rest of the coast.

There are occasional weather patterns that, while not common, are distinctive enough that the visitor to the coast should be aware of them. When a high-pressure system builds up inland, a reversal of the usual winds can occur, the direction then being from the interior. These winds, called Santa Anas or foehns, are hot, dry, and sometimes quite strong. During the summer they can contribute to the destructive spread of brush fires. In the winter they are even stronger, sometimes of hurricane strength, but are not as warm. Santa Anas are markedly affected by local topography; for example, the winds are frequently stronger in canyons.

In the summer, tropical air may intrude even to Northern California, bringing the possibility of thunderstorms, which are usually rare along the coast. Even more uncommon are full-fledged tropical storms, which may bring heavy rains or showers to Southern California in the summer.

With moderate temperatures and dry summers, California has ideal weather for visiting the coast. Because of seasonal and geographical differences in the state, visitors should check local conditions before departing on a trip to the coast.

Russian River Estuary

History of the Coast

In 1510, the Spanish author Garcí Rodríguez Ordóñez de Montalvo wrote of a mythical island, named California, that he described as being rich in gold, located "on the right hand of the Indies...very near the Terrestrial Paradise," and inhabited only by women and man-eating "griffins." Montalvo's fictitious island, and other rumors of lands of great wealth, are thought to have led many early Spanish explorers to believe that such a land might actually exist along the uncharted areas of the west coast of the New World. One of the earliest Spanish explorers, Hernando Cortés, is thought to have been the first to apply the name California (probably taken from Montalvo's writings) to the area now called Baja California, which he explored from 1535 to 1537.

At the time of Cortés's expeditions, California was inhabited by as many as 200 to 300 thousand Native Americans; California was then one of the most densely populated Native American areas in North America, north of the present boundary of Mexico. Scattered tribes, many of which were settled along the coast, were usually peaceful and isolated from one another. More than 100 dialects of 21 distinct languages were spoken. Shell mounds indicate the presence of California Native Americans at least three to four thousand years ago; their relatively simple nomadic culture is believed to have remained unchanged until their encounters with the Spanish.

During their early explorations, the Spanish referred to all of the Spanish-claimed land from what is now Baja California in Mexico to Alaska as "California." The first European to explore the coast of California north of the state's present boundary with Mexico was Juan Rodríguez Cabrillo, a Portuguese navigator hired by the Spanish. Cabrillo sailed up the coast in 1542, charting the landmarks now known as San Diego Bay, Santa Catalina Island, the Channel Islands, Point Conception, Point Piños, Monterey Bay, Point Año Nuevo, and Point Reyes. Because of bad weather, Cabrillo's pilot, Bartolomé Ferrelo, eventually continued the voyage north and passed Cape Mendocino, the westernmost point on the California coast; it is believed he named the cape after the Viceroy of New Spain, Don Antonio de Mendoza.

Although Cabrillo found no evidence of the riches and wealthy cities rumored to exist in California, Spain nonetheless laid

Wood block prints by Mallette Dean, *Monterey Peninsula*, Federal Writers' Project © 1941.

claim to the entire coast. The English, however, were bitter rivals of the Spanish and refused to recognize Spanish-claimed lands and rights of navigation. The English government, therefore, tacitly encouraged private adventurers to attack Spanish ships and settlements. The most famous of these adventurers was Francis Drake, who sailed up the coast of South and Central America in 1579, attacking and plundering numerous Spanish ships along the way. Drake continued northward along the California coast in a futile search for the northern passage that was thought to exist between the Pacific and Atlantic Oceans. Drake stopped along the coast to refit his vessel at an area he named Nova Albion; although the actual landing site is unknown, he may have landed at what is now called Drakes Bay off Point Reyes.

Spanish trading galleons traveling from Manila to Acapulco along the northern trade routes visited the coast in the late 1500s, in search of protective harbors where their crews could rest and gather supplies. Pedro de Unamuno anchored a galleon at Morro Bay in 1587, and Sebastián Rodríguez Cermeño landed at Drakes Bay in 1595, where his galleon was destroyed by storms. Remarkably, Cermeño and his crew traveled the remaining distance to Acapulco in a small launch, completing one of the first detailed surveys of the coast.

The Spanish again tried to find a harbor for their trading galleons in 1602, when the Spanish king directed Sebastián Vizcaíno to explore the California coast and to confirm whether there actually existed a North American passage to the Atlantic. Vizcaíno found no such passage, but he did chart a major portion of the coast and discovered what he described to be a suitably protective harbor at Monterey Bay. Many of the place names Vizcaíno gave to California coastal landmarks still exist today.

Between 1697 and 1767, the Spanish established a number of Jesuit missions in Baja California (now part of Mexico) for the purpose of converting the Native Americans to Catholicism and to strengthen the Spanish claim to the coast. In 1767, Juan Gaspar de Portolá carried out a royal Spanish decree banishing the Jesuits; the missions were then turned over to the Franciscans, led by Padre Junípero Serra.

Portolá and Serra were directed to expand the mission and presidio system northward, and in 1769 they traveled to San Diego Bay, where they established the first Alta (or upper) California mission. Portolá then continued northward on a land expedition in search of the Monterey Bay, which Vizcaíno had so favorably described over a century earlier. Portolá actually camped at Monterey Bay on his first expedition, but failing to recognize it as the bay that Vizcaíno had charted, he continued north and eventually sighted San Francisco Bay. Portolá returned to San Diego and organized a second expedition, which returned to Monterey Bay in 1770. Portolá was joined by Padre Serra, and they proceeded to establish the Monterey Mission and Presidio; this settlement became the first official capital of Alta California in 1775.

Between 1769 and 1823, Padre Serra, followed by Padre Fermin Francisco de Lasuén, settled a total of 21 missions on or near the California coast. The missions were usually located in the areas of the greatest Native American populations, since the missionaries' goal was to convert the resident inhabitants to the Catholic faith. However, contact with the Spanish missionaries was devastating to the Native Americans, who suffered horribly from diseases inadvertently introduced by the Europeans. The Native Americans were also forced to work on the missions' agricultural and ranch lands. Many of the Native Americans rebelled, and hostile outbreaks periodically occurred at the

missions. A visiting Frenchman at Monterey in 1786, Jean François Galaup de la Pérouse, described the missionaries' treatment of the Native Americans as being worse than American plantation slavery in the southeast.

Mexico won independence from Spain in 1821, and California was formally declared a territory of the Republic of Mexico in 1825. In 1833, the Mexican Governor of California, José Figueroa, ordered the secularization of the missions, requiring that their lands be distributed among the mission Native Americans and the resident "Californios," most of whom were of Hispanic heritage. However, Figueroa died in 1835, and a majority of the Native American-held land was subsequently taken by the Californios, who managed these huge land grants as livestock "ranchos."

San Francisco Bay, before and during the gold rush.
San Francisco Archives, San Francisco Public Library

After secularization many of the Native Americans attempted to return to their tribes and former lifestyles; those who did not were put to work as unpaid ranch hands on the ranchos. The ranchos produced hides and tallow, much of which were sold to American traders from Boston who arrived by ship. San Diego was once a principal area for collecting and curing hides obtained from the ranchos.

Americans began arriving on the coast in the late 1700s in search of animal furs and whales. The abundance of sea ot-

ters and fur seals created a profitable coastal fur trade, which also attracted the Russians, who built a north coast settlement at Fort Ross in 1812. New England whalers took large numbers of migrating California gray whales, which were cut and processed at land-based whaling stations. Whaling stations were established on the coast at Ballast Point in San Diego Bay, Portuguese Bend near Palos Verdes, San Simeon, Moss Landing, Monterey Bay, and at Bolinas near Point Reyes. So many whales were taken (in the 1840s the patios and sidewalks at the Monterey station were paved with whale vertebrae) that conservation measures were necessary in the early 1900s to preserve the remaining whale herds.

The United States did not want to be limited to land east of the Rocky Mountains, and therefore viewed California, with its desirable coastal ports, as important territory. Increasing numbers of American settlers in California, and fear that Mexico might cede its California territory to England for payment of debts, prompted the U. S. government to begin attempts to purchase the land from Mexico. When negotiations broke down, partly due to the war between the U. S. and Mexico, the U.S. helped to initiate a revolution within California.

In 1846, a small band of American settlers (who were later joined by a small army detachment led by Captain John C. Frémont) easily forced a ranking Mexican general, Mariano G. Vallejo, to surrender at Sonoma, where the Americans first raised the California Republic "Bear Flag." Commander John Sloat shortly thereafter raised the American flag at Monterey, the ex-capital of Mexican California. Skirmishes between the Californios and the Americans ended in 1847 with the surrender of the Californios. In 1848, California became a United States territory. The state's constitution was drafted at Monterey in 1849, and in 1850 California was admitted as the thirty-first state of the Union.

A discovery of gold in 1848 triggered the California gold rush, when thousands of Americans, Europeans, and Chinese flocked to the gold fields of the Sierra Nevada and the state's northern ranges. Most of the immigrants entered the state by

sea, through the Golden Gate and the port of San Francisco. San Francisco grew rapidly during the 1850s, serving as an important marketplace and hub of the gold rush period.

The gold rush also initiated the development of the state's north coast, which was rich in the timber necessary to the gold mines and the state's rapidly growing towns and cities. In the 1850s, shipping timber by sea was easier than shipping it overland, and hundreds of mills and small settlements were established in the north adjacent to nearly every bay and cove that could serve as a port. The timber was mostly brought to San Francisco, Los Angeles, and through the Sacramento River Delta to the Central Valley.

The city of Los Angeles was incorporated in 1850, and many of the neighboring ranchos were gradually subdivided into small home and farm sites. In 1876, the Southern Pacific Railroad completed a new line, which ran between San Francisco and Los Angeles, providing a link with the transcontinental railroad system that had previously ended in the San Francisco Bay Area. In 1885, the Santa Fe railway to Los Angeles was completed, creating intense competition between the two railroads, which quickly dropped fares from the midwest to Los Angeles to as low as $1 per person. Thousands of people began arriving in Los Angeles and land speculation rapidly reached a frenzied state, initiating the first of several Southern California population booms.

As Southern California developed, many areas along the coast became popular vacation spots. Beach resort communities were established at Santa Monica in 1872 (where a small hotel advertised that "a week spent at the beach will add 10 years to your life") and at Venice in 1904. Huge luxury hotels were also constructed along the coast, including the Hotel Del Coronado near San Diego, the Redondo Beach Hotel near Los Angeles, the Potter Hotel in Santa Barbara, and the Hotel Del Monte in Monterey.

Between 1860 and 1880, the Chinese formed the first marine fishing industry in California when they established on the

coast a number of fishing villages from which they exported millions of dollars worth of fish (mostly abalone and shrimp) to China. Prior to 1860 the only fishermen on the coast were the Native Americans, who took fish primarily for their own consumption. After 1890 the Chinese were gradually driven from the lucrative industry by resident Italians, Japanese, Portuguese, and Yugoslavs.

In the late 1800s, Southern California experienced an oil boom, considered to be as economically important as the earlier Northern California gold rush. Exploration for oil began in the 1860s, but it was not until the 1890s (following improvements in refining, storage, and drilling) that the industry greatly expanded. In succeeding years, Southern California established itself as a producer of approximately a quarter of the world's oil and gas supply. The San Pedro Bay and Long Beach area in Los Angeles soon became, and still is, a major distribution point for the oil industry. Oil development expanded to offshore leases of the Outer Continental Shelf in the 1950s.

During the 1900s, especially during and immediately following World War II, the military built a number of coastal defense bases, including those at San Diego, Camp Pendleton, Point Mugu, Port Hueneme, Vandenberg, Fort Ord, and along the Marin County coastline. Residential development also continued along the coast; today, approximately 80 percent of the state's population lives within 30 miles of the ocean. This development created a demand for preservation of coastal public lands; the first coastal park, Big Basin Redwoods in Santa Cruz County, was purchased in 1902, and the State Park System was formed in 1927. There are currently over one hundred state beaches, parks, reserves, preserves, and monuments on or near the coast.

Santa Monica c. 1870s

Long Beach c. 1890

Historic Landmarks

There are virtually thousands of historic monuments, landmarks, and points of interest along the California coast; included here are descriptions of a few of the more famous and interesting historic spots.

Carson Mansion, Humboldt County: The Carson Mansion in Eureka is a famous example of Victorian architecture, built by lumber magnate William Carson in the 1880s. Having made his fortune from logging the redwood trees around Humboldt Bay, he built his mansion from choice, clear-grain redwood. Carved panels, onyx fireplaces, and stained glass windows decorate the interior. Cupolas, porches, and "gingerbread" detail are part of the Queen Anne-style exterior. The 18-room, three-story, ornamented mansion was built for Carson by dozens of carpenters and artists, and is now a private club. Although it is not open to the public, the mansion can be viewed from Second Street in Eureka.

Fort Humboldt State Historic Park, Humboldt County: Fort Humboldt was built on a hilltop overlooking Eureka to protect the settlers from the Native Americans. The fort was occupied from 1853 to 1870; only minor battles with the Native Americans occurred during that time. There were over 14 buildings on the site, including officers quarters, a hospital, a powder magazine, a blacksmith shop, a stable, and a bakehouse. The fort's most noted officer was Ulysses S. Grant, who spent six unhappy months stationed there in 1854. The commissary building still stands, and contains a small museum; there is also an exhibit of early logging tools and machinery on the fort grounds, located off Highland Avenue in Eureka.

Mendocino, Mendocino County: The town of Mendocino was first called Meiggsville after Harry Meiggs, who bought the land and opened a timber mill at the mouth of the Big River in the 1850s. Although Meiggs (and probably the name Meiggsville) departed in 1854, the timber mill operated until 1937. Around the turn of the century, Mendocino was a bustling port, boasting at least 21 saloons and 8 hotels. Because most of the residents were transplanted New Englanders, the buildings reflected the New England salt box style and early Gothic style with steep roofs and clapboard walls. Some restored homes are now operated as bed and breakfast facilities.

After the closing of the mill, Mendocino's population declined until the 1950s, when numerous artists were lured to the town by its remoteness and inexpensive homes. The influx of artists and other people who value the charm and beauty of Mendocino continues today. Some of the notable buildings in Mendocino include the MacCallum House (1882), the Presbyterian Church (1868), the Mendocino Hotel (1878), the Joss House (1855), and the Masonic Hall (1872).

Carson Mansion, Eureka

Mendocino Hotel

Fort Ross, Sonoma County: In 1812 Captain Kuskov established a permanent Russian settlement at Fort Ross, 12 miles north of the Russian River. The settlement, one of several along the Sonoma coast, was the center of Russian activities in California. Fur harvesting and agricultural cultivation were quite successful; the surplus products were either traded to the Spanish in California, shipped up to Alaska to support Russian settlements there, or sent to Russia. The Fort Ross settlers, probably the first to commercially harvest redwood timber in California, began shipping the lumber to the Hawaiian Islands in the 1820s.

The fort's population, which never exceeded 400, comprised both Russians and Aleuts (natives of the Aleutian Islands). The fort included two blockades along the palisades, a chapel, and the commander's headquarters. Outside the fort were approximately 50 buildings, including the blacksmith, carpenter, and cooper shops, and a stable for cows. Because of depleted resources, the Russians gave up their colony at Fort Ross in 1841. For a short time, it was owned by John Sutter. The fort is now owned by the State Department of Parks and Recreation, and is partially restored and open to the public.

Drakes Beach, Marin County: One of the first Europeans to explore the coast was the English sailor Francis Drake, who anchored and repaired his vessel, the *Golden Hinde*, in a Northern California harbor in 1579. Drake's journal of his voyage leaves the exact location of his landing unclear; however, historians believe it was most likely either Drakes Bay at Point Reyes, nearby Bodega Bay, or along the Marin County shores of San Francisco Bay. Monuments commemorating Drake are located at Drakes Beach near the ranger's office on Drakes Beach Road, and at Drakes Estero.

Drake stayed at his landing site for 36 days, and named the land Nova Albion (New England), partly because the eroded white cliffs (such as those at Drakes Beach) reminded him of the white banks of Dover in England. Drake's crew built a stone fort at the landing site, and found the Native Americans to be extremely friendly; so friendly, in fact, that Drake reported that the Native Americans crowned him as their king and regarded the English as gods. Drake eventually returned to England, becoming the first ship's captain to circumnavigate the globe.

Golden Gate Bridge, San Francisco and Marin counties: The entrance to San Francisco Bay was named *chrysopylae* (golden gate) by John C. Frémont in 1846. He chose this name because he expected the riches of the Orient to be shipped through the Golden Gate. Several years later, the gold riches from the Sierra foothills gave the Golden Gate a new meaning. Although the idea for a bridge spanning the bay entrance was first publicly mentioned in 1869, it was not seriously considered until 1916. Numerous problems delayed the building of the bridge, the most significant of which were design and engineering problems encountered because of severe currents caused by the tides through the gate.

In May 1937 the twin-tower, single-suspension bridge built by Joseph Strauss was opened. The bridge took four years to build, and although the construction claimed several lives, it was a major engineering success. The 8,981-foot-long bridge is a color called international orange; the acrylic paint is touched up on a continuous basis, to combat the corrosive effects of the salt air. The Golden Gate Bridge is one of the few

Golden Gate Bridge

San Francisco bay area bridges that pedestrians can walk across. On May 27, 1937, when the bridge was first opened to pedestrians, 202,000 people walked across it. The fiftieth anniversary of the bridge was celebrated on May 24, 1987 by many, including an estimated 800,000 pedestrians on the bridge and its approaches. Two observation areas at each end of the bridge provide panoramic views of the bridge, the bay, and San Francisco.

Cliff House, San Francisco County: The Cliff House, which has a restaurant and bar and is perched on the cliffs overlooking the Pacific Ocean, Seal Rocks, and the long strand of Ocean Beach, is the third building on the site to be called the "Cliff House." The first Cliff House was built in 1863 and was a popular destination for San Franciscans. Adolf Sutro bought the building in 1883 and dispatched the rough clientele that habitually gathered there. Unfortunately, the Cliff House burned down on Christmas Day in 1894. Two years later, Sutro rebuilt the Cliff House, seven stories high, complete with spires and towers, only to have it burn down again in 1907. The present building, made of stucco, was built following the second fire and is still a popular destination point for both resident San Franciscans and visitors. The National Park Service, which operates the Cliff House as part of the Golden Gate National Recreation Area, has undertaken a major renovation of the facility, to be completed in 2004.

City of Monterey, Monterey County: Numerous adobe buildings in downtown Monterey, dating from the 1800s, recall the town's historical importance in the development of the state. Many of these buildings are part of Monterey State Historic Park. The oldest government building on the Pacific Coast, the Custom House, was built intermittently from about 1827 to 1846. It was here that American and British traders paid duty to the Mexicans, and it is also here that the United States flag was first officially raised in California, on July 7, 1846. Colton Hall was the first American public building in California, and was the site of the state constitutional convention in 1849. Walter Colton, for whom the building is named, and Robert Semple started the first newspaper in California.

There are several adobe houses that typify styles popular at the time. The Larkin House, built in the 1830s, shows the typical Monterey-style architecture with two stories, balconies, and verandas. This was the home of Thomas Oliver Larkin, who was the first and only U.S. Consul to Mexican California; he held this position from 1843 to 1846, and was also a key figure in initiating the American revolution in California. The Casa Gutierrez, built in the 1840s, is a home typical of average citizens during the Mexican period. Other historic buildings in the area include private houses, California's first theater, public buildings, and a whaling station.

Hearst Castle

Hearst Castle, San Luis Obispo County: La Casa Grande in San Simeon was designed by William Randolph Hearst and his architect, Julia Morgan; construction began in 1922 and continued until 1951, when Hearst died. The house by that time had 100 rooms; 38 bedrooms, 31 bathrooms, 14 sitting rooms, a kitchen, a movie theater, 2 libraries, a billiard room, and a dining room were included in this Spanish/Moorish Revival mansion. Guests who stayed with Hearst were bound by four rules: they were required to come to the Great Hall in La Casa Grande every evening; they had to attend the nightly movie in his private theater; no liquor was permitted in the guests' suites; and no one was to mention the word death in Hearst's presence.

The grounds around La Casa Grande include 123 acres of gardens, terraces, pools, guesthouses, a small airstrip, and tennis courts. The three guesthouses are Mediterranean-style mansions; the gardens contain exotic plants from all over the world. The Roman Pool is thought to have been inspired by a first-century Roman mausoleum. Hearst also had a collection of wild animals, including zebras, tahr goats, aoudad (Barbary sheep), monkeys, cheetahs, lions, leopards, panthers, and polar bears; the descendents of these animals still graze on the grounds. The property is now Hearst San Simeon State Historic Museum; visits are by guided tour only.

Santa Barbara Mission, Santa Barbara County: Mission Santa Barbara was originally founded in 1786; the present structure on Los Olivos Street in the city of Santa Barbara was dedicated in 1820. The mission's well-maintained features, early Spanish Renaissance architecture, massive six-foot-thick walls, and the fact that it has been in continuous use for religious services since its dedication make this one of the most distinctive and attractive California missions. The graves of approximately 4,000 Native Americans are contained within the mission's scenic old cemetery.

Santa Catalina Island, Los Angeles County: Juan Rodríguez Cabrillo originally named the island San Salvador, after his ship, when he landed here in 1542. This was the first recorded landing by a European on the island. In 1602, Sebastián Vizcaíno landed on the island and renamed it Santa Caterina after a Christian martyr, the patron saint of spinsterhood. After the missions were established on the mainland, the island's Native Americans were exposed to European diseases as a result of trading with the mainland villages, and many died; measles alone killed 200 Native Americans in the early 1800s.

Until Mexico gained its independence from Spain, Santa Catalina Island was a base for unlawful trade with the mainland. The numerous coves hid ships that were avoiding duty payments on trade goods. During the next few decades, Catalina had various residents; in 1887, Avalon was named by Mrs. E.J. Whitney after a Celtic paradise. In the early 1900s, Judge Joseph B. Banning, one of the island's owners, built a wharf and house on the Isthmus. This became a popular spot for filming movies; *Mutiny on the Bounty, Treasure Island*, and *The Ten Commandments* were filmed here. William Wrigley, Jr. bought the island in 1919 and converted it into the resort it is today. He planted palm trees, cleaned the beaches, and built the famous Casino on a rocky promontory in Avalon Bay.

Dana Point, Orange County: Dana Point was named after author Richard Henry Dana, who described this point in his book *Two Years Before the Mast* after he had anchored his ship here in 1835. At that time, the harbor was a major port for square-rigged ships and a loading area for cattle hides brought in from the Californios' ranchos. The cliffs were steep and hard to climb, so the hides were thrown down to the harbor from the cliff tops, then carried by the sailors on their heads out to the ships. As Dana described, "It was really a picturesque sight: the great height, the scaling of the hides, and the continual walking to and fro of the men, who looked like mites, on the beach. This was the romance of hide droghing!" A new harbor with artificial breakwaters and room for 2,500 boats was completed in 1970.

Cabrillo National Monument, San Diego County: Located near the tip of Point Loma, overlooking San Diego Bay, this monument commemorates Juan Rodríguez Cabrillo, the first European to land in Alta California. Cabrillo sailed into San Diego Bay (which he named San Miguel) in 1542, and probably landed at what is now known as Ballast Point, originally called La Punta de Guijarros ("The Point of Cobblestones") by Sebastián Vizcaíno in 1602; the point is so named because the rocks there were used as ballast for Boston trading ships in the 1800s. Ballast Point was also the location of a Spanish fort from 1797 to 1838 (the fort was involved in the "Battle of San Diego" in 1803, a dispute between American fur traders and the Spanish), and two New England-based whaling stations were established there in the mid-1800s. The old Point

Loma Lighthouse, located on the hill above the Cabrillo statue and the museum, was built in 1854 but abandoned in 1891 because its elevation obscured it during periods of fog. The old lighthouse building is now open to visitors.

Hotel Del Coronado, San Diego County: This Coronado hotel, built during the 1880s, was designed by Stanford White, and is the largest wooden structure west of the Mississippi. All building materials had to be ferried to the peninsula from San Diego, then carried three miles by train to the building site. The beach resort hotel, located on 26 acres on Orange Avenue in Coronado, includes a two-acre interior courtyard, guest rooms in the original Victorian building, as well as newer accommodations. Hotel guests have included Charles Lindbergh, Frank Sinatra, and Marilyn Monroe.

Cabrillo National Monument

Hotel Del Coronado

How to Use the Guide

The Guide is divided into fifteen sections by county, from north to south, and into subsections by local area. Each subsection includes a map, descriptions of the accessways, and a grid that contains fifteen categories for facilities and seven for environment. The categories are defined as follows:

FACILITIES

ENTRANCE/PARKING FEE: Indicates that there is an entrance or parking fee for the area. These fees frequently apply only to summer use, and in some cases, apply only for overnight stays and not for day use.

PARKING: Indicates that there is adjacent on-street parking or a parking lot.

RESTROOMS: Noted where either flush, chemical, or pit toilets are provided.

LIFEGUARD: Lifeguards, where present, are usually on duty only in the summer.

CAMPGROUND: Indicates either a private campground or one that is part of a local, state, or federal park unit.

SHOWERS: Noted where there are hot showers or outdoor cold showers.

FIREPITS: Indicates any designated cooking area, including barbecues, grills, and fire rings.

STAIRS TO BEACH: Used to indicate a stairway which leads to the beach; in certain cases, there may also be an alternate way to get to the beach.

PATH TO BEACH: Indicates that it is necessary to take a path or trail to get to the beach, as opposed to an area where access to the beach is directly from the parking area.

BIKE PATH: Indicates a paved bikeway separate from the road. Bike lanes or recommended routes are not included.

HIKING TRAILS: Indicates designated hiking trails, not simply areas where hiking is possible.

FACILITIES FOR THE DISABLED: Includes wheelchair ramps, modified restrooms, or other specific modifications that enable persons with disabilities to use the facility.

BOATING FACILITIES: Includes boat launch ramps, hoists, slips, docks, moorings, etc.

FISHING: Noted where a fishing pier or fish cleaning facility exists; also noted for any area commonly used for fishing.

EQUESTRIAN TRAIL: Noted for any area that has a designated equestrian trail.

ENVIRONMENT

SANDY BEACH/DUNES/ROCKY SHORE: Noted where the environment of the accessway includes a sandy beach, dunes, or a rocky shore; these categories are not mutually exclusive and may refer to only a portion of the area.

UPLAND FROM BEACH: Used when all or part of an accessway is upland from the shoreline. Some coastal accessways have both nearshore and upland portions.

STREAM CORRIDOR: Indicates a stream in the vicinity; sometimes the accessway is actually along the stream.

BLUFF: Used when the environment includes a bluff, although the accessway itself may not be directly on the bluff, but below or nearby.

WETLAND: Indicates a wetland in the vicinity; sometimes the accessway is actually located alongside the wetland.

CALIFORNIA
COASTAL ACCESS
· G U I D E ·

Del Norte Coast Redwoods State Park

Del Norte County

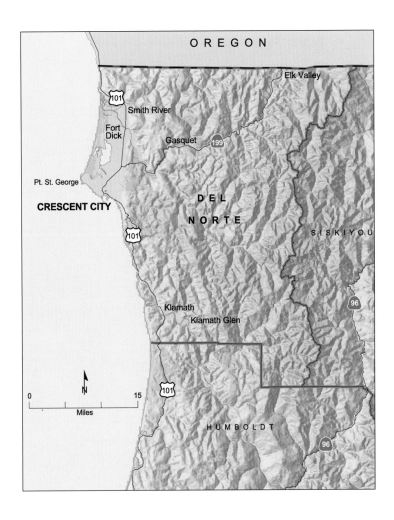

Del Norte County's two principal resource-based industries reflect its predominant natural features; the timber industry utilizes its extensive redwood forests, and commercial fishing depends on the anadromous fish rivers. Both rivers and forests provide recreational opportunities as well. The fertile Smith River plain supports dairy and Easter lily farming, in addition to providing a home to two-thirds of the county's population.

Until 1850, the Del Norte coast was home only to the Yurok and Tolowa Indians. The Yurok, who lived along the banks of the Klamath River, are recognized today as having had one of the most complex Indian civilizations in California. In 1828, members of Jedediah Smith's expedition came overland to the coast, but permanent settlers did not arrive until 1850; in that year the Klamath area was flooded with white men seeking gold. By 1855 the Yurok had been relocated to an Indian reservation along the Klamath. Today the Hoopa (originally Hupa) Reservation on the Trinity River is the home for both Yurok and Hupa, as well as several other tribes.

Klamath recalls its Indian past each summer during the salmon festival, when ceremonial dances are performed by local Indians, in addition to logging contests and general community fair festivities. The Klamath River is famous for salmon and steelhead trout fishing; summer and fall bring hundreds of anglers to its shores.

Crescent City was also settled in the 1850s, and became an important port for the transfer of gold from the strikes in southern Oregon, and, later, from those in the nearby Smith River watershed. At one point, all of Crescent Beach was staked for gold claims. By the 1870s, gold fever had subsided and lumber became Crescent City's major export. The timber industry in Del Norte had its most recent peak in the 1950s; now, with only a few stands of old-growth redwood remaining, cutting is slowing down to reach the rate at which new trees reach maturity.

The average temperature difference during the year along the Del Norte coast is only 13 degrees; however, coastal fog is prevalent in summer, and heavy rains in winter. In the past, huge storms have capsized many boats offshore, at times leaving no survivors. In 1964, as a result of an earthquake off Alaska, a tsunami hit Crescent City and destroyed the business district. That same year, a storm flooded the Klamath River, washing away the town of Klamath.

The towns of Smith River, Crescent City, and Klamath are now popular tourist areas, as is Redwood National Park, which draws visitors to see the world's tallest trees and large old-growth redwood groves. Within its boundaries are two state parks: Del Norte Coast Redwoods and Jedediah Smith Redwoods State Parks.

Headquarters for Redwood National Park are in Crescent City, 1111 2nd Street 95531; call: (707) 464-6101. For more information on Del Norte County, write: Del Norte Chamber of Commerce, 1001 Front Street, Crescent City 95531, or call: (707) 464-3174. For information about the Public Bus write: Del Norte Senior Center, 810 H Street, Crescent City 95331, or call: (707) 464-3069. For information on DIAL-A-RIDE, a service that carries passengers for an 8-10 mile area around Crescent City, call (707) 464-9314 or (707) 464-4314.

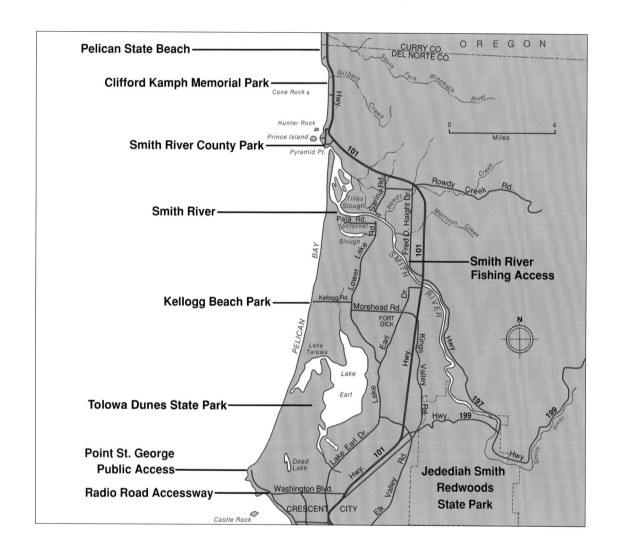

Pelican State Beach

Clifford Kamph Memorial Park

Smith River County Park

Smith River

Smith River Fishing Access

Kellogg Beach Park

Tolowa Dunes State Park

Point St. George Public Access

Radio Road Accessway

OREGON

CURRY CO.
DEL NORTE CO.

South Fork Winchuck River

Gilbert Creek

Cone Rock

Hunter Rock

Prince Island

Pyramid Pt.

101

Rowdy Creek Rd.

Morrison Creek

Tillas Slough

Pala Rd.
Yontocket
Slough

Carina Rd.

Rowdy Rd.

Fred D. Haight Dr.

SMITH RIVER

Lower Lake Rd.

Kellogg Rd.

Morehead Rd.

FORT DICK

Earl Dr.

Lake Earl

Lake Talawa

PELICAN BAY

Kings Valley Rd.

Hwy.

197

Hwy. 199

199

Smith River

Hwy.

Jedediah Smith Redwoods State Park

Dead Lake

Lake Earl Dr.

101 Hwy.

Washington Blvd.

CRESCENT CITY

Castle Rock

Elk Valley Rd.

0 Miles 4

N

Smith River

Smith River County Park

48

Del Norte County

OREGON BORDER TO CRESCENT CITY

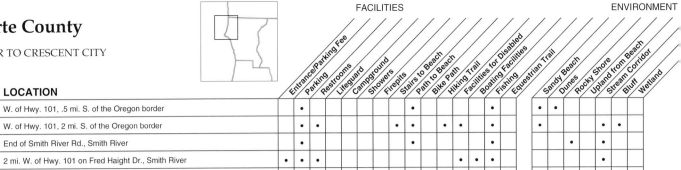

NAME	LOCATION	Entrance/Parking Fee	Parking	Restrooms	Lifeguard	Campground	Showers	Firepits	Stairs to Beach	Path to Beach	Bike Path	Hiking Trail	Facilities for Disabled	Boating Facilities	Fishing	Equestrian Trail	Sandy Beach	Dunes	Rocky Shore	Upland from Beach	Stream Corridor	Bluff	Wetland
Pelican State Beach	W. of Hwy. 101, .5 mi. S. of the Oregon border		•							•					•		•	•					
Clifford Kamph Memorial Park	W. of Hwy. 101, 2 mi. S. of the Oregon border		•	•		•		•		•	•				•		•				•	•	
Smith River County Park	End of Smith River Rd., Smith River		•							•					•				•	•			
Smith River Fishing Access	2 mi. W. of Hwy. 101 on Fred Haight Dr., Smith River	•	•	•									•	•	•					•			
Smith River	W. of Hwy. 101, 3.5 mi. S. of the town of Smith River		•					•						•	•					•			
Kellogg Beach Park	W. end of Kellogg Rd., 5.5 mi. S. of the Smith River mouth		•												•		•	•	•				
Lake Earl Wildlife Area / Tolowa Dunes State Park	W. of Lower Lake Rd. and Lake Earl Dr., Crescent City		•	•		•	•			•		•	•	•	•			•					•
Jedediah Smith Redwoods State Park	4241 Kings Valley Rd., Crescent City		•	•		•						•	•		•					•	•		
Point St. George Public Access	End of Radio Rd., Crescent City		•							•					•		•		•				
Radio Road Accessway	Off Radio Rd., 1 mi. S. of Point St. George, Crescent City		•						•	•					•		•		•				

PELICAN STATE BEACH: A road across from the Pelican Beach Motel leads to the sandy beach; five acres of undeveloped shoreline, grassy dunes, and abundant driftwood.

CLIFFORD KAMPH MEMORIAL PARK: Blufftop picnic tables and benches. Paths lead to the beach; several small bridges cross a creek that empties into the ocean. Restrooms are wheelchair accessible.

SMITH RIVER COUNTY PARK: Pebble beach located at the mouth of the Smith River; fishing and bird-watching from Pyramid Point.

SMITH RIVER FISHING ACCESS: Facilities include picnic tables, a concrete one-lane boat ramp, and wheelchair-accessible restrooms. Groceries, fishing supplies, raft rentals, and bait and tackle available.

SMITH RIVER: Smith River is part of California's Wild and Scenic River System, and one of seven stabilized river deltas in California. This year-round fishing river includes king salmon and silver salmon runs, steelhead trout and cutthroat trout, smelt, perch, and flounder. Upstream areas are safe for swimming. The Smith River National Recreation Area has over 65 miles of trails, four campgrounds – Big Flat, Grassy Flat, Panther Flat, and Patrick Creek; for reservations, call: 1-800-280-2267; interpretive center and headquarters: 707-457-3131.

Accommodations, guided boat trips, water-skiing equipment, and fishing and boating supplies are available off Hwy. 101 mainly around the town of Smith River. Trailer campgrounds include Ship Ashore Resort, Salmon Harbor Resort, and Valley View Camper-Trailer Park. Easter in July Festival celebrates Smith River as the "Easter Lily Capital of the World."

KELLOGG BEACH PARK: Wide, sandy beach, administered by the county, backed by extensive dune fields and marshes. Facilities include parking and picnic tables. Call: (707) 464-7237.

LAKE EARL WILDLIFE AREA / TOLOWA DUNES STATE PARK: The 5,500-acre wildlife area and adjacent 5,000-acre park includes Lake Earl and the smaller Lake Talawa; both are estuarine freshwater lagoons that are periodically open to the Pacific Ocean. The lakes are surrounded by an ancient dune complex that has evolved into many different ecological communities, providing habitat for a diverse array of birds, animals, and plants. The lakes are connected and contain salmon and trout. Motorboats are prohibited during waterfowl hunting season. Public boat access is located at the Pacific Shores subdivision south of Kellogg Rd. and on Lakeview Drive. The lakes are very shallow and can accommodate only small, shallow-draft boats. There is hiking through sand dunes, wooded ridges, pond and lake areas, wetlands, marshes, and flowered meadows; do not trespass on adjacent private property.

Facilities include a trail to a beach off Sand Hill Rd., approximately a one-mile walk through the dunes; Brush Creek Trail (1.5 miles long) with parking off Lake Earl Drive; a small parking area at Teal Point on Lake Earl, accessible by a public road in the Pacific Shores subdivision south of Kellogg Rd.; and parking and a trail to the beach at Crooked Creek, west of Lake Earl. North of the lakes, west of Yontocket Slough off Pala Rd., there is a parking area with restrooms and picnic tables. There is a Wildlife Center at 2591 Old Mill Rd. at the Department of Fish and Game Headquarters, which features a .82-mile-long self-guided nature walk, and access to other hiking and bicycle trails. Wheelchair accessible restrooms are available. For information, call: (707) 464-2523.

Tolowa Dunes State Park units are adjacent to the wildlife area; call: (707) 464-6101. Six walk-in environmental campsites are located ½-mile from the beach off Kellogg Rd.; no water available. There are fire rings, stoves, and chemical toilets. Fee charged; pay at Mill Creek or Jedediah Smith State Parks during the summer or 4241 Kings Valley Rd. at other times. For information, call: (707) 464-6101.

JEDEDIAH SMITH REDWOODS STATE PARK: Part of Redwood National Park, with thousands of acres of old-growth coast redwood forest. Hiking, kayaking, swimming, fishing, and gold panning; 20 miles of trails wind through groves of virgin coastal redwoods. Trails are closed to bicycles, although a nearby scenic seven-mile-long gravel road winds through the forest. 106 family campsites are available; for reservations, call: 1-800-444-7275. Park Visitor Center is located at the campfire center. Wheelchair-accessible restrooms summer only. For information, call: (707) 464-6101 ext. 5113.

POINT ST. GEORGE PUBLIC ACCESS: A long trail leads to the beach and rocky shore from the bluffs. Beds of littleneck, razor, and Washington clams; rock fishing and smelt netting. The St. George light can be seen at the end of St. George reef, seven miles offshore.

RADIO ROAD ACCESSWAY: An elevated wooden walkway leads to stairs that lead to the beach. Adjacent private property; do not trespass.

Tsunami

The Japanese word *tsunami* (harbor wave) refers to a series of waves caused by a sudden movement along a large portion of the ocean floor. Tsunamis are typically caused by distant underwater earthquakes, local earthquakes, large submarine slides, volcanic eruptions or meteorites.

The wavelength of a tsunami may be several hundred miles, with periods ranging from 15 minutes to an hour or more. A tsunami wave can travel at speeds in excess of 400 miles per hour. In the open ocean a tsunami is hardly noticeable. However, as the waves approach a shoreline, their height increases significantly; tsunami waves 75 feet high and greater have been observed.

Throughout history, coastal villages and towns located in the path of a tsunami have been completely destroyed. In California, the likelihood of a devastating tsunami with 75-foot waves is remote. Smaller tsunamis occur rather frequently. From 1950 to 2002, 36 tsunamis of varying intensities struck the West Coast of the U.S.

However, lesser tsunami waves are capable of causing considerable damage to developed portions of the California coastline. For example, tsunami waves generated by the 1964 Anchorage, Alaska earthquake made landfall at the Crescent City area in Del Norte County. A series of waves about 12 feet high washed inland as far as 1,600 feet. Twelve people drowned and the city's central business district was destroyed. It has since been rebuilt further inland.

NAME	LOCATION	Entrance/Parking Fee	Parking	Restrooms	Lifeguard	Campground	Showers	Firepits	Stairs to Beach	Path to Beach	Bike Path	Hiking Trail	Facilities for Disabled	Boating Facilities	Fishing	Equestrian Trail	Sandy Beach	Dunes	Rocky Shore	Upland from Beach	Stream Corridor	Bluff	Wetland
Pebble Beach Public Fishing Access	W. of Pebble Beach Dr., Crescent City	•	•						•	•					•		•		•			•	
Preston Island	Condor St. and Pebble Beach Dr., Crescent City	•					•			•					•				•				
Brother Jonathan Park/Vista Point	W. end of 9th St., Crescent City	•	•									•							•			•	
Street Ends Leading to the Beach	W. ends of 3rd, 5th, and 6th streets, Crescent City	•	•						•	•					•				•			•	
Crescent Lighthouse at Battery Point	S. end of "A" St., Crescent City	•	•							•					•				•			•	
Beach Front Park	W. of Howe Dr., Crescent City	•	•				•			•	•				•		•			•			
Shoreline Campground Accessway	W. of Sunset Circle Way, Crescent City	•	•	•		•	•	•		•					•		•				•		
Crescent City Harbor	W. of Hwy. 101 and Citizens Dock Rd., Crescent City	•	•	•						•				•	•								

PEBBLE BEACH PUBLIC FISHING ACCESS: Pebble Beach Dr. parallels the shore on a bluff overlooking the beach. There are several pull-outs for parking and viewing, and three stairways to the beach. A picnic table and historical marker of a Tolowa Indian settlement are located at the southern end.

PRESTON ISLAND: Rocky spit at the north end of Crescent City. A paved road leads down to the beach; picnic tables are available.

BROTHER JONATHAN PARK/VISTA POINT: Grassy blufftop park and cemetery memorializes the 1865 shipwreck of the steamer *Brother Jonathan*. Across the street is a vista point with a picnic table and bench.

STREET ENDS LEADING TO THE BEACH: A ¾ mile rocky beach can be reached by paths or stairways from the west ends of Third, Fifth, and Sixth streets. All accessways are signed, and there is parking at Third and Fifth streets.

CRESCENT LIGHTHOUSE AT BATTERY POINT: The museum in this offshore 1856 lighthouse is accessible and open, tides permitting, Apr.-Sept., Wed.-Sun.; fee charged. The museum contains a historical maritime collection; call: (707) 464-3089. There are picnic tables on the bluff and a view of Crescent City Harbor; on a clear day, St. George Reef Lighthouse is visible in the distance.

BEACH FRONT PARK: The Northcoast Marine Mammal Center on Howe Drive provides a rescue and rehabilitation service for the northernmost counties in California, along with educational materials and gift shop; call: (707) 465-6265. Adjacent Fred Endert Park has a public indoor swimming pool, playground, picnic tables, shuffleboard courts, and a putting green. B St. Pier, an 800-foot-long public fishing pier, is on B St. between Beach Front Park and Crescent Lighthouse at Battery Point. Del Norte County Visitors Center and Redwood National Park Information Center and Headquarters are both located just east of the park. Visitors Center: (707) 464-3174. Redwood National Park: (707) 464-6101.

SHORELINE CAMPGROUND ACCESSWAY: A public path runs along the levee at Elk Creek, through Shoreline Campgrounds from Sunset Circle Dr. to the beach. Picnic tables, campsites, and 189 trailer campsites with hookups are available. Call: (707) 464-2473.

CRESCENT CITY HARBOR: Now one of the safest harbors on the north coast, protected by large rock jetties reinforced with concrete. The harbor includes lumber docks for freighters; the Small Boat Basin with boat ramps, docks, and guest slips; and the Citizens Dock. This dock is a public wharf built in 1950 and rebuilt in 1987. The commercial fishing fleet, three fish processing plants, and a Coast Guard Station are based here; other facilities include seafood restaurants, diving and marine supplies, groceries, engine and hull maintenance areas, ramp, hoist, and fuel dock.

The southern part of Crescent City Harbor, Chart Room Marina, has a public two-lane boat ramp, sportsboat marina, two trailer parks, a sewage pump-out station, snack bar, and picnic tables. Access to the southern jetty via Whaler Island, where there is a view area.

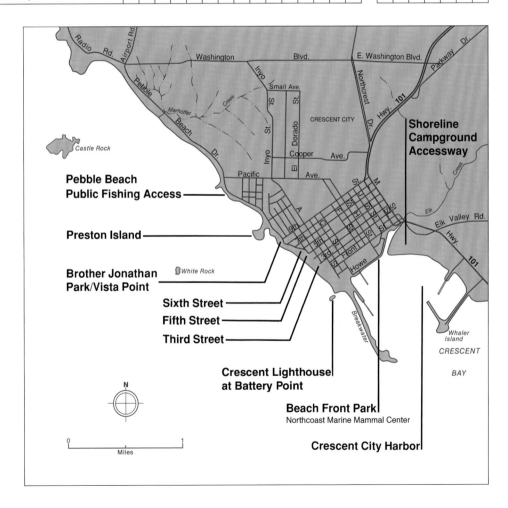

Wilson Creek Beach

Mill Creek Campgrounds

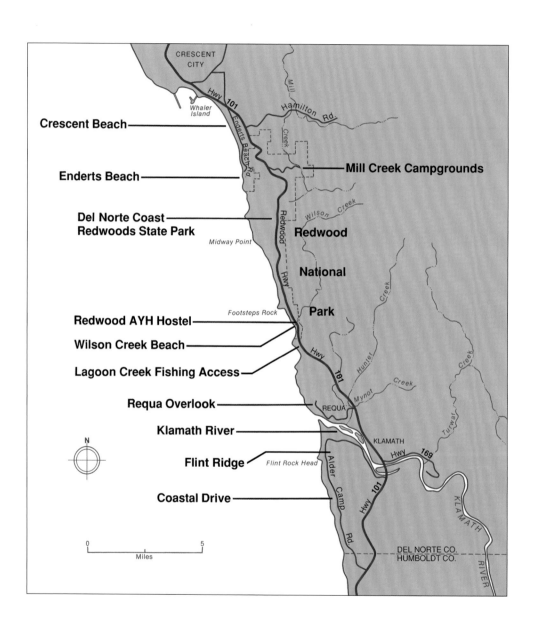

Crescent Beach

Enderts Beach

Del Norte Coast
Redwoods State Park

Mill Creek Campgrounds

Redwood AYH Hostel

Wilson Creek Beach

Lagoon Creek Fishing Access

Requa Overlook

Klamath River

Flint Ridge

Coastal Drive

Del Norte County

REDWOOD NATIONAL PARK / KLAMATH RIVER

NAME	LOCATION	Entrance/Parking Fee	Parking	Restrooms	Lifeguard	Campground	Showers	Firepits	Stairs to Beach	Path to Beach	Bike Path	Hiking Trail	Facilities for Disabled	Boating Facilities	Fishing	Equestrian Trail	Sandy Beach	Dunes	Rocky Shore	Upland from Beach	Stream Corridor	Bluff	Wetland
Redwood National Park	Coastal and inland areas between Crescent City and Orick		•	•		•		•		•	•		•	•	•		•		•	•	•	•	•
Crescent Beach	Off Enderts Beach Rd., 2 mi. S. of Crescent City		•	•				•					•		•		•						
Enderts Beach	End of Enderts Beach Rd., S. of Crescent City		•	•				•		•		•					•		•				
Del Norte Coast Redwoods State Park	7 mi. S. of Crescent City		•	•		•	•	•		•		•	•		•			•		•			
Mill Creek Campgrounds	E. of Hwy. 101, 5 mi. S. of Crescent City	•	•	•		•	•	•				•	•		•					•	•		
Redwood AYH Hostel-DeMartin House	14480 Hwy. 101 at Wilson Creek Rd., Klamath	•	•	•				•					•		•		•		•				
Wilson Creek Beach	W. of Hwy. 101, 5.5 mi. N. of Klamath		•	•				•							•		•		•				
Lagoon Creek Fishing Access	W. of Hwy. 101, 5 mi. N. of Klamath		•	•				•						•	•						•		•
Requa Overlook	Off Patrick Murphy Memorial Dr., 1.5 mi. from Requa Rd.		•											•								•	
Klamath River	W. of Hwy. 101 between Requa and Klamath Townsite		•				•								•	•						•	
Flint Ridge	Off Alder Camp Rd., S. of the Klamath River		•	•		•		•				•								•		•	
Coastal Drive	Alder Camp Rd., S. of the Klamath River		•																•			•	

REDWOOD NATIONAL PARK: Established in 1968 and expanded in 1978, the park preserves coastal redwoods, including the world's tallest trees, in virgin stands as well as new-growth forests. Within its boundaries are three state parks: Del Norte Coast Redwoods; Jedediah Smith Redwoods, which is inland from Crescent City; and Prairie Creek Redwoods in Humboldt County. The diverse habitats of the coastal area redwood forests, riparian woodlands, and grassy prairies support a comparable diversity of wildlife.

The fifty-mile-long park extends through both Humboldt and Del Norte counties. An extensive trail system throughout the park provides access to the park's natural features and recreational facilities. Headquarters for the park are in Crescent City, at Second and K streets: (707) 464-6101 ext. 5064. Check here for information on the new equestrian trails and horse camping, and for specific trail conditions, which may change due to heavy rains.

Redwood National Park in Del Norte County includes Crescent Beach, Crescent Beach Overlook, Enderts Beach, Lagoon Creek Fishing Access, Coastal Trail, Requa Overlook, and Coastal Drive. Del Norte Coast and Jedediah Smith Redwoods State Parks also lie within Redwood National Park's boundaries.

CRESCENT BEACH: Picnic tables, some modified for wheelchair use, are in a grassy area between the parking lot and the beach. There are also wheelchair-accessible restrooms. Crescent Beach Overlook, on a bluff off the west side of Hwy. 101, has parking and a picnic area.

ENDERTS BEACH: A 1/2-mile walk to primitive campsites and the beach; there are picnic tables and grills at the campsites. Daily tidepool walks are offered during the summer at low tide. Call: (707) 464-6101.

DEL NORTE COAST REDWOODS STATE PARK: Contains over 6,000 acres of redwoods, beaches, and rain forests. There is camping at Mill Creek east of Hwy. 101, and a trail system leading both inland and to the coast. Wilson Creek Beach is a day-use area at the southern end of the park. For information, call: (707) 464-6101 ext. 5064.

MILL CREEK CAMPGROUNDS: The campgrounds contain 145 campsites in a valley that is sheltered from ocean winds and fog. Some sites can accommodate trailers although there are no hookups; there

are also hike-in/bicycle sites. Hot showers, restrooms, and a campfire center are provided, as well as fire rings, picnic tables, and cupboards at each site. There are hiking and nature trails, and fishing in Mill Creek. A new interpretive program with campfires, a junior rangers program, and nature hikes are also available. Wilson Creek Beach is six miles from the campground. For reservations, call: 1-800-444-7275.

REDWOOD AYH HOSTEL-DEMARTIN HOUSE: The hostel has 30 beds; fee charged. Wheelchair-accessible; open year-round 7:30 AM-10:00 AM, 4:30 PM-9:30 PM. There is access to Wilson Creek Beach, 100 yards from the hostel. Information: (707) 482-8265.

Klamath River Mouth

WILSON CREEK BEACH: Sandy beach with abundant driftwood and tidepools; picnic tables. There is fishing in Wilson Creek.

LAGOON CREEK FISHING ACCESS: A freshwater lagoon with nearby picnic areas and a self-guided Yurok Loop nature trail. The lagoon is stocked with trout and is also popular for canoeing. This is the northern trailhead for the Coastal Trail, a four-mile trail that follows the bluffs south to Requa Overlook. Interpretive walks are conducted in the summer.

REQUA OVERLOOK: Views of the Klamath River valley and coastline; picnic tables and information displays. There is a trailhead to the Coastal Trail, which runs north along the cliffs to Lagoon Creek.

KLAMATH RIVER: California's second largest river is one of the best-known fishing areas in Northern California. From Klamath Townsite to the mouth of the river, motels and trailer parks, restaurants, fishing supplies, and boat rentals are available on both sides of the river.

FLINT RIDGE: Eleven primitive campsites; off Coastal Drive. To the south is High Bluff overlook and picnic area.

COASTAL DRIVE: An eight-mile drive that winds along the bluffs between the Klamath River and Prairie Creek Redwoods State Park in Humboldt County. The road is narrow and unpaved in parts and is not recommended for trailers.

Patrick's Point State Park

Humboldt County

Humboldt County in Northern California can claim many superlatives: the world's tallest tree, a redwood (*Sequoia sempervirens*) found near Orick; the westernmost point in California, Cape Mendocino in southern Humboldt; the second largest enclosed bay in California, Humboldt Bay; and perhaps one of the wettest places in California, Honeydew, near Humboldt Redwoods State Park. The annual Cross-Country Kinetic Sculpture Race, from Arcata to Ferndale, is also unique to Humboldt County. All of these can be found on or near the coast.

The coastal area of Humboldt can be divided generally into three areas. In the north, steep cliffs and bluffs, and the forests of Redwood National Park and commercial timberland dominate the landscape. From south of Trinidad to the Eel River, the coastal area consists mainly of low-lying fertile river deltas and bays; the more populated areas around the Mad River, Humboldt Bay, and the Eel River are centers of dairying, fishing, and timber processing. The southern portion of the county is characterized by steep ridges rising several thousand feet from the ocean, especially in the King Range National Conservation Area, where grass, brush, and Douglas-fir are the predominant vegetation.

Humboldt County was not widely settled by Europeans until 1850, following gold strikes on the Trinity River. Since it was easier to sail up the coast from San Francisco than to travel overland, port towns on the Humboldt coast vied for dominance. Trinidad, Arcata, and Eureka have each been the main port at different times.

Timber became an important industry early in the area's development; five months after the first European settlers arrived at Humboldt Bay, a sawmill was operating in Eureka. By 1870, most of the trees around Humboldt Bay were cut. Eventually, timber and fishing industries superseded mining, and Eureka ultimately became the main port.

Today timber and fishing, in addition to tourism, remain the most important industries. Eureka's fishing industry provides about 90% of the state's catch of Pacific Ocean shrimp and Dungeness crab. Chinook and silver salmon, sea and surf perch, several kinds of clams, oysters, rock and ling cod, cabezon, and rockfish are commonly found in the area. For the amateur angler, the numerous bays and rivers provide a fine opportunity for fishing.

The weather along the Humboldt coast, as in Del Norte County, is foggy in the summer and rainy in the winter; the temperature rarely drops below freezing. Spring and fall have clear skies and warm temperatures.

For visitor information, write the Eureka/Humboldt County Convention and Visitors Bureau at 1034 Second Street, Eureka 95501, call (707) 443-5097 or 1-800-346-3482, or see www.redwoodvisitor.org. The headquarters for Redwood National Park is at 1111 Second Street, Crescent City, CA 95531. Call (707) 464-6101 ext. 5064 or see www.nps.gov/redw/ for information on Redwood National Park and several state redwood parks that are managed jointly. For local fishing information, call (707) 444-8041. For Humboldt Transit Authority bus service information, including service to many of the sites listed in this guide, write: Humboldt Transit Authority, 6700A Highway 101 North, Eureka 95501, call: (707) 443-0826, or see www.hta.org.

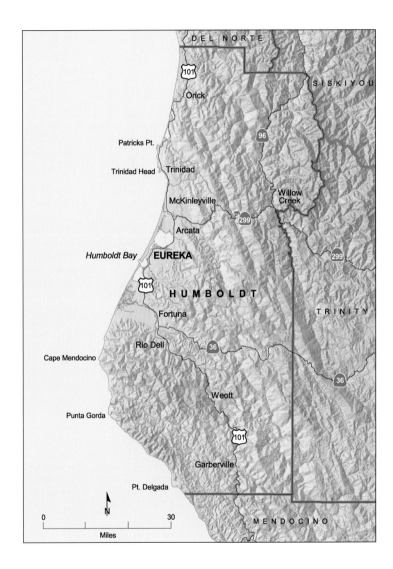

Redwoods

The magnificent coast redwoods *Sequoia sempervirens* grow in forests along the northern and central California coast, ranging from the Oregon border to the Santa Lucia Mountains south of Monterey. The trees are the tallest on earth, with many individuals achieving heights exceeding 300 feet, and trunks measuring up to 20 feet in diameter at their base. The coast redwood is named for its dark, red-colored wood, which is highly desirable for use as lumber. The trees can be identified by their great size, stranded cinnamon-brown bark, and flat, half-inch-long sharply pointed needles.

Coast redwoods grow only along the narrow strip of Pacific coastline, which is frequently covered by fog. The fog, by condensing and dripping off the redwoods' leaves, may supply up to ten inches of precipitation annually; this added moisture allows the redwoods to survive in an often arid environment. Within its range, the redwood is extremely successful and nearly always dominates other trees that grow nearby, such as Douglas-fir, Sitka spruce, tanbark oak, and madrone. The reasons for the coast redwood's dominance are its prolific regeneration through seed and trunk sprouting, and the facts that it is both highly resistant to insects and disease and well adapted to deal with fires and periodic flooding.

Although coast redwoods produce large amounts of seed, most regeneration in redwood forests is attributed to the trees' ability to sprout from the trunk or root system. Sprouting, a redwood reproductive method that is unique among cone-bearing trees, allows coast redwoods to dominate areas where seed regeneration would not normally occur, such as forests with leaf litter thick enough to prevent seeds from reaching mineral soil. These root and trunk sprouts are also unusually hardy and fast growing because they draw water and nutrients from the already well-developed root system of the parent tree.

Another unique survival mechanism of the coast redwoods is the ability to send out new vertical roots following silt deposition from periodic flooding. Since other trees that compete with redwoods for resources of soil and sun cannot tolerate the buildup of silts on their root systems, forests of pure redwood tend to dominate along stream floodplain areas.

Low intensity ground fires, which were once common in undisturbed natural forests, also favor coast redwoods. Redwoods have an unusually thick and non-resinous bark that is more resistant to fires than that of competing hardwoods and Douglas-fir. When periodic small fires occur, the effect is to eliminate trees other than redwoods; fire scars at the base of redwoods are evidence of previous fires that left these trees still standing.

The first commercial logging of redwoods occurred in the Oakland hills during the 1820s and in Monterey County in the 1830s. In 1848, the California gold rush began, resulting in an incredible demand for redwood lumber and providing the first impetus for settlement of the timber-rich north coast. The first lumber mill began operating in Eureka in 1850, and within ten years approximately 300 mills had been built along the coast, the majority being in Mendocino and Humboldt counties.

Excessive and often poorly managed logging prompted the establishment of park lands to preserve some of the most scenic redwood forests. Big Basin Redwoods State Park was purchased in 1902 to save one of the last remaining forests of old-growth redwood in the Santa Cruz Mountains. In 1906, President Theodore Roosevelt established the Monterey Forest Reserve, and designated Muir Woods in Marin County as a National Monument.

The largest park acquisition occurred in 1968 when Congress established the Redwood National Park in Humboldt and Del Norte counties. This 58,000-acre park near the coast encompassed the Mill Creek drainage area, and the Jedediah Smith, Del Norte Coast, and Prairie Creek Redwoods state parks. In 1978, the Redwood Creek drainage area near Orick was added to Redwood National Park, and later additions have increased the park's size to 112,000 acres.

Humboldt County

NORTHERN HUMBOLDT COAST

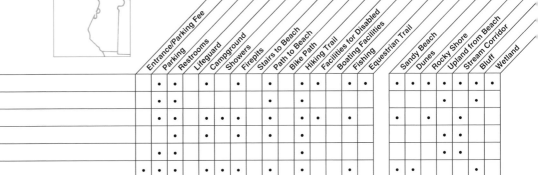

NAME	LOCATION	Entrance/Parking Fee	Parking	Restrooms	Lifeguard	Campground	Showers	Firepits	Stairs to Beach	Path to Beach	Bike Path	Hiking Trail	Facilities for Disabled	Boating Facilities	Fishing	Equestrian Trail	Sandy Beach	Dunes	Rocky Shore	Upland from Beach	Stream Corridor	Bluff	Wetland
Redwood National Park	Off Hwy. 101, N. of Orick		•	•		•		•		•		•	•		•	•	•	•	•	•	•	•	•
Coastal Drive	Off Alder Camp Rd., 10 mi. N. of Orick		•	•				•		•									•		•		
Prairie Creek Redwoods State Park	W. of Hwy. 101, 6.5 mi. N. of Orick		•	•		•	•	•		•		•	•		•			•		•		•	
Butler Creek Backpack Camp	Off West Ridge Rd., 9 mi. N. of Orick			•		•						•						•	•				
Fern Canyon	End of Davidson Rd.		•	•						•							•	•				
Gold Bluffs Campground	Off Davidson Rd., 5 mi. W. of Hwy. 101	•	•	•		•	•	•		•			•			•	•				•	
Elk Prairie Campground	W. of Hwy. 101, 6.5 mi. N. of Orick	•	•	•		•		•				•	•					•	•			

REDWOOD NATIONAL PARK: Over 112,000 acres of redwood forests near the coast in Humboldt and Del Norte counties. Hiking trails and scenic drives provide access to beaches, streamside redwood groves, and overnight campsites. The world's tallest tree, a redwood towering 367.8 feet above Redwood Creek, is located near Orick. The Redwood Information Center and Shuttle Bus Terminal at the southern end of the park is in Orick: (707) 464-6101 ex. 5265; park headquarters can be reached at (707) 464-6101 ex. 5064. For more information on Redwood National Park, see Del Norte County.

Redwood National Park in Humboldt County includes Coastal Drive, Carruthers Cove, Prairie Creek Redwoods State Park, and Gold Bluffs Beach.

COASTAL DRIVE: An eight-mile drive along coastal bluffs from the Klamath River in Del Norte County to the northern end of Prairie Creek Redwoods State Park; trailers not recommended. A guide to the drive is available at park information offices. The Carruthers Cove Trail leads from the south end of Coastal Drive north to Carruthers Cove Beach.

PRAIRIE CREEK REDWOODS STATE PARK: A 14,500-acre park with three campgrounds: Elk Prairie along Hwy. 101, Gold Bluffs Beach, and Butler Creek Backpack Camp; environmental campsites are available. Ossagon Trail, just north of West Ridge Rd., leads to the north end of Gold Bluffs Beach. The park features hiking and nature trails, including a guided nature trail for the blind and disabled, picnic areas, fishing streams, a herd of Roosevelt elk, and interpretive talks. Fees for camping. Call: (707) 464-6101 ex. 5300.

BUTLER CREEK BACKPACK CAMP: Located on the Coastal Trail. The camp is 7.5 miles from park headquarters, 2.6 miles from Fern Canyon, and is located on dunes in an alder meadows habitat near the ocean. There are three campsites with fire rings, large enough for small groups. All backpackers must park their vehicles and register at park headquarters. Call: (707) 464-6101.

FERN CANYON: Part of Prairie Creek Redwoods State Park. A short trail leads through the narrow canyon whose steep walls are covered with many species of ferns, including five-finger ferns, lady ferns, deer ferns, and chain ferns. Footbridges are installed after the heavy rains have passed each year, and are removed in the fall. The trail connects with the James Irvine Trail, which continues for four miles to the Park Headquarters. Call: (707) 464-6101.

GOLD BLUFFS CAMPGROUND: The sand below the high cliffs of Gold Bluffs Beach was mined for gold in the 1850s; area is a refuge for the Roosevelt elk. There are 25 campsites, each with a picnic table and cupboard; water nearby. Three hiker/bicyclist sites are at the east end of the campground. Tables and fire rings are provided; restrooms and cold showers in the adjacent campground. Campsites have direct access to the beach; camping fee. Call: (707) 464-6101.

ELK PRAIRIE CAMPGROUND: 75 campsites, each with a picnic table and cupboard; two wheelchair-accessible sites. Hiker/bicyclist campsites also available. Restrooms and nature trails adjacent to the campground are wheelchair accessible. Campfire programs in summer; camping fee. Call: (707) 464-6101.

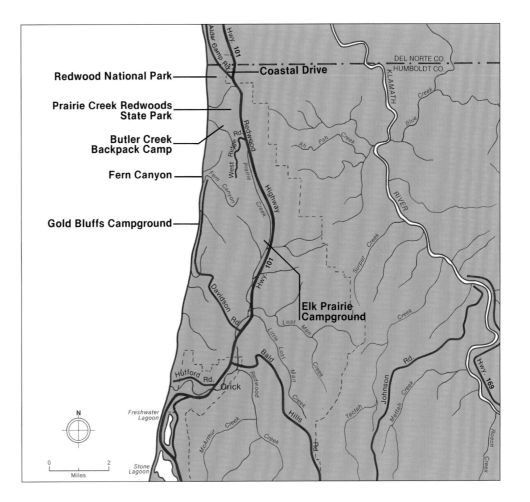

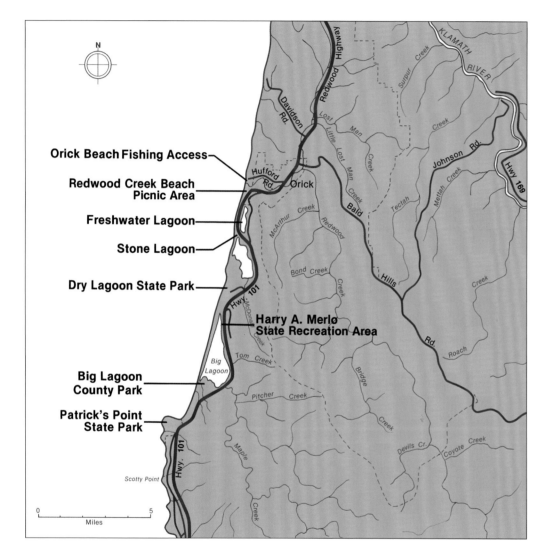

Orick Beach Fishing Access

Redwood Creek Beach Picnic Area

Freshwater Lagoon

Stone Lagoon

Dry Lagoon State Park

Harry A. Merlo State Recreation Area

Big Lagoon County Park

Patrick's Point State Park

Dry Lagoon State Park

Patrick's Point State Park

Humboldt County

ORICK TO PATRICK'S POINT

NAME	LOCATION	Entrance/Parking Fee	Parking	Restrooms	Lifeguard	Campground	Showers	Firepits	Stairs to Beach	Path to Beach	Bike Path	Hiking Trail	Facilities for Disabled	Boating Facilities	Fishing	Equestrian Trail	Sandy Beach	Dunes	Rocky Shore	Upland from Beach	Stream Corridor	Bluff	Wetland
Orick Beach Fishing Access	End of Hufford Rd., 1 mi. W. of Orick		•												•		•			•			
Redwood Creek Beach Picnic Area (Redwood National Park)	W. of Hwy. 101, 2 mi. S. of Orick		•	•									•				•			•			
Freshwater Lagoon	E. of Hwy. 101, 3 mi. S. of Orick		•	•		•								•	•		•						•
Stone Lagoon	Off Hwy. 101, 4.5 mi. S. of Orick	•	•	•		•								•	•		•	•					•
Dry Lagoon State Park	Off Hwy. 101, 6 mi. S. of Orick		•	•		•									•		•	•	•	•			•
Harry A. Merlo State Recreation Area	East Shore of Big Lagoon, off Hwy. 101		•												•					•			•
Big Lagoon County Park	Off Hwy. 101, 12 mi. S. of Orick	•	•	•		•		•	•					•	•	•	•	•					•
Patrick's Point State Park	W. of Hwy. 101, 5 mi. N. of Trinidad	•	•	•		•	•	•	•	•		•	•				•		•	•	•	•	

ORICK BEACH FISHING ACCESS: At the mouth of Redwood Creek within Redwood National Park on the north side; take the left fork when Hufford Rd. splits. The road ends on a dike at the creek mouth. The road is rough and narrow, and not recommended for campers or trailers. Fishing for salmon and steelhead trout.

REDWOOD CREEK BEACH PICNIC AREA: A developed area of Redwood National Park, located south of Redwood Creek, with beach access, picnic tables, and restrooms. Adjacent is the Redwood Information Center and Shuttle Bus Terminal, which provides information on local, state, and national parks in the area, and shuttle bus tickets. Entire facility is wheelchair accessible, including a boardwalk that leads from the center to Redwood Creek Estuary. Summer interpretive programs. Call: (707) 464-6101.

FRESHWATER LAGOON: Boat ramp at the north end; a small road off the east side of Hwy. 101 leads to the ramp and beach. The lagoon is stocked with trout. Camping is allowed west of Hwy. 101; first-come, first-served spaces can be obtained through the Redwood Information Center. Access to a sandy beach with driftwood; popular for surf fishing.

STONE LAGOON: Boat-in camping area of Dry Lagoon State Park. A short road at the north end of Stone Lagoon leads over a ridge to a spit separating the lagoon from the ocean; small boats may be launched at the end of this road. Six campsites, first-come, first-served, are accessible by boat from the east shore of the lagoon, where there is a boat ramp and parking area adjacent to the old Little Red Hen building. Camping fee. For information, call: (707) 488-2071.

DRY LAGOON STATE PARK: Includes the spit between Stone Lagoon and the ocean, the marshy Dry Lagoon, and the spit between Big Lagoon and the ocean. A day-use area just south of Dry Lagoon has four picnic sites. Migratory waterfowl and shorebirds feed at the lagoons, and anadromous fish enter the lagoons when the spits are breached during heavy rains. There are five environmental campsites; to reserve, call: (707) 488-2071.

HARRY A. MERLO STATE RECREATION AREA: Undeveloped. Provides day-use parking and access for sport fishing and waterfowl hunting in Big Lagoon. A nature trail that leads up Stagecoach Hill through wild azaleas is popular in spring. Call: (707) 488-2071.

BIG LAGOON COUNTY PARK: Located on a marsh that is adjacent to the ocean. Two-lane concrete boat ramp, with a 40-foot ten-ton limit. Facilities include restrooms, picnic tables, and 30 adjacent first-come, first-served primitive camping sites; camping and day-use fee. Steelhead fishing and swimming. Access from the park south to Agate Beach at Patrick's Point State Park, and to Big Lagoon Spit on the west side of Big Lagoon. Restrooms are wheelchair accessible. Call: (707) 445-7652.

PATRICK'S POINT STATE PARK: A 640-acre park of forests and meadows located on bluffs that are rimmed by rocky headlands and a sandy beach. There are 123 campsites including group and hiker/bicyclist sites, 16 picnic sites, and a campfire center. Nature and hiking trails include the two-mile Rim Trail, which follows an old Indian trail along the bluffs, and a self-guided nature trail through the octopus tree grove. Some trails and restrooms are wheelchair accessible. A small museum features natural and Indian history; nearby is a Yurok Indian village reconstruction and the Native American Plant Garden. Shorebirds and songbirds are abundant, and migrating gray whales can be observed from the bluffs. The three-mile coastline includes Agate Beach, where small pieces of agate and jade can be found, and rocky areas with tidepools. Sea lions can be seen offshore. Camping and day-use fee. Call: (707) 677-3570.

Stone Lagoon, Dry Lagoon State Park

Mad River Beach County Park

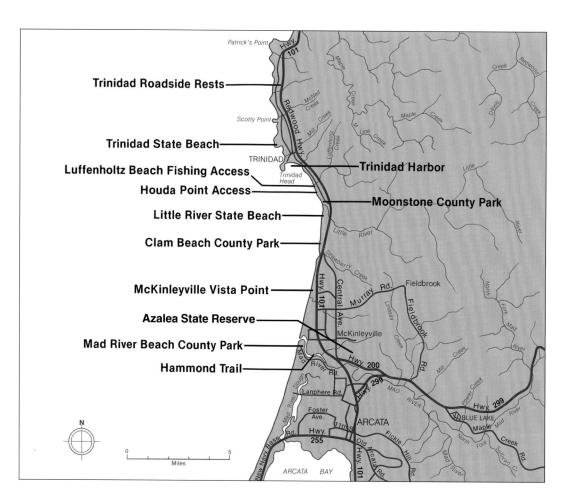

Trinidad Roadside Rests

Trinidad State Beach

Luffenholtz Beach Fishing Access

Houda Point Access

Little River State Beach

Clam Beach County Park

McKinleyville Vista Point

Azalea State Reserve

Mad River Beach County Park

Hammond Trail

Trinidad Harbor

Moonstone County Park

Humboldt County

TRINIDAD TO THE MAD RIVER

NAME	LOCATION	Entrance/Parking Fee	Parking	Restrooms	Lifeguard	Campground	Showers	Firepits	Stairs to Beach	Path to Beach	Bike Path	Hiking Trail	Facilities for Disabled	Boating Facilities	Fishing	Equestrian Trail	Sandy Beach	Dunes	Rocky Shore	Upland from Beach	Stream Corridor	Bluff	Wetland
Trinidad Roadside Rests	Hwy. 101, 2 and 4 mi. N. of Trinidad	•	•										•	•					•				
Trinidad State Beach	Off Trinity St., N. of Main St., Trinidad	•	•				•		•				•				•		•			•	
Trinidad Harbor	W. end of Edwards St., E. of Trinidad Head	•	•							•				•	•		•		•				
Luffenholtz Beach County Fishing Access	W. of Scenic Dr., 2 mi. S. of Trinidad	•	•				•	•	•					•			•		•				•
Houda Point Access	W. of Scenic Dr., 2.5 mi. S. of Trinidad	•								•							•		•				•
Moonstone County Park	W. of Scenic Dr., 3 mi. S. of Trinidad	•															•						
Little River State Beach	W. of Hwy. 101, 4 mi. S. of Trinidad	•								•					•		•	•				•	
Clam Beach County Park	W. of Hwy. 101, 3.5 mi. N. of McKinleyville	•	•			•				•				•	•	•	•	•				•	
McKinleyville Vista Point	Hwy. 101, 2 mi. N. of McKinleyville	•																	•			•	
Azalea State Reserve	E. of Hwy. 101, on North Bank Road	•	•									•								•			
Mad River Beach County Park	End of Mad River Rd., W. of Hwy. 101	•	•				•		•				•	•	•	•	•	•			•		
Hammond Trail	W. of Hwy. 101, from Clam Beach S. to the Mad River											•			•	•					•	•	

TRINIDAD ROADSIDE RESTS: Northbound and southbound areas. Dog runs, nature exhibits, paths, picnic tables, and benches; trailer sanitation station.

TRINIDAD STATE BEACH: A day-use park on the north side of Trinidad Point, adjacent to the town of Trinidad. The northern unpaved parking lot off Stagecoach Rd. provides trail access to the headlands and to College Cove, a sandy beach. Facilities include a paved parking area, restrooms, picnic tables, and grills at the southern end of the park, adjacent to Mill Creek; trails lead to the creek, and down the bluff to the beach. Trails from the southern beach lead to the harbor area and along Trinidad Bay. Beach access is by a footpath leading from the Humboldt State University Marine Laboratory, which has a wheelchair-accessible visitors area with marine exhibits and an aquarium. Open 9 AM-5 PM Monday-Friday, 10 AM-5 PM weekends, except in summer; call: (707) 826-3671.

The Humboldt North Coast Land Trust, a nonprofit land trust based in the Trinidad area, owns and operates a number of coastal trails and access points in the Trinidad area and other parts of Humboldt County. For information, send a self-addressed, stamped envelope to: Humboldt North Coast Land Trust, P.O. Box 457, Trinidad 95570.

TRINIDAD HARBOR: Restaurant, moorings, charter boats, and fuel dock at Bob's Boat Basin. The marine railway, used for launching small boats, is open May-October. For information, call: (707) 677-3625. Trails lead from Edwards St., Wagner St., and the end of Parker Creek Rd. to the harbor and bay.

LUFFENHOLTZ BEACH COUNTY FISHING ACCESS: Turn south off Main St. onto Trinidad Scenic Dr., just west of Hwy. 101 in Trinidad. Scenic overlook with picnic tables and restrooms. Access to the beach is along steep, rocky cliffs.

HOUDA POINT ACCESS: Roadside parking area with benches; trails lead to two pocket beaches south of Luffenholtz Beach. Owned and operated by Humboldt North Coast Land Trust.

MOONSTONE COUNTY PARK: Broad, sandy beach, near Luffenholtz Beach County Fishing Access; good clamming

LITTLE RIVER STATE BEACH: North end of Clam Beach County Park. Broad sandy beach, with access to the Little River delta at the north end.

CLAM BEACH COUNTY PARK: Overnight roadside stops are permitted; facilities include restrooms and picnic tables. Good clamming area; ponds east of Hwy. 101 on the frontage road are stocked with trout. Camping fee.

McKINLEYVILLE VISTA POINT: Picnic table; panoramic view north to Trinidad.

AZALEA STATE RESERVE: Thirty-acre park provides restrooms, picnic areas, and self-guided nature trails. Call: (707) 677-3570.

MAD RIVER BEACH COUNTY PARK: Take the Janes Rd. exit off Hwy. 101, turn right on Heindon Rd., left on Iverson Rd., and right on Mad River Road. Do not trespass or allow dogs onto the adjacent private farm lands. Two parking areas; one leads to the river and boat ramp, the other is near the ocean beach. Picnic table and restrooms are located near the boat ramp.

HAMMOND TRAIL: Multi-use recreational trail that runs near the coast between Clam Beach and the Mad River. Includes an equestrian/hiking loop trail from Clam Beach to McKinleyville with a bridge across the Mad River, ½-mile east of Mad River Beach County Park; connects bicycle lanes between Clam Beach through McKinleyville to Arcata.

Trinidad State Beach

Arcata Marsh

North Spit, Samoa Public Access

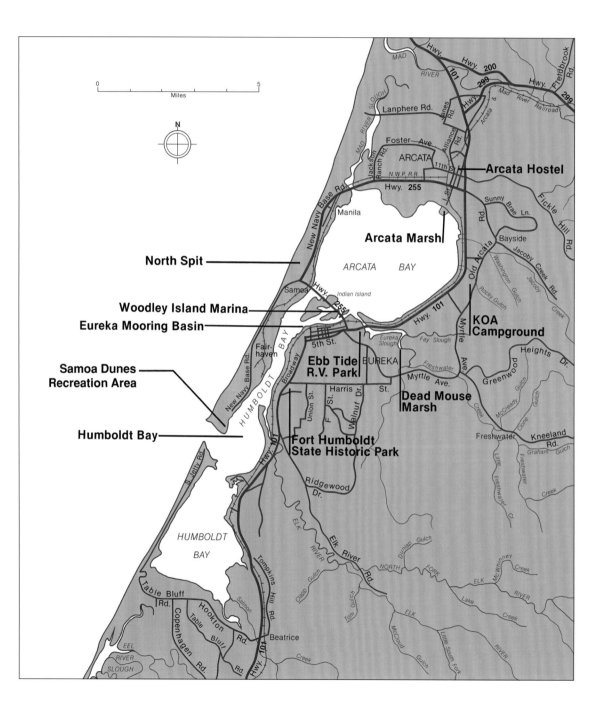

Humboldt County

HUMBOLDT BAY

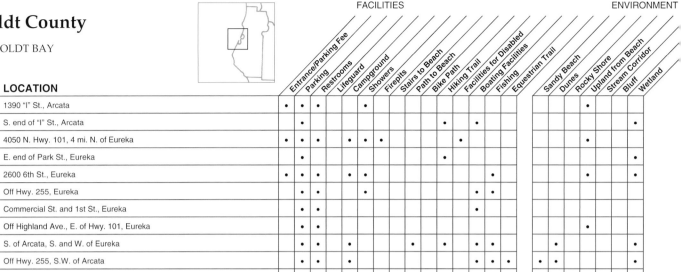

NAME	LOCATION	Entrance/Parking Fee	Parking	Restrooms	Lifeguard	Campground	Showers	Firepits	Stairs to Beach	Path to Beach	Bike Path	Hiking Trail	Facilities for Disabled	Boating Facilities	Fishing	Equestrian Trail	Sandy Beach	Dunes	Rocky Shore	Upland from Beach	Stream Corridor	Bluff	Wetland
Arcata Hostel	1390 "I" St., Arcata	•	•	•			•													•			
Arcata Marsh	S. end of "I" St., Arcata		•									•		•									•
KOA Campground	4050 N. Hwy. 101, 4 mi. N. of Eureka	•	•	•		•	•	•					•							•			
Dead Mouse Marsh	E. end of Park St., Eureka		•									•											•
Ebb Tide R.V. Park	2600 6th St., Eureka	•	•	•		•	•								•					•			•
Woodley Island Marina	Off Hwy. 255, Eureka		•	•			•							•	•								
Eureka Mooring Basin	Commercial St. and 1st St., Eureka		•	•										•									
Fort Humboldt State Historic Park	Off Highland Ave., E. of Hwy. 101, Eureka		•	•																		•	
Humboldt Bay	S. of Arcata, S. and W. of Eureka		•	•		•				•		•		•	•		•						•
North Spit	Off Hwy. 255, S.W. of Arcata		•	•		•								•	•	•	•	•					•
Samoa Dunes Recreation Area	Off Hwy. 255, end of North Spit		•	•		•		•										•					•

Humboldt Bay

ARCATA HOSTEL: Open mid-June to mid-August. Facilities include 15 beds, showers, kitchen, and bike storage; reservations are suggested. Fee. For information, write or call: 1390 "I" Street, Arcata 95521; (707) 822-9995.

ARCATA MARSH: Restored freshwater marsh and wildlife sanctuary. Trails through the area are used for walking, jogging, and bird-watching. Bird blinds have been constructed, and Audubon members lead Saturday morning bird-watching tours. Call: (707) 826-7031. The Arcata boat ramp, a two-lane concrete ramp, is adjacent to the marsh area and has a picnic table. Another trail leads through the marsh along the dikes, just west of the oxidation pond; the parking lot is on South G Street, adjacent to the Jacoby Creek Unit of the Humboldt Bay National Wildlife Refuge.

KOA CAMPGROUND: 165 sites; 29 sites are specifically for tent campers; 2 of the restrooms are wheelchair accessible. Facilities include: picnic tables, barbeques, laundry, store, snack bar, and recreation hall; fee. For more information, call: (707) 822-4243 or 1-800-562-3136.

DEAD MOUSE MARSH: Trails along the dikes lead into the marsh area; good bird-watching.

EBB TIDE R.V. PARK: 81 trailer sites. Fee; weekly and monthly rates are available. Laundry facilities; paths to nearby fishing areas. For more information, call: (707) 445-2273.

WOODLEY ISLAND MARINA: 237-slip marina. Two-ton hoist, work area, and sidewalk area with benches; no boat ramp. Hot pay showers and laundry facilities for tenants only; coffee shop. The rest of the island is a wildlife reserve; no public access. For more information, call: (707) 443-0801.

EUREKA MOORING BASIN: A small boat harbor with a one-lane concrete boat ramp, fuel docks, pumpout station, and berths. Picnic tables and the Humboldt Yacht Club are located in the harbor area. Adjacent 1st and 2nd streets are part of the ongoing restoration of the Victorian buildings and businesses on the waterfront, including mini-parks and walkways; several cross-streets end at the bay. The Humboldt Bay Harbor Cruise leaves from the foot of C Street, mid-April through October; information: (707) 445-1910.

FORT HUMBOLDT STATE HISTORIC PARK: Logging exhibit and military and logging museums with picnic areas and views of Humboldt Bay. Open 8:30 AM-4 PM daily. For information, call: (707) 445-6567.

HUMBOLDT BAY: A number of access points exist around Humboldt Bay. The Humboldt Bay National Wildlife Refuge includes most of the south bay and some mudflats, marshes, and islands in the north bay; the snowy egret and the great blue heron are commonly sighted birds. For information, call: (707) 733-5406. The Lanphere Dunes unit of the National Wildlife Refuge is accessible by permit only. Guided nature walks are led by Friends of the Dunes on Saturdays. Call: (707) 444-1397.

The Eureka Waterfront Boardwalk, centered on F Street, connects the bayfront to the historic downtown district of Eureka. The City has also made improvements to Waterfront Dr. to provide vehicular and pedestrian access from Washington St. to F St. and from L St. to T St. The City's Adorni Memorial Recreational Center is located on the bay at the foot of J St. This facility includes a boardwalk access along the shore. Other bay front access points include the ends of Commercial St. and a pier at Del Norte Street. The City of Eureka is conducting habitat restoration between the Bayshore Mall and Del Norte Street, and there is a pedestrian trail along the marsh; for information, call (707) 441-4207. Camping for R.V.'s only is available at the nearby Redwood Acres Fairground, 3750 Harris St.; for information, call: (707) 445-3037. The Maritime Museum is located at 423 First St., open Tuesday through Saturday 12 noon-4 PM. For information, call: (707) 444-9440.

NORTH SPIT: A long spit separates Arcata Bay from the ocean. On the ocean side, the sandy beach can be reached from several parking areas off Hwy. 255 and New Navy Base Road. Dunes are between the road and the beach. The Samoa Public Access on the bay has a ten-ton 40-foot capacity concrete boat ramp, and camping is allowed. Fee.

Manila Community Park, at Peninsula Dr. and Victor Blvd. in Manila, has a playground, baseball diamond and playing fields, picnic tables, and access to the bay and marshes. Horses are allowed on trails near the bay. The Coast Guard Station at the south end of New Navy Base Rd. provides access to the jetty at the end of the spit (May-Sept. only), to the southern dune area, and to small coves along the bay. The jetty is a popular fishing area; waves can be hazardous.

SAMOA DUNES RECREATION AREA: An off-road vehicle recreation area located adjacent to the Coast Guard station and immediately north of the mouth of Humboldt Bay. Facilities include trails for off-road vehicles and motorcycles, campgrounds, picnic tables, fire rings, and a parking area; the facilities are managed by the Bureau of Land Management. Call: (707) 825-2300. Samoa Boat Ramp County Park with a boat ramp and 20 site campground is adjacent. Fee. Call: (707) 445-7651.

Salmon and Steelhead Trout

Steelhead trout, king salmon, and silver salmon are three of the most important Northern California fish species; all three are anadromous, meaning that they migrate from the ocean into rivers to spawn.

King (chinook) salmon are found in the ocean from Monterey northward, and run upstream from mid-July through October. Average weight at spawning is 20 lbs. To spawn, females dig nests in the gravel bottoms of cool streams and lay their eggs, which are immediately fertilized by males. All Pacific salmon die shortly after spawning.

King salmon eggs hatch in 50-60 days, and several months later the young migrate to the ocean; when mature, usually at age three or four, they return to the same stream to spawn. Most king salmon migrate into fresh water in the fall and spawn from October-January. There are, however, some spring-run king salmon in certain rivers, and even some winter-run fish in the Sacramento River.

Silver salmon are mostly found north of Monterey and migrate to fresh water in fall and early winter, just before they spawn. Their habits are similar to those of the king salmon except that silver salmon usually weigh 7-12 lbs. at spawning, prefer smaller streams than king salmon, and their young remain in fresh water a year (occasionally two years) before migrating to sea in spring. Most silvers mature at age three.

King salmon are caught by anglers in the ocean from Avila Beach (in San Luis Obispo County) north to Oregon, and also in the mouths and tidewater areas of northern rivers such as the Klamath, Smith, and Eel. Silver salmon have been designated as a threatened species in most of California's coastal waters, where they may not be taken by anglers; check with the Department of Fish and Game for current regulations.

Steelhead trout are found from San Luis Obispo County north to Alaska and are rarely caught at sea; they are found in smaller streams than those that salmon enter as well as in the larger rivers. Steelhead enter streams from June-November to spawn and, unlike salmon, do not necessarily die after spawning; steelhead may live up to eight years and spawn as many as five times.

After hatching, steelhead young stay in fresh water for one or two seasons and then migrate to the sea. Steelhead usually weigh 2-10 lbs. at maturity, and resemble rainbow trout when in fresh water, but change coloration, as do salmon, when entering salt water.

According to the California Department of Fish and Game, the habitats of these anadromous fish have been seriously degraded and hundreds of thousands of fish have died in recent years due to such things as the loss of stream beds to gravel and gold dredging operations; the building of dams and unscreened irrigation diversions; improper logging and road building operations; untreated industrial and domestic wastes; increased amounts of fertilizers from return irrigation waters; higher water temperatures; and drought conditions.

In an effort to mitigate the effects of these human activities, the Department of Fish and Game has built fish ladders to enable migrating fish to reach spawning areas upstream from dams, has aided federal agencies in constructing and operating fish hatcheries, and is currently working with the logging and lumber industry to remove barriers and pollutants from spawning streams.

Anglers should consult the current regulations published by Fish and Game regarding salmon and steelhead fishing. The *California Ocean Sport Fishing Regulations* are available from:

Department of Fish and Game
Marine Region Headquarters
20 Lower Ragsdale Drive
Monterey, CA 93940   (831) 649-2870

The *California Freshwater Sport Fishing Regulations* are available from:

Department of Fish and Game
619 Second Street
Eureka, CA 95501   (707) 445-6493

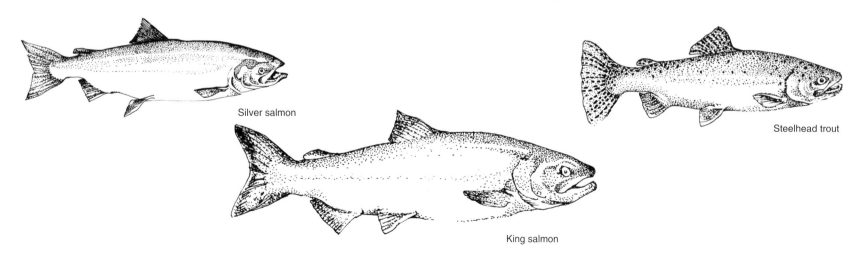

Silver salmon

Steelhead trout

King salmon

Humboldt County

HUMBOLDT BAY / EEL RIVER

NAME	LOCATION	Entrance/Parking Fee	Parking	Restrooms	Lifeguard	Campground	Showers	Firepits	Stairs to Beach	Path to Beach	Bike Path	Hiking Trail	Facilities for Disabled	Boating Facilities	Fishing	Equestrian Trail	Sandy Beach	Dunes	Rocky Shore	Upland from Beach	Stream Corridor	Bluff	Wetland
King Salmon Resort	Off Buhne Dr., W. of Hwy. 101, 1 mi. N. of Fields Landing	•	•	•		•								•	•		•						•
Fields Landing County Boat Launch	Foot of Railroad Ave., Fields Landing		•	•										•	•								•
South Spit and Jetty	S. Jetty Rd., N. of Table Bluff Rd.		•												•		•	•	•				•
Table Bluff County Park	Table Bluff Rd. off Hookton Rd., W. of Hwy. 101 at Beatrice		•																			•	•
Crab County Park	W. end of Cannibal Island Rd., W. of Hwy. 101, Loleta	•	•						•						•		•				•		•
Eel River	W. of Hwy. 101 at Fernbridge		•												•		•				•		
Humboldt County Fairgrounds Campground	W. of Main St., Ferndale	•	•			•	•	•										•					
Centerville Beach County Park	W. end of Centerville Rd., 5 mi. W. of Ferndale	•	•				•							•	•		•	•					
Headwaters Forest Reserve	South end of Eel River Road	•	•									•								•			

KING SALMON RESORT: Take the King Salmon Dr. exit off Highway 101. Sport fishing charter boats, trailer camping, fuel dock and hoists, and fishing and boating equipment are available at Johnny's Marina and R.V. Park, (707) 442-2284, and E-Z Landing R.V. Park and Marina, (707) 442-1118. A picnic area with restrooms is adjacent to the PG&E power plant on King Salmon Dr.

FIELDS LANDING COUNTY BOAT LAUNCH: Public two-lane concrete boat ramp on Humboldt Bay. The annual Cross-Country Kinetic Sculpture Race begins the bay crossing portion here.

SOUTH SPIT AND JETTY: Day use managed by the Bureau of Land Management (BLM). Public fishing all year. Dangerously large waves; exercise caution when walking on the jetty. For information, call: (707) 825-2300.

TABLE BLUFF COUNTY PARK: The park provides views of southern Humboldt Bay, including the south spit and jetty, and the Humboldt Bay National Wildlife Refuge.

CRAB COUNTY PARK: Take Loleta Dr. to Cannibal Island Rd. to get to the park; fishing area with access to the ocean, dunes, and river delta.

EEL RIVER: Within the Eel River delta, there are many roads that end at gravel beds along the river. Adjacent land is private property; do not trespass. Starting from the mouth of the river and moving east, the following streets end at the river: Camp Weott Rd., Dillon Rd., Tappendorf Lane, Sage Rd., and Singley Rd. near Fernbridge, and Sandy Prairie Rd. near Fortuna. One must cross through a farm gate on Sage Rd. to get to the river. Pedrazzini County Park is off Cannibal Island Rd. at the bridge over Eel River on Cock Robin Island Road.

HUMBOLDT COUNTY FAIRGROUNDS CAMPGROUND: Camping; reservations necessary during the fair in August. Fee; hookups, dump station, restrooms and showers are available. For more information, call: (707) 786-9511. The campground is in Ferndale, noted for its many ornate Victorian buildings, which were constructed during the late 1800s. Ferndale also hosts two kinetic sculpture races in the spring.

CENTERVILLE BEACH COUNTY PARK: Four acres of park with access to the beach, reaching from the mouth of the Eel River to False Cape in the south. Private farmlands to the east; do not trespass. Four-wheel drive vehicles allowed on the beach; dangerous rip currents.

HEADWATERS FOREST RESERVE: North access is at the end of Elk River Road. A 5 ½ mile trail follows an abandoned logging road to the old-growth forest. Limited access to the southern part of the reserve is available May-Nov. 15. For reservations, call the Bureau of Land Management (BLM): (707) 825-2300.

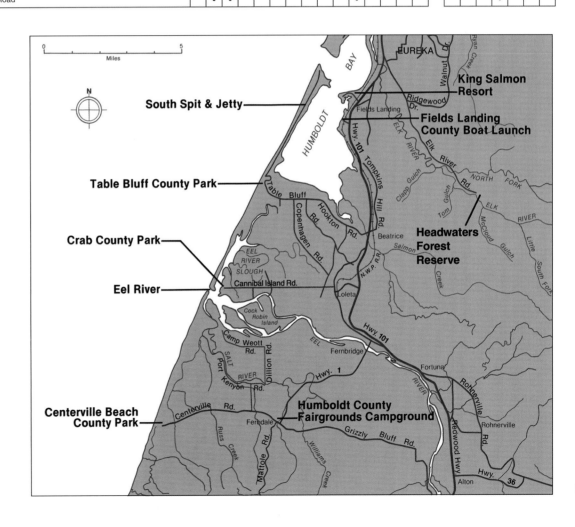

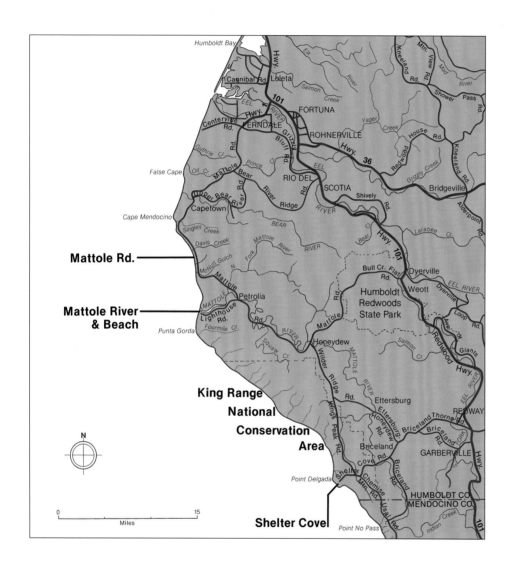

Mattole Rd.

Mattole River & Beach

King Range National Conservation Area

Shelter Cove

N

0   15
Miles

Humboldt Bay

Cannibal Rd.   Loleta

Hwy. 1

101

FORTUNA

Centerville Rd.   FERNDALE

ROHNERVILLE

False Cape

Oil Cr.

RIO DEL   Hwy. 36

Cape Mendocino   Capetown

Singley Creek

Davis Creek

McNulty Gulch

Petrolia

Lighthouse Rd.

Punta Gorda   Fourmile Cr.

Honeydew

Ettersburg

Briceland

Briceland Cove Rd.

Point Delgada

HUMBOLDT CO.
MENDOCINO CO.

Point No Pass

Elk River

Salmon Creek

Kneeland Rd.   Mtn. View Rd.

Mad River

Shower Pass Rd.

Yager Creek

House Rd.

Redwood Rd.   Kneeland Rd.

SCOTIA   Shively Rd.   Bridgeville

Alderpoint Rd.

BEAR RIVER   Bear Cr.   Larabee Cr.   Hwy. 101

Mattole River

Bull Cr. Flat Rd.   Dyerville   EEL RIVER

Humboldt Redwoods State Park   Weott   Dyerville Loop Rd.

Salmon Cr.   Giants   Redwood Hwy.   EEL RIVER

REDWAY

Thorne Rd.   Briceland Rd.

GARBERVILLE   Hwy. 101

Chemise Mtn. Rd.   Usal Rd.   Indian Creek

Falls below Kings Peak, King Range National Conservation Area

Humboldt County

SOUTHERN HUMBOLDT COAST

NAME	LOCATION	Entrance/Parking Fee	Parking	Restrooms	Lifeguard	Campground	Showers	Firepits	Stairs to Beach	Path to Beach	Bike Path	Hiking Trail	Facilities for Disabled	Boating Facilities	Fishing	Equestrian Trail	Sandy Beach	Dunes	Rocky Shore	Upland from Beach	Stream Corridor	Bluff	Wetland
Mattole Road	Between Singley Creek and McNutt Gulch									•							•	•	•			•	
Mattole River and Beach	W. end of Lighthouse Rd., 5 mi. W. of Petrolia	•	•								•				•	•	•	•			•		•
King Range National Conservation Area	Honeydew to Point No Pass in Mendocino County	•	•	•		•					•			•			•	•	•	•			
Shelter Cove	W. end of Shelter Cove Rd.	•	•		•		•	•		•		•	•			•	•	•	•		•		

MATTOLE ROAD: The Mattole Rd. between Cape Town and Petrolia includes an 8-mile stretch that runs along the coast at ocean level. The beach is accessible at 4 locations between Singley Creek and McNutt Gulch: at mile marker 25.02, at mile marker 24.02 near Devils Gate, 500 feet south of mile marker 23.07, and on the north side of the Russell Chambers Bridge across McNutt Gulch, just before the road turns inland, where there is a path to the beach through a gap in the fence. The northern paths are 50 feet wide, and the path at McNutt Gulch is 20 feet wide. Private property is adjacent to these paths; do not trespass. Shoulder parking at each location.

MATTOLE RIVER AND BEACH: Lighthouse Rd. from Petrolia follows the Mattole River down to the beach. A trail leads south from the beach into the King Range, past an abandoned Coast Guard Lighthouse at Punta Gorda. It is possible to hike along the beach south to Shelter Cove; the distance is about 24 miles. Watch for rattlesnakes in driftwood and rocky areas.
Transit: Lost Coast hikers shuttle offers service to the mouth of the Mattole; call: (707) 986-9909.

KING RANGE NATIONAL CONSERVATION AREA: 62,200 acres of steep coastal mountains and shoreline administered by the Bureau of Land Management (BLM). This primitive area, which constitutes part of the "Lost Coast," has very few roads; 6,700 acres of the land is still privately held. BLM operates six camping areas year round. Facilities include picnic tables, fire rings, and restrooms. Water is available at Wailaki, Nadelos, and Mattole Camping areas. The Lost Coast Trail begins at the Mattole River in the King Range NCA and follows the coastline south to Black Sands Beach in Shelter Cove; the trail then climbs Chemise Mountain and enters the Sinkyone Wilderness State Park, ending at Usal Beach. In all, eighty miles of hiking trails traverse the King Range NCA. Permits are not required for individual backpackers and groups of six or fewer, but permits are required for outfitters and organized groups. Campfire permits are required all year, and fire restrictions may be in effect during the summer months; check with BLM. All hikers are encouraged to sign trail registers near trailheads.

There are two beaches accessible by road within the King Range, in addition to Shelter Cove: Mattole Beach and Black Sand Beach off Beach Dr. north of Shelter Cove. Since road conditions vary during the year, check with BLM for the best routes into the King Range. Write: King Range National Conservation Area Project Office, P.O. Box 189, Whitethorn, CA 95589; call: (707) 986-5400; or see www.ca.blm.gov/arcata.

SHELTER COVE: Take Briceland-Thorne Rd. from Redway to Shelter Cove Rd.; Shelter Cove is a privately owned community within the King Range National Conservation Area. The main cove at the end of Machi Rd. has a sandy beach and a public boat launch managed by the Humboldt County Harbor District. Six coastal access facilities are managed by the BLM: Mal Coombs Park at Point Delgada with parking, restrooms, and stairs to a rocky intertidal zone; Seal Rock and Abalone Point; two smaller, unnamed parcels of public land along Lower Pacific Dr. (look for the split rail fences); and Black Sands Beach Trailhead and Parking Area, with restrooms and drinking water. The historic Cape Mendocino Lighthouse, moved in 1998 from its original location, now overlooks the sea at Mal Coombs Park. Motels, an R.V. park, and campground provide accommodations. There is also a golf course and a small daylight airstrip. The Shelter Cove Campground and Deli offers information on shuttle services for backpackers; for information, call: (707) 986-7474. Mario's Marina provides boat launching and dry storage facilities and fuel, in addition to a motel and restaurant; for information, call: (707) 986-1145.

Vista, North of Shelter Cove, King Range National Conservation Area

Mattole River Mouth

Near Point Arena

Mendocino County

Over 130 miles long, Mendocino County's coast is noted for its dramatically eroded sea cliffs and numerous small pocket beaches. Undeveloped terraces typically extend for miles along the coast between Highway 1 and the Pacific Ocean, interrupted only occasionally by small towns and villages, and the deep ravines of creek, stream, and river inlets.

The county's shoreline is extremely rugged, characterized by offshore sea stacks and abundant tidepool areas. Sandy beaches are usually found at the mouths of freshwater stream channels; access is often steep and difficult, but the beaches are generally secluded and protected from winds by the surrounding sea cliffs.

The rocky shores and clear waters of the coast are popular for diving. Although many people dive simply for the enjoyment of exploring undersea features such as the kelp forest off the bay at Van Damme State Park, a majority of divers are attracted by the abalone and rockfish that are prevalent here. During abalone season, which is from April to November in Northern California, beaches and coastal campgrounds are often crowded with divers.

Unique to California's coastline is the northern Mendocino County area known as the "Lost Coast." At Rockport, Highway 1 heads eastward, leaving the coast north of here accessible only by Usal Road, a narrow, dirt logging road with steep grades and numerous hairpin turns, which extends for 30 miles to the Humboldt County line. Although the southern half of the Lost Coast is mostly private, Sinkyone Wilderness State Park and the King Range National Conservation Area in the north provide some of the last remaining true wilderness areas in California that are adjacent to the coast.

Within the Lost Coast area the abandoned port towns of Rockport, Usal, and Bear Harbor serve as reminders of Mendocino's more populated era during the 19th century. Between 1850 and 1900, nearly every bay or anchorage along the coast large enough to harbor a ship was used as a shipping point for the valuable redwood and Douglas-fir timber of the north coast, which were in high demand during and after the California Gold Rush period.

The demand for timberland resulted in the government's consolidation in 1856 of the Pomo, Yuki, and Sinkyone Indian tribes, the original settlers of the coast, into the Mendocino Indian Reservation, located between the Noyo and Ten Mile rivers. The military established Fort Bragg in 1857 to oversee the reservation. The government relocated the Indians inland to the Round Valley around 1864 due to the timber industry's interests in the reservation's forest lands. Fort Bragg was abandoned and the community subsequently became a lumber and port town. Easily harvestable timber along the coast became scarce after 1900 and many of the small towns of the north county were abandoned; remaining lumber production was eventually consolidated at Fort Bragg.

Mendocino, a former mill site established in 1852, is now an artists' community and a popular tourist destination; the central town section has been designated as a historical preservation district, ensuring that its 19th-century New England-style architecture will be maintained.

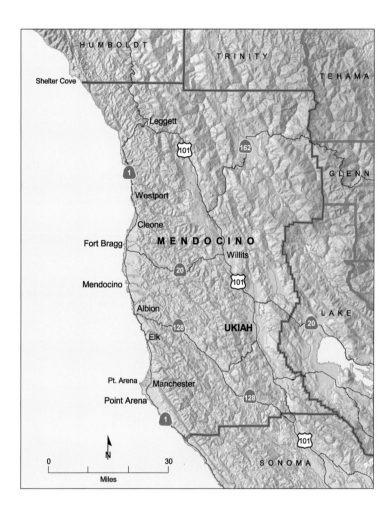

For additional information on Mendocino County's coast, contact the Mendocino Coast Chamber of Commerce, 332 N. Main Street (P.O. Box 1141), Fort Bragg 95437, (707) 961-6300, www.mendocinocoast.com. For transit information on service between Ft. Bragg and the Navarro River, contact the Mendocino Stage, Box 1177, Mendocino 95460, (707) 964-0167. For service south of the Navarro River, contact the Mendocino Transit Authority (MTA), 1-800-696-4682, www.4mta.org. Both the MTA and the Mendocino Stage buses may be boarded at regular stops or flagged at any safe location.

Coastal Landforms

Waves, wind, subsidence, and uplift are the major physical forces that act on the shoreline and that produce, over a long period of time, a variety of coastal landforms.

Perhaps the most obvious coastal landform is the beach, an area adjacent to the surf zone consisting of a veneer of sand and gravel over a bedrock base. In California, most beach sediments are carried to coastal areas by rivers and are distributed and deposited along the shoreline by waves and currents. The width and shape of the beach depend on the prevailing wave conditions and the type of sediment present at a particular location; for example, beaches that are made of large sand grains usually are steeper than beaches that are made of fine sand.

Sand spits and sand bars are formed when waves and currents deposit sand, gravel, or other sediments on the sea floor. Spits are narrow ridges or points of sand that project into the water from a point on the shore. Bars are submerged mounds or embankments of sediments that build up on the sea floor, often at the mouths of bays and estuaries. The San Francisco Bar, located offshore from the Golden Gate, is one example of a large sand bar system; this bar is formed by sediments carried from the Sacramento and San Joaquin rivers, and if not dredged periodically, it would prevent ship movement to and from San Francisco Bay.

The erosive power of waves also contributes significantly to the creation of various coastal landforms. Breaking waves cut into the shoreline and form the bluffs and cliffs so common to the California coast. Sea caves are created when waves first remove the least resistant material, or attack the weakest areas, in a cliff face. As waves continue to advance landward they plane, or level, the shoreline area, forming a wave-cut bench, or platform, just below the water's surface. Before the wave-cut bench is completed, the more resistant portions of

sea cliffs may project above sea level; these projections, which appear as offshore rocks or tiny islands, are called sea stacks.

The evolution of coastal landforms is demonstrated dramatically along the Mendocino coast. Today, the shoreline is characterized by countless sea stacks located offshore adjacent to a series of promontories, or headlands, that project into the ocean. This rugged shoreline actually is the beginning of the youngest in a series of wave-cut terraces, also called marine terraces, that extend inland more than two miles.

Over the last half-million years, the Mendocino coastal area has risen in response to the forces that have been building the coast mountain ranges. Thus, a terrace that was cut by the ocean more than 500,000 years ago is now elevated about 650 feet above sea level. From this oldest terrace, four more terraces step down in elevation and age as they near the present shoreline; each terrace is approximately 100 feet lower and 100,000 years younger than the preceding one.

Sea Stacks, Mendocino Coast

South Kibesillah Gulch View Area

Mendocino County

NORTHERN MENDOCINO COAST

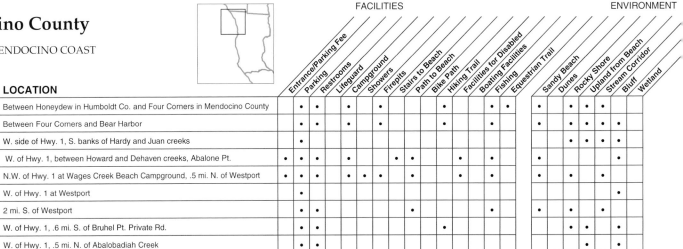

NAME	LOCATION	Entrance/Parking Fee	Parking	Restrooms	Lifeguard	Campground	Showers	Firepits	Stairs to Beach	Path to Beach	Bike Path	Hiking Trail	Facilities for Disabled	Boating Facilities	Fishing	Equestrian Trail	Sandy Beach	Dunes	Rocky Shore	Upland from Beach	Stream Corridor	Bluff	Wetland
King Range National Conservation Area	Between Honeydew in Humboldt Co. and Four Corners in Mendocino County		•	•		•		•				•		•	•	•	•		•	•	•		
Sinkyone Wilderness State Park	Between Four Corners and Bear Harbor		•	•		•						•			•		•		•	•	•	•	
Overlooks	W. side of Hwy. 1, S. banks of Hardy and Juan creeks		•																•	•	•	•	
Westport-Union Landing State Beach	W. of Hwy. 1, between Howard and Dehaven creeks, Abalone Pt.	•	•	•		•			•	•			•		•		•						
Wages Creek Beach	N.W. of Hwy. 1 at Wages Creek Beach Campground, .5 mi. N. of Westport	•	•	•		•	•	•		•			•		•		•						
Westport Headlands	W. of Hwy. 1 at Westport		•																			•	
Chadbourne Gulch	2 mi. S. of Westport		•	•					•						•		•		•		•		
Bruhel Point Bluff	W. of Hwy. 1, .6 mi. S. of Bruhel Pt. Private Rd.		•	•						•							•		•		•		
South Kibesillah Gulch View Area	W. of Hwy. 1, .5 mi. N. of Abalobadiah Creek		•	•															•		•		
Seaside Creek Beach	W. of Hwy. 1, .75 mi. N. of Ten Mile River		•														•			•			

KING RANGE NATIONAL CONSERVATION AREA: Comprises 62,200 total acres; 1,620 are in Mendocino County. Access is from Shelter Cove Rd. and Kings Peak Rd. in Humboldt County.

SINKYONE WILDERNESS STATE PARK: Consists of 7,367 acres of undeveloped beaches, bluffs, and coastal mountains extending from the King Range National Conservation Area to Bear Harbor, including dense redwood forests and spectacular coastal views. Unpaved Briceland Rd. (Route 435) provides the best access to the north end of the park. Usal Rd. (Route 431) from Rockport is unpaved but passable year-round to Usal Beach Camp, seasonally passable north. Both roads are unsuitable for trailers or R.V.'s. The rugged Lost Coast Trail (Wilderness Permit required) runs the length of the park (16.7 mi.) from Orchard Creek to Usal Creek; most of the park's campsites are located along the trail. Trails also lead north from Needle Rock, including access to the Low Gap Trail, which crosses the park east to west. Trails at Whale Gulch and Bear Harbor lead from Briceland Rd. to narrow, black sandy beaches. There are 15 campsites accessible by 2-wheel drive vehicles at Usal Beach Campground; tables and fire rings available; no piped water at campsites. The park has 32 primitive hike-in campsites at scattered locations, including 10 sites along the Lost Coast Trail; each site accommodates up to 8 people. There is parking at Usal Beach Campground, Needle Rock, Low Gap Campground, and Orchard Creek trailhead (seasonal only). Campers are asked to self-register at Usal Campground or Needle Rock Ranch House. For trail maps and information, contact: State Parks, Eel River District Office, P.O. Box 245, Whitethorn 95489; (707) 986-7711.

WESTPORT-UNION LANDING STATE BEACH: Blufftop day-use areas are located along the two miles of frontage road just west of Hwy. 1 at Abalone Point. Facilities include 2 picnic areas and 130 primitive sites in 7 camping areas. There is blufftop wheelchair access at Howard Creek; no wheelchair access to the beach. Fee charged for camping; call: (707) 937-5804.

WAGES CREEK BEACH: The sandy ocean beach at the mouth of Wages Creek is accessible from the privately run campground south of the creek, just off Hwy. 1; R.V. sites with hookups are adjacent to the beach. Surf net rentals and monthly trailer storage available. Fee charged for camping or day use. Beach, walkways, and restrooms are wheelchair accessible. Open all year;2 call ahead for camping and flood conditions during December-February: (707) 964-2964.

WESTPORT HEADLANDS: Blufftop ocean views, west of the village of Westport.

CHADBOURNE GULCH: Shoulder parking; the dirt road off Hwy. 1 on the north side of the gulch, which is closed to motor vehicles, leads to the mile-long sandy beach. Popular for surf fishing and surfing. Harbor seals are often sighted offshore.

BRUHEL POINT BLUFF: Several trails along the grassy blufftops west of Hwy. 1 provide scenic views of the headlands and rocky shore below. Access to trails is from a paved vista point near a stand of Monterey pines and from a nearby unpaved pull-out west of the highway.

SOUTH KIBESILLAH GULCH VIEW AREA: Blufftop overlook and parking area with a good view south to Ft. Bragg. No beach access. For information, call: (707) 463-4267.

SEASIDE CREEK BEACH: Small, undeveloped ocean beach directly off Hwy. 1.

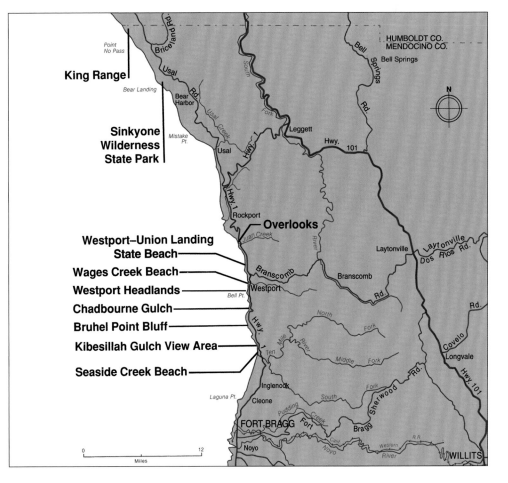

71

Mendocino Pygmy Forests

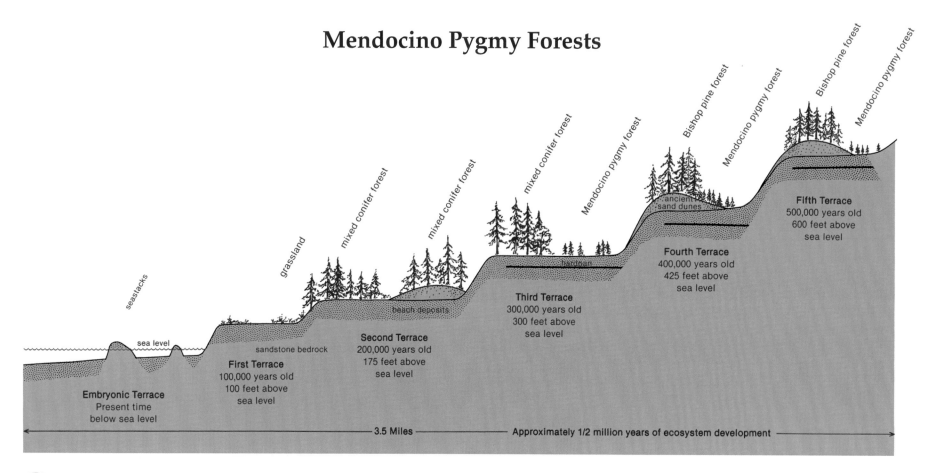

Several small forested areas known as pygmy forests are located along the Mendocino County coast between Fort Bragg and the Navarro River. These forests contain unusually dwarfed trees and shrubs that grow very slowly, typically reaching heights of only a few feet, although they may live to be over 100 years old.

The pygmy forests of Mendocino grow on landforms known as marine terraces. These terraces were created in the Pleistocene era by the periodic uplifting of the wave-eroded Mendocino coastline, resulting in a staircase-type landform with five distinct terraces. The pygmy forests are found along the upper three older terraces where a soil condition, called podzol, has formed. Podzol soils generally only evolve where the terraces are flat and rainwater collects and eventually drains through the sandy soils. This water is very acidic, primarily due to the decomposition of certain leaves and pine needles; as the water drains through the soil, it carries with it important plant nutrients and minerals, leaving the upper soil layers impoverished. Some of these minerals

combine with the soil a few feet below the surface to form a concrete-like hardpan layer.

The iron-rich hardpan layer prevents plant roots from penetrating to the more fertile soils below. It also raises near-surface groundwater levels, causing water to flood often and persist on the surface for long periods of time during the winter. In the spring and summer, however, water slowly drains past the hardpan, leaving the soils above it, where most plant roots are located, devoid of moisture. These extreme conditions of flooding in the winter, drought in the summer, lack of important nutrients, and high soil acid levels all combine to severely limit plant growth. All the plants within the pygmy forests are capable of growing to more normal heights in fertile soils.

The extremely slender and sometimes twisted and gnarled trees of the pygmy forest are a remarkable contrast to the surrounding tall forests of redwood, pine, and fir. Common pygmy forest species include the pygmy cypress, *Cupressus*

pygmaea; Bishop pine, *Pinus muricata*; Bolander pine, *P. contorta* spp. *bolanderi*; and the dwarf (Fort Bragg) manzanita, *Arctostaphylos nummularia*. Except for the Bishop pine, all of these are endemic to the Mendocino pygmy forest.

Portions of the Mendocino pygmy forest have been formally preserved within the Jug Handle State Reserve (which contains the former Pygmy Forest Reserve), in Van Damme State Park, and in a small parcel owned by the University of California. The pygmy forests in Jug Handle State Reserve and Van Damme State Park are accessible by self-guided nature trails.

To obtain maps or literature on Jug Handle State Reserve pygmy forests, contact:

Jug Handle Creek Farm
P.O. Box 17
Caspar, CA 95420   (707) 964-4630

Mendocino County

FORT BRAGG / CASPAR

NAME	LOCATION	Entrance/Parking Fee	Parking	Restrooms	Lifeguard	Campground	Showers	Firepits	Stairs to Beach	Path to Beach	Bike Path	Hiking Trail	Facilities for Disabled	Boating Facilities	Fishing	Equestrian Trail	Sandy Beach	Dunes	Rocky Shore	Upland from Beach	Stream Corridor	Bluff	Wetland
MacKerricher State Park	Runs from Ten Mile River S. to Pudding Creek, Cleone	•	•	•		•	•	•		•		•	•	•	•	•	•	•	•		•	•	•
Glass Beach	W. of Hwy. 1, End of Elm St., Fort Bragg		•							•							•						
Noyo Harbor	End of Noyo Harbor Dr. at the Noyo River	•	•				•							•	•	•	•						
Mendocino Coast Botanical Gardens	18220 N. Hwy. 1, 2 mi. S. of Fort Bragg	•	•									•	•								•		
Path to Beach	Ocean Dr. near Hwy. 1, N. of Caspar									•							•						
Jefferson Way	W. of Hwy. 1, within Jug Handle State Reserve	•	•							•							•		•			•	
Jug Handle State Reserve	E. and W. of Hwy. 1, 1.5 mi. N. of Caspar	•	•		•					•		•	•				•		•	•	•	•	

MACKERRICHER STATE PARK: Six miles of sandy beach and dunes, a 143-site fee campground, and an offshore underwater park for divers. Beach access from Hwy. 1 is at Pudding Creek, and at the ends of Ward Ave., Mill Creek Dr., and the main park entrance road. Some campsites and restrooms are wheelchair accessible.

A logging truck haul road, for day use only, runs along the shoreline from Pudding Creek to Ten Mile River. Access to road is just north of Pudding Creek trestle, at Mack Ave., or at Ward Ave. in Cleone; pedestrians/bicyclists only from two miles north of Pudding Creek. A six-mile-long equestrian trail runs along the beaches and dunes west of the haul road. Hiking trails along the Laguna Point headlands and around Lake Cleone; a wheelchair-accessible redwood path extends part way around the lake. Another path leads to a seal-watching station at the tip of Laguna Point. 15-acre Lake Cleone, at end of Mill Creek Rd., is popular for fishing and non-power boating; a boat ramp adjoins the parking and picnic area. Call: (707) 937-5804.

GLASS BEACH: A short dirt trail leads from a blufftop parking area at the end of Elm St. down to several sandy pocket beaches. Call: (707) 937-5804.

NOYO HARBOR: Sandy beach at Noyo River mouth beneath Hwy. 1 bridge accessible from North Harbor Drive. Boat charter and fishing equipment rentals available at the harbor village east of the bridge. Boat ramp on south bank of the river at end of Basin St.; rock jetty on river's north bank has parking, launch ramp, outdoor shower, and wheelchair-accessible restrooms. Call: (707) 964-4719.

MENDOCINO COAST BOTANICAL GARDENS: The 47-acre horticultural gardens are located on a coastal terrace above the ocean. Nature trails lead through the lush native and exotic vegetation of the gardens to the adjoining headlands and sandy beach. Restrooms and some paths are wheelchair accessible; electric carts are available for disabled persons. Entrance fee for gardens; free access to beach and bluffs. Call: (707) 964-4352.

PATH TO BEACH: West of the Pine Beach Inn sign, a short path leads through groves of Monterey pine to sea cliffs and a small sandy beach, part of Jug Handle State Reserve.

JEFFERSON WAY: Two pocket beaches within Jug Handle State Reserve are accessible at the west end of Jefferson Way.

JUG HANDLE STATE RESERVE: Five miles of interpretive nature trails traverse the headlands and inland coastal terraces adjacent to Jug Handle Creek. Brochures, available at the picnic area next to the parking lot and rangers' headquarters off Hwy. 1, describe the formation of local landforms along the self-guided "ecological staircase" trail. A sandy beach at the mouth of Jug Handle Creek is accessible from the trails near the south side of the Hwy. 1 overpass.

The privately owned Jug Handle Farm and Nature Center adjoins the reserve, east of Hwy. 1 and south of Jug Handle Creek. Facilities, all wheelchair accessible, include overnight accommodations in a renovated farmhouse, small campground (tent only), a self-guided nature trail, and library. Group educational programs available. Overnight fee; minor contribution of work requested. Reservations required; call: (707) 937-5804.

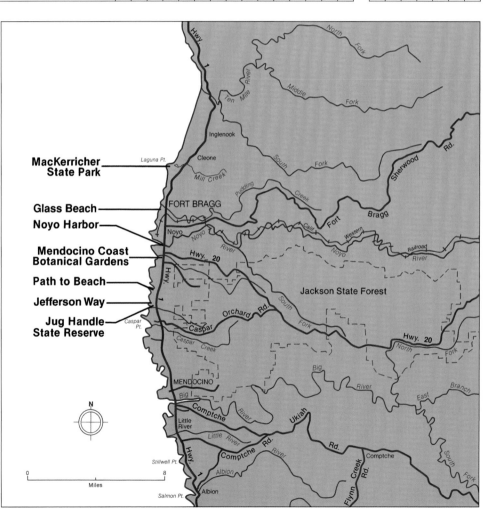

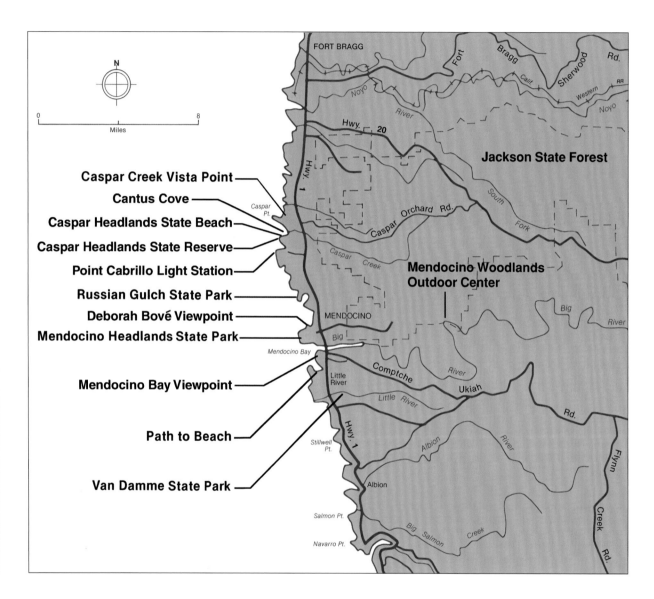

FORT BRAGG

Jackson State Forest

Caspar Creek Vista Point
Cantus Cove
Caspar Headlands State Beach
Caspar Headlands State Reserve
Point Cabrillo Light Station
Russian Gulch State Park
Deborah Bové Viewpoint
Mendocino Headlands State Park

Mendocino Bay Viewpoint

Path to Beach

Van Damme State Park

Mendocino Woodlands Outdoor Center

MENDOCINO

Hwy. 20

Caspar Pt.

Caspar Orchard Rd.

Caspar Creek

Noyo River

Fort Bragg Sherwood Rd.

Calif Western RR

Noyo

South Fork

Big River

Mendocino Bay

Big

Comptche

Little River

Ukiah

Little River

Albion River

Flynn Creek Rd.

Rd.

Stillwell Pt.

Albion

Salmon Pt.

Big Salmon Creek

Navarro Pt.

Hwy. 1

N

0   8
Miles

Mendocino Headlands State Park

Mendocino County

CASPAR / MENDOCINO

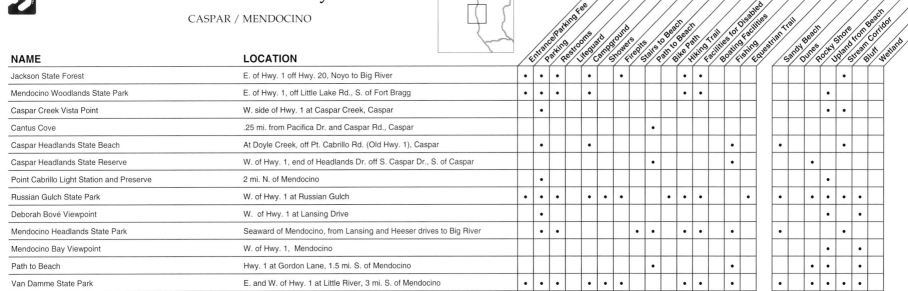

NAME	LOCATION	Entrance/Parking Fee	Parking	Restrooms	Lifeguard	Campground	Showers	Firepits	Stairs to Beach	Path to Beach	Bike Path	Hiking Trail	Facilities for Disabled	Boating Facilities	Fishing	Equestrian Trail	Sandy Beach	Dunes	Rocky Shore	Upland from Beach	Stream Corridor	Bluff	Wetland
Jackson State Forest	E. of Hwy. 1 off Hwy. 20, Noyo to Big River	•	•	•		•		•				•	•								•		
Mendocino Woodlands State Park	E. of Hwy. 1, off Little Lake Rd., S. of Fort Bragg	•	•	•		•						•	•							•			
Caspar Creek Vista Point	W. side of Hwy. 1 at Caspar Creek, Caspar		•																•	•			
Cantus Cove	.25 mi. from Pacifica Dr. and Caspar Rd., Caspar								•														
Caspar Headlands State Beach	At Doyle Creek, off Pt. Cabrillo Rd. (Old Hwy. 1), Caspar		•			•							•				•					•	
Caspar Headlands State Reserve	W. of Hwy. 1, end of Headlands Dr. off S. Caspar Dr., S. of Caspar								•				•						•				
Point Cabrillo Light Station and Preserve	2 mi. N. of Mendocino		•																•				
Russian Gulch State Park	W. of Hwy. 1 at Russian Gulch	•	•	•		•		•		•	•	•			•		•		•			•	
Deborah Bové Viewpoint	W. of Hwy. 1 at Lansing Drive		•																•	•			
Mendocino Headlands State Park	Seaward of Mendocino, from Lansing and Heeser drives to Big River	•	•					•	•		•	•		•	•		•		•		•		
Mendocino Bay Viewpoint	W. of Hwy. 1, Mendocino																		•	•			
Path to Beach	Hwy. 1 at Gordon Lane, 1.5 mi. S. of Mendocino								•										•				
Van Damme State Park	E. and W. of Hwy. 1 at Little River, 3 mi. S. of Mendocino	•	•	•		•		•				•	•		•		•		•	•	•		

JACKSON STATE FOREST: A 50,200-acre demonstration forest of 'forest management' practices, featuring a self-guided demonstration/nature trail, picnic areas, hiking (mostly along old logging roads), and 44 primitive campsites. Use permits are required, and must be obtained in advance from Jackson State Forest, 802 N. Main St., Fort Bragg 95437; (707) 964-5674.

MENDOCINO WOODLANDS STATE PARK: A reservation-only camping facility for groups of between 30 and 200 people, near the southwest edge of Jackson State Forest. Year-round and seasonal camps are available with cabins, tent cabins, and kitchen/dining halls. There are also trails within the woodlands, with connections to State Forest trails. Reservations can be made up to two years in advance: Mendocino Woodlands Camping Association, P.O. Box 267, Mendocino 95460; (707) 937-5755.

CASPAR CREEK VISTA POINT: Overlooks the Pacific Ocean and heavily wooded Caspar Creek.

CANTUS COVE: A trail leads to the bluff from Pacifica Dr., approximately ¼ mile from the intersection of Pacifica Dr. and Caspar Rd. in the village of Caspar.

CASPAR HEADLANDS STATE BEACH: The small sandy beach is located at the mouth of Doyle Creek where Point Cabrillo Dr. (old Hwy. 1) runs adjacent to the ocean. Call: (707) 937-5804. Caspar Beach R.V. Campground, located at 45201 Point Cabrillo Dr., offers full hookups and tent sites. Facilities include a convenience store, laundry, showers, and wheelchair-accessible restrooms.

CASPAR HEADLANDS STATE RESERVE: Blufftop parcels totaling five acres are interspersed with private property. Pedestrian access is from Headlands Dr., off South Caspar Drive. A use pass is required, available at no charge from the State Parks district office, located three miles south on the east side of Hwy. 1 across from Russian Gulch State Park. Call: (707) 937-5804.

PT. CABRILLO LIGHT STATION AND PRESERVE: The 270-acre preserve offers access to rocky headlands; open 9 AM-6 PM daily. The lighthouse, with an operating Fresnel lens, has been fully restored and is open March through October, Friday through Monday 11 AM-4 PM; call: (707) 937-0816.

RUSSIAN GULCH STATE PARK: Thirty campsites, including one that is wheelchair accessible, are situated in the redwood-forested valley of Russian Gulch, below Highway 1. Additional camping facilities include a 40-person group campsite, a hiker/bicyclist area for up to 12 persons, and 4 primitive equestrian campsites for up to 4 persons. Reservations for equestrian sites are available through State Parks district headquarters only; for other camping reservations, call: 1-800-444-7275. A sandy beach beneath the arched Highway 1 bridge is accessible from the road leading past the park's recreation hall. There is an outdoor shower located near the beach.

Twelve miles of hiking trails include paths along the headlands that pass a 200-foot-diameter ocean blowhole, and vista points of Mendocino to the south. Some trails are available to bicyclists and horses. Fees are charged for day use and camping; summertime reservations strongly recommended. The offshore area is an underwater park for divers. Call: (707) 937-5804.

DEBORAH BOVÉ VIEWPOINT: The blufftop west of Lansing Drive, just south of the intersection with Highway 1, offers vistas of the rocky shore from a hand carved bench; shoulder parking.

MENDOCINO HEADLANDS STATE PARK: Two miles of spectacular sea cliffs, dotted with numerous wave tunnels, rim the promontory of the Mendocino headlands adjacent to the town of Mendocino and the Big River. The headlands are accessible by blufftop trails along the southern terraces or from the parking lots off Heeser Dr. to the north and west. Paths and a stairway lead down the bluffs to the beach. A visitor center at the historic Ford House has maps, brochures, interpretive displays, and picnic tables. A sandy beach at the mouth of the Big River is adjacent to the southern headlands and is accessible from North Big River Road. Some 7,300 acres of the Big River watershed have been added to Mendocino Headlands State Park, including stretches of river frontage. For State Park information, call: (707) 937-5804.

Privately run Mendocino Campground with 60 sites is located just south of the Big River at Comptche-Ukiah Rd. and Hwy. 1. Call: (707) 937-3130. Canoes can be rented hourly and daily at Catch-A-Canoe, 44900 Comptche-Ukiah Rd., just off Hwy. 1, on the south bank of the Big River. Call: (707) 937-0273.

MENDOCINO BAY VIEWPOINT: A vista of the historic town of Mendocino with its bay and rocky headland is available from the bluff just south of the Big River Bridge; turn west off Highway 1 on Brewery Gulch Road. Shoulder parking. A second vista point is located 1/2 mile south where Brewery Gulch Rd. rejoins Highway 1.

PATH TO BEACH: Located south of Mendocino Bay, a 1/2 mile-long path leads through an undeveloped part of Van Damme State Park to bluffs and a small rocky beach; steep and hazardous shoreline access down the bluffs.

VAN DAMME STATE PARK: A small sandy beach with outdoor showers, fire rings, restrooms, and a paved parking lot is west of Hwy. 1, adjacent to the Little River mouth. The park campground contains 74 sites, including 10 environmental sites and one group site for up to 75 people; restrooms are wheelchair accessible. Twenty-four enroute campsites are available in this lot if the regular campground is full. Reservations are strongly recommended during summer; call: 1-800-444-7275. Fees charged for camping and park day use; no fee for beach day use. The visitor center and museum has several dioramas and other interpretive materials, and is open on weekends year-round.

The park includes an offshore underwater park for divers, the forested valley of the Little River, and the Pygmy Forest self-guided nature trail, accessible from hiking trails at the eastern end of the park; from the entrance to Little River Airport Rd. south of the park entrance; or from the parking lot at the intersection of Little River-Airport Rd. and Albion-Little River Rd., about three miles east of Hwy. 1. There is a short wheelchair-accessible redwood walk through the forest. The Charlotte Hook Reserve, a small ten-acre parcel of pygmy forest, is accessible east of Hwy. 1 on the Comptche-Ukiah Road. The Pygmy Forest is so named because of the unusually stunted mature cone-bearing trees that grow on the hardpan soil in this area. For State Park information, call: (707) 937-5804.

Albion Flat

Albion River, Albion

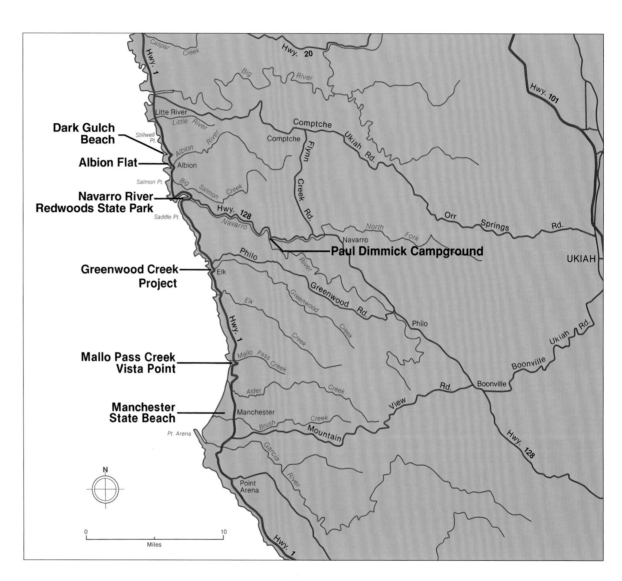

Dark Gulch Beach

Albion Flat

Navarro River Redwoods State Park

Paul Dimmick Campground

Greenwood Creek Project

Mallo Pass Creek Vista Point

Manchester State Beach

Mendocino County

ALBION / ELK / MANCHESTER

NAME	LOCATION	Entrance/Parking Fee	Parking	Restrooms	Lifeguard	Campground	Showers	Firepits	Stairs to Beach	Path to Beach	Bike Path	Hiking Trail	Facilities for Disabled	Boating Facilities	Fishing	Equestrian Trail	Sandy Beach	Dunes	Rocky Shore	Upland from Beach	Stream Corridor	Bluff	Wetland
Dark Gulch Beach	W. of Hwy. 1, Dark Gulch		•														•						
Albion Flat	N. bank of the Albion River, Albion	•	•	•		•	•	•		•				•	•		•				•		
Navarro River Redwoods State Park	W. end of Navarro Bluff Rd.		•	•		•									•		•				•		
Paul Dimmick Campground	8 mi. E. of Hwy. 1 on Hwy. 128 (Navarro Rd.)	•	•	•		•							•		•					•	•		
Greenwood Creek Project	W. of Hwy. 1, N. of Greenwood Creek, Elk		•				•	•							•		•		•	•	•	•	
Mallo Pass Creek Vista Point	W. of Hwy. 1, S. of Mallo Pass Creek, 4.25 mi. N. of Manchester		•																•	•	•		
Manchester State Beach	Runs 3.5 mi. S. from Alder Creek to just N. of Pt. Arena, Manchester	•	•	•		•				•		•			•		•	•			•		

DARK GULCH BEACH: A path to a cove beach leads through the Heritage House grounds; limited roadside parking just south of the Heritage House entrance.

ALBION FLAT: Boat harbor and private campground at the end of Albion River Rd. provides access to a sandy ocean beach at the mouth of the Albion River. A boat ramp and dock, bait and tackle, and canoe rentals are available. Tent or R.V. camping, with or without hookups; day-use and camping fees charged. No fee required for pedestrian access to the beach. For information and reservations, contact: Albion River Campground, P.O. Box 217, Albion 95410; (707) 937-0606. Just east of Albion Flat is Schooner's Landing, with R.V. camping, boat launch and dock, hiking trails, picnicking, and swimming. For reservations and information, contact: Schooner's Landing, P.O. Box 218, Albion 95410; (707) 937-5707.

NAVARRO RIVER REDWOODS STATE PARK: A broad, sandy ocean beach is at the end of Navarro Bluff Rd., which runs along the south bank of the Navarro River. A trail through the woods on the south side of the road runs from the beach to the county right-of-way. Ten primitive campsites with toilets; no water. For information, call: (707) 895-3141.

PAUL DIMMICK CAMPGROUND: Twenty-eight campsites in a redwood grove between Route 128 and the north fork of the Navarro River, 6 miles east of Highway 1. All sites have picnic tables, fire rings, and access to running water; wheelchair-accessible restrooms are available. The campground is a popular spot for kayaking, canoeing, and winter steelhead fishing. Campground host in summer only. Call: (707) 937-5804.

GREENWOOD CREEK PROJECT: An unpaved parking lot is located in the town of Elk, on the west side of Hwy. 1. A fairly steep, broad path leads south from the parking lot down to a wide sand and pebble beach. Scenic sea stacks and sea caves are directly offshore. Paths north of the parking lot criss-cross the blufftop. Picnic tables and fire rings are located on the blufftop and at the beach. No vehicles are allowed seaward of the parking lot. No camping is allowed on the beach or in the parking lot. For information, call: (707) 937-5804.

MALLO PASS CREEK VISTA POINT: A large paved parking lot is located on the high bluff above the south bank of Mallo Pass Creek, overlooking the rocky ocean shore to the west and the lush, riparian forests of Mallo Pass Creek canyon to the north and east.

MANCHESTER STATE BEACH: 760 acres of beach and dunes west of Manchester. Alder Creek Rd., Kinney Rd., and Stoneboro Rd. all provide access from Hwy. 1 to small parking areas at the edge of the dunes; trails lead from the road ends to a long sandy beach scattered with driftwood. The offshore area is an underwater park for diving, and includes the Point Arena Underwater State Reserve. A 46-site primitive campground is off Kinney Rd., ¼-mile from the beach. The campground includes one hiker/bicyclist site and one 40-person group camp; east of the main park entrance are 9 primitive walk-in sites (about 1 mile walk). Call: (707) 937-5804.

East of Manchester State Beach is a privately run KOA campground off Kinney Road. Facilities include a store, laundry, game room, swimming pool, hot tub, and showers. Overnight fee; additional fee for hookups. Summer reservations recommended: Manchester Beach KOA, P. O. Box 266, Manchester 95459; (707) 882-2375.

Manchester State Beach

Fish Rock Beach

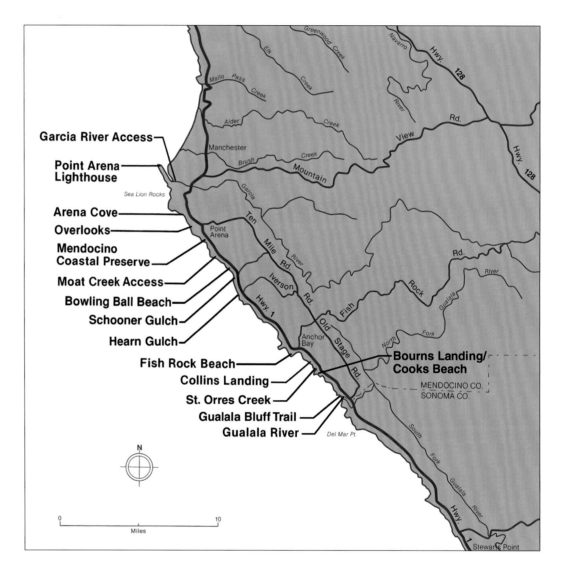

Garcia River Access

Point Arena
Lighthouse

Arena Cove

Overlooks

Mendocino
Coastal Preserve

Moat Creek Access

Bowling Ball Beach

Schooner Gulch

Hearn Gulch

Fish Rock Beach

Collins Landing

St. Orres Creek

Gualala Bluff Trail

Gualala River

Bourns Landing/
Cooks Beach

MENDOCINO CO.
SONOMA CO.

N

0   10
Miles

Mendocino County

SOUTHERN MENDOCINO COAST

FACILITIES · ENVIRONMENT

NAME	LOCATION	Entrance/Parking Fee	Parking	Restrooms	Lifeguard	Campground	Showers	Firepits	Stairs to Beach	Path to Beach	Bike Path	Hiking Trail	Facilities for Disabled	Boating Facilities	Fishing	Equestrian Trail	Sandy Beach	Dunes	Rocky Shore	Upland from Beach	Stream Corridor	Bluff	Wetland
Garcia River Access	End of Miner Hole Rd., 2 mi. N. of town of Pt. Arena					•									•						•		•
Point Arena Lighthouse	End of Lighthouse Rd., 1.25 mi. S. of town of Pt. Arena	•	•										•						•	•		•	
Arena Cove	W. end of Port Rd., Pt. Arena	•	•				•						•	•	•				•				
Overlooks	Along W. side of Hwy. 1, from Schooner Gulch to .75 mi. S.		•																•	•			
Mendocino Coastal Preserve	W. of Hwy. 1, 1 mi. S. of town of Pt. Arena		•																•	•		•	
Moat Creek Access	W. of Hwy. 1, 2.5 mi. S. of town of Pt. Arena		•							•									•			•	
Bowling Ball Beach	W. of Hwy. 1, 3.5 mi. S. of town of Pt. Arena		•							•							•		•			•	
Schooner Gulch	W. of Hwy. 1 at Schooner Gulch Rd., 3.5 mi. S. of Pt. Arena		•							•							•		•			•	
Hearn Gulch	W. of Hwy. 1, N. of Iverson Road		•							•					•		•				•	•	
Fish Rock Beach	At Anchor Bay Campground, .2 mi. N. of town of Anchor Bay	•	•	•		•	•			•					•		•		•				
Collins Landing	At Serenisea Lodge, 1.75 mi. S. of town of Anchor Bay								•	•					•				•				
St. Orres Creek	W. of Hwy. 1 at St. Orres Creek		•							•					•		•				•		
Bourns Landing / Cooks Beach	W. of Hwy. 1, 2 mi. N. of Gualala		•							•					•		•					•	
Gualala Bluff Trail	W. of Hwy. 1 at Gualala		•																			•	
Gualala River	W. of Hwy. 1 at Gualala		•							•					•		•				•		•

GARCIA RIVER ACCESS: Miner Hole Rd., a gravel road off Hwy. 1, leads across private lands to the south bank of the Garcia River, which is a popular fishing access. The area is noted as a wintering habitat for tundra swans. No dogs, camping, or hunting permitted; do not trespass on adjacent private property.

POINT ARENA LIGHTHOUSE: The 115-foot-high lighthouse, built in 1870, has one of the most powerful lights on the coast. Open to the public 11 AM-3:30 PM daily, 10 AM-4:30 PM April to October; fee for entrance. Docent tours of the lighthouse and small museum are available. Restrooms and ground level tours are wheelchair accessible; no paved paths. Three former lighthouse keepers' houses on the site are available for rental year-round; reservations required; call: (877) 725-4448. For lighthouse information, call: (707) 882-2777.

ARENA COVE: A small rocky beach is at the end of Port Rd. within the Point Arena Cove area. The new 322-foot-long steel and concrete pier is completely wheelchair accessible. Facilities include boat launch and hoist (5 ton cap.), open 7 AM-4 PM daily all year (weather permitting), fish-cleaning tables, outdoor showers, restrooms, and offshore moorings. No overnight tie-up to pier. Arena Cove is the only launch facility between Bodega Bay and Fort Bragg, except the high-tide-only ramp at Albion Flat. Pier supervisor: (707) 882-2583.

MENDOCINO COASTAL PRESERVE: Access to this 80-acre blufftop preserve, managed by the California Institute of Environmental Studies, is restricted to educational groups and for scientific purposes by appointment only. Noted as a natural coastal grassland habitat. For information, call: (916) 278-6620.

MOAT CREEK ACCESS: Formerly known as the "Whiskey Shoals Access." There is a small dirt parking area and trail to the small rocky beach and tidepools; no dogs allowed. A trail leads across the bluffs from Moat Creek south to Ross Beach; do not trespass on adjacent agricultural lands.

BOWLING BALL BEACH: A short trail leads across the blufftop and down to the beach, named for the large, wave-rounded bowling-ball-like rocks visible along the surfline at low tide. Another trail forks off from the Bowling Ball Beach trail and leads through a redwood-forested ravine to Schooner Gulch, a wide beach with tidepools and unique geologic formations. Park on the Hwy. 1 shoulder across from Schooner Gulch Rd. on the west side of the highway. Call: (707) 937-5804.

SCHOONER GULCH : Over 32 acres of undeveloped state park land. Short trails lead through a forested ravine, across the blufftops, and down the bluffs to Schooner Gulch Beach, a wide, sandy beach featuring tidepools, driftwood, and wave-carved cliff formations. At minus tides, one can hike north along the beach to Bowling Ball Beach.

HEARN GULCH: Blufftop trail and beach access managed by the Redwood Coast Land Conservancy; call: (707) 785-3327.

FISH ROCK BEACH: Privately run Anchor Bay Campground, directly off Hwy. 1 and adjacent to the ocean, provides access to a mile-long sandy beach and tidepools; picnic tables, wheelchair-accessible restrooms, hot showers, and fish-cleaning facilities are available. Popular fishing and diving access; car-top and inflatable boats can be launched from the beach. A 76-site campground with partial hookups for trailers is open all year. Fees charged for parking, camping, and day use. For information and reservations, contact: Anchor Bay Campground, Gualala 95445; (707) 884-4222.

COLLINS LANDING: Rocky cove and tidepools accessible by a steep trail and stairs beginning at the north end of the Serenisea Lodge facilities; the trailhead and stairway overlook Fish Rock Island to the north, which is used as a haul-out by seals and sea lions. Use of the accessway requires the permission of the Serenisea Lodge management; the lodge office, open 9 AM-9 PM daily, is adjacent to the accessway. Call: (707) 884-3836.

ST. ORRES CREEK: Opposite onion-domed St. Orres Inn; parking on seaward shoulder of Hwy. 1. Beach access trail managed by the Redwood Coast Land Conservancy; call: (707) 785-3327.

BOURNS LANDING / COOKS BEACH: Two trails lead through opening in fence, near northern end loop of Bourns Landing Rd., off Hwy. 1 two miles north of Gualala; one path to blufftop overlook, second to Cooks Beach. Call: (707) 785-3327.

GUALALA BLUFF TRAIL: Views of the river mouth and ocean from the bluff, seaward of Gualala's commercial district; trail managed by the Redwood Coast Land Conservancy; call: (707) 785-3327.

GUALALA RIVER: West of Hwy. 1, the Gualala River runs parallel to the Pacific shoreline for two miles. Sandy beaches along the river provide freshwater swimming and views of the coastal sandspit of Gualala Point Regional Park. River access is from the dirt roads just north of the Hwy. 1 Gualala River bridge.

Sonoma Coast State Beaches

Sonoma County

The Sonoma County coast, with its steep cliffs, wave-cut marine terraces, and occasional small towns, is one of the few remaining undeveloped coastal areas in California. The northern third of the coast is mostly private land with limited public access, including the ten-mile-long Sea Ranch development; certain other private lands in the north county are accessible for a fee. Salt Point State Park, 5,676 acres, 20 miles north of Jenner is a popular area for camping, hiking, and whale watching.

Historic Fort Ross, located on a coastal bluff eight miles north of the Russian River, was occupied from 1812 to 1841 by the Russian American Fur Company, who used the fort as a base for seal and sea otter hunting and for farming. At one time there were as many as 300 people occupying the fort. After decimating the seal and sea otter populations, however, and having little success at farming, the Russians left in 1841. Today the site is a state park with a museum and visitors center.

The Russian River, which flows into the Pacific Ocean at Jenner, is noted for its spring and summer canoeing and swimming, and its winter steelhead fishing. It's a popular summer resort area, without the persistent fog found near the ocean, and there are many campgrounds and recreational facilities with river access along Highway 116, which leads inland from the coast.

The southern third of the coast is almost exclusively State Beach; the Sonoma Coast State Beaches comprise over a dozen sandy coves between Goat Rock and Bodega Bay, with headquarters at Salmon Creek Beach. These coves lead to pocket beaches that are excellent for beachcombing. This part of the coast is also popular for fishing; however, the ocean is too rough for swimming, and strong rip currents and cold water temperatures make even walking in shallow water dangerous. Lifeguards are not provided, and visitors should be very careful when walking on the beaches.

Along Highway 1 just south of these beaches is Bodega Bay, Sonoma's southernmost coastal town. The Bodega Harbor, in the north part of the bay, is a center for commercial fishing, recreation, and sportfishing and is particularly crowded during salmon season. Bodega Bay also has piers, berths, and boat tours, and holds a Fishermen's Festival in April.

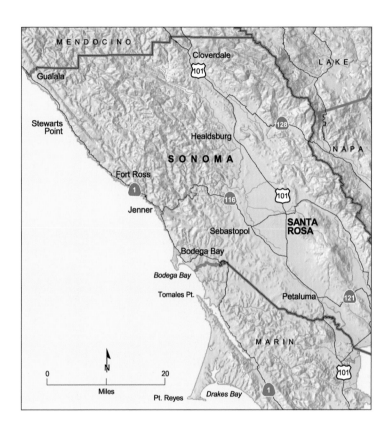

For information on the Bodega Bay Fishermen's Festival, contact the Bodega Bay Chamber of Commerce at Highway 1, P.O. Box 146, Bodega Bay 94923, www.bodegabay.com, (707) 875-3422 or (707) 875-3866. For information on Sonoma Coast State Beaches and current State Beach fee policy, call: (707) 875-3483. Surf and weather information is available anytime from the Bodega Bay Surf Shack at: (707) 875-3944 (recorded information when the shop is closed).

The Mendocino Transit Authority (MTA) provides transit service between Pt. Arena and Santa Rosa, including Highway 1 in Sonoma County north of the Russian River. For information, contact the MTA at 241 Plant Road, Ukiah 95482, 1-800-696-4682, or see www.mcn.org/a/mta/. Transit service between San Francisco and Sonoma County is available via Golden Gate Transit; call 1-800-735-2922 or 1-800-735-2929 (TDD), or see www.goldengate.org for information.

Stump Beach, Salt Point State Park

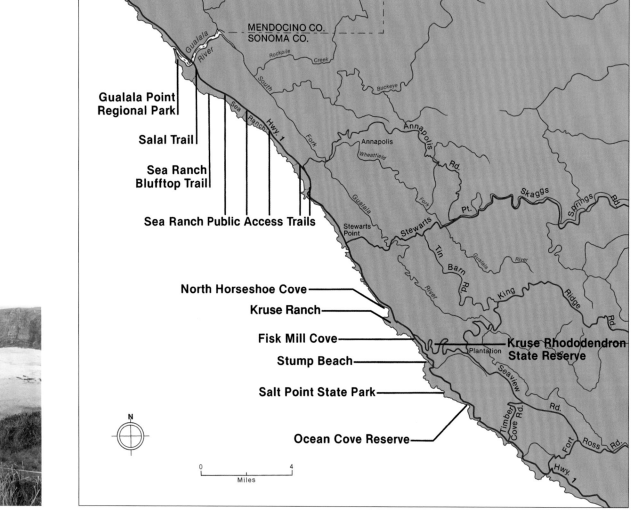

Gualala Point Regional Park

Salal Trail

Sea Ranch Blufftop Trail

Sea Ranch Public Access Trails

North Horseshoe Cove

Kruse Ranch

Fisk Mill Cove

Stump Beach

Salt Point State Park

Ocean Cove Reserve

Kruse Rhododendron State Reserve

MENDOCINO CO.
SONOMA CO.

Gualala River

Rockpile Creek

South

Buckeye

Sea Ranch

Hwy. 1

Fork

Annapolis

Annapolis

Wheatfield

Rd.

Gualala

Fork

Pt.

Skaggs

Springs Rd.

Stewarts Point

Stewarts

Tin Barn Rd.

Gualala River

King

Ridge Rd.

Plantation

Seaview Rd.

Rd.

Timber Cove Rd.

Fort Ross Rd.

Hwy. 1

N

0   4
Miles

![logo]

Sonoma County

NORTHERN SONOMA COAST

FACILITIES ENVIRONMENT

NAME	LOCATION	Entrance/Parking Fee	Parking	Restrooms	Lifeguard	Campground	Showers	Firepits	Stairs to Beach	Path to Beach	Bike Path	Hiking Trail	Facilities for Disabled	Boating Facilities	Fishing	Equestrian Trail	Sandy Beach	Dunes	Rocky Shore	Upland from Beach	Stream Corridor	Bluff	Wetland
Gualala Point Regional Park	Hwy. 1, 1 mi. S. of Gualala	•	•	•		•		•	•	•	•	•	•		•	•	•		•	•	•	•	
Salal Trail	.6-mi. trail begins at Gualala Point Regional Park		•	•				•		•							•		•	•	•	•	
Sea Ranch Blufftop Trail	Between Gualala Point Regional Park and Walk-On Beach		•					•		•							•		•	•	•	•	
Sea Ranch Public Access Trails	Hwy. 1 at Walk-On, Shell, Stengel, Pebble, and Black Point beaches		•	•				•		•							•		•			•	
North Horseshoe Cove	Hwy. 1, 1.5 mi. S. of Rocky Point		•					•		•			•					•				•	
Kruse Ranch	Hwy. 1, 2 mi. S. of Rocky Point		•					•		•								•			•		
Kruse Rhododendron State Reserve	E. of Hwy. 1, off Kruse Ranch Rd.		•	•				•												•	•		
Fisk Mill Cove (Unit of Salt Point State Park)	Hwy. 1, just S. of Kruse Ranch Rd.		•	•		•	•	•		•			•		•		•		•			•	
Stump Beach (Unit of Salt Point State Park)	1 mi. N. of main entrance to Salt Point State Park		•	•		•	•	•		•			•			•		•			•		
Salt Point State Park	Hwy. 1, 20 mi. N. of Jenner		•	•	•	•	•	•		•		•	•	•	•	•	•		•			•	
Ocean Cove Reserve	12 mi. S. of Stewarts Point	•	•	•		•		•		•			•			•		•			•		

GUALALA POINT REGIONAL PARK: The 195-acre park encompasses both uplands and shore. Day-use parking fee; facilities include a paved, wheelchair-accessible trail to the beach, restrooms, and a volunteer-staffed visitor center with changing exhibits, open weekends from Memorial Day to Labor Day. The camping area is in a redwood forest along the Gualala River, east of Highway 1. Facilities include picnic tables, firepits, and a trailer sanitation facility. Camping fees are charged for improved and hiker/bicyclist sites. Information: (707) 785-2377 or www.sonoma-county.org.

Coastal access is provided at seven locations within the Sea Ranch subdivision via the Salal Trail, the Sea Ranch Blufftop Trail, and five Sea Ranch Public Access Trails.

SALAL TRAIL: The trail follows a creek through the Sea Ranch to a small cove. Respect adjacent private property. A fee is charged for parking at the trailhead in Gualala Point Regional Park.

SEA RANCH BLUFFTOP TRAIL: Three-mile-long trail along the bluff-top from Gualala Point Regional Park to Walk-On Beach. Private property is adjacent to the trail; do not trespass.

SEA RANCH PUBLIC ACCESS TRAILS: There are five public parking areas west of Hwy. 1 with trails leading to the beach:

Walk-On Beach: Accessible from the Blufftop Trail or from a trailhead just north of the Leeward Spur Road near Highway 1 milepost 56.53; wheelchair accessible, with a ramp to the beach.

Shell Beach: Trail starts just south of Whalebone Reach near Highway 1 milepost 55.20.

Stengel Beach: Trail begins just north of the stables, near Highway 1 milepost 53.96; there is a bridge and a stairway to the beach.

Pebble Beach: Trail is south of Navigator's Reach, near Highway 1 milepost 52.32.

Black Point Beach: Trail starts near Highway 1 milepost 50.80, just north of the Sea Ranch Lodge.

Wheelchair-accessible restrooms are available at parking areas for all trailheads; fee for parking. Public access rights extend to the base of the bluff or the first line of vegetation on the beaches and throughout the trail and parking easements. All roads in the Sea Ranch are private; do not trespass. Observe signs. Information: (707) 565-2041.

NORTH HORSESHOE COVE: The cove can be reached by trails from any of the Kruse Ranch trailheads. Call: (707) 847-3221.

KRUSE RANCH: 1,350 acres of undeveloped state park land, on both sides of Highway 1. Trails provide access to bluffs and coves, including North Horseshoe Cove and Fisk Mill Cove. Call: (707) 847-3221.

KRUSE RHODODENDRON STATE RESERVE: 300-acre reserve with five miles of hiking trails, restrooms, and parking. No picnic facilities are provided. Rhododendrons bloom from April to June. Please remain on trails. For information, call: (707) 847-3221.

FISK MILL COVE: A unit of Salt Point State Park, with trails to and along bluffs and beaches. Facilities include a parking lot, picnic tables, and wheelchair-accessible restrooms.

STUMP BEACH: A unit of Salt Point State Park. Facilities include picnic tables, firepits, and stairs to the beach.

SALT POINT STATE PARK: A 5,676-acre park that includes both upland and coastal land. Gerstle Cove Campground has 30 improved family sites, a 40-person group camp, and a trailer sanitation facility. Woodside Campground has 79 improved sites, 10 walk-in sites, and 10 unimproved hiker/bicyclist sites; for camping reservations, call: 1-800-444-7275.

The park includes marked hiking and riding trails, picnic areas, small beaches, and tidepools; camping fees are charged. A paved, level wheelchair-accessible path leads from the parking lot in the Gerstle Cove day-use area (west of Hwy. 1) to Salt Point. All restrooms are wheelchair accessible. There is an underwater ecological preserve within the park; a small boat launch is available. Call: (707) 847-3221.

OCEAN COVE RESERVE: Privately owned bluff and beach. Pay camping and day-use fees at the grocery store. Fishing and diving area.

83

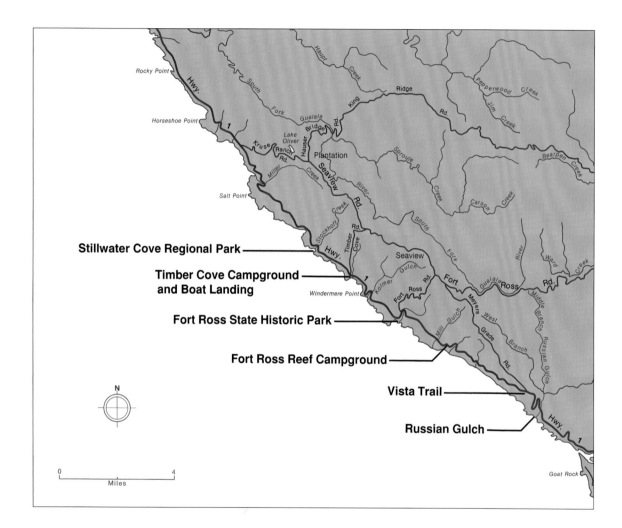

Stillwater Cove Regional Park

Timber Cove Campground and Boat Landing

Fort Ross State Historic Park

Fort Ross Reef Campground

Vista Trail

Russian Gulch

Fort Ross State Historic Park

Fort Ross State Historic Park

Sonoma County

STILLWATER COVE TO RUSSIAN RIVER

NAME	LOCATION	Entrance/Parking Fee	Parking	Restrooms	Lifeguard	Campground	Showers	Firepits	Stairs to Beach	Path to Beach	Bike Path	Hiking Trail	Facilities for Disabled	Boating Facilities	Fishing	Equestrian Trail	Sandy Beach	Dunes	Rocky Shore	Upland from Beach	Stream Corridor	Bluff	Wetland
Stillwater Cove Regional Park	Hwy. 1, 3 mi. N. of Fort Ross	•	•	•		•	•	•	•	•		•	•				•		•	•	•	•	
Timber Cove Campground and Boat Landing	Hwy. 1, 12 mi. N. of Jenner	•	•	•		•	•	•		•				•	•		•		•	•		•	
Fort Ross State Historic Park	Hwy. 1, 11 mi. N. of Jenner	•	•	•			•		•			•	•		•		•		•	•		•	
Fort Ross Reef Campground	Hwy. 1, 8 mi. N. of Jenner	•	•	•		•			•			•	•		•		•		•	•	•	•	
Vista Trail	Hwy. 1, 9 mi. N. of Jenner		•	•								•	•						•			•	
Russian Gulch	Hwy. 1, 2.5 mi. N. of Jenner		•	•					•								•		•	•	•		

STILLWATER COVE REGIONAL PARK: A long stairway leads to the beach. Picnic tables, restrooms, and a path to the beach are wheelchair accessible. For park information, call: (707) 847-3245; for campsite reservations, call: (707) 565-2267.

TIMBER COVE CAMPGROUND AND BOAT LANDING: Privately owned facilities. Day use and camping with hook-ups; use of hot tub included in fee. Boat rentals and sales, guided boat tours, boat launch, and scuba rentals and sales available; air station, fishing licenses, bait and tackle, and propane. Home and cabin rentals. For information: 21350 Highway 1, Jenner 95450; (707) 847-3278.

FORT ROSS STATE HISTORIC PARK: An American outpost for Russian fur traders in the 19th century. Many of the original buildings have been restored. The Fort Ross Living History Program is an annual event that recreates a typical day at the fort in 1836; authentically costumed volunteers portray Russian colonists and foreign visitors to the fort; demonstrations of threshing, musket and cannon drills, cooking, and other activities are presented, and visitors are encouraged to participate. Day-use fee for museum and visitors center, picnic tables, hiking trails, and beach. Guided tours are available daily. There is an offshore underwater park for divers adjacent to the park. For information, call: (707) 847-3286.

FORT ROSS REEF CAMPGROUND: Fee for day use and 20 campsites. Trails lead to beach at Fort Ross in the north, and to two coves. The southern portion includes a trail to the beach at the bottom of Timber Gulch. No dogs. No reservations taken for campsites.

VISTA TRAIL: Blufftop views from a paved wheelchair-accessible loop-trail, picnic tables, restrooms, and a parking lot north of Russian Gulch. Call (707) 847-3286.

RUSSIAN GULCH: A Sonoma Coast State Beach. Parking lot north of gulch with restroom and trail to beach; fee. Call: (707) 847-3286.

Russian Gulch

Mouth of the Russian River

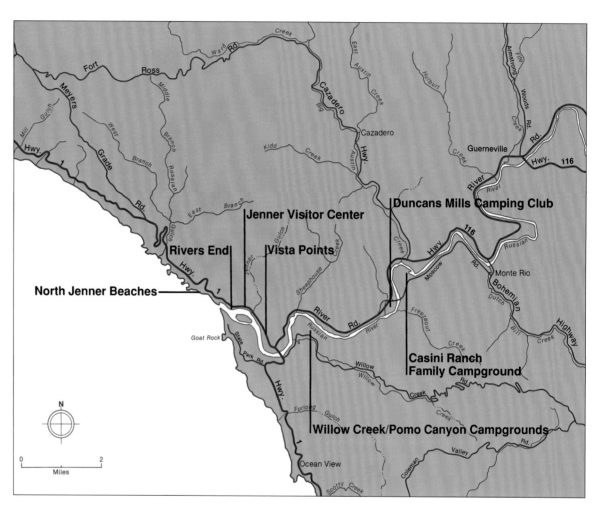

Sonoma County

RUSSIAN RIVER

NAME	LOCATION	Entrance/Parking Fee	Parking	Restrooms	Lifeguard	Campground	Showers	Firepits	Stairs to Beach	Path to Beach	Bike Path	Hiking Trail	Facilities for Disabled	Boating Facilities	Fishing	Equestrian Trail	Sandy Beach	Dunes	Rocky Shore	Upland from Beach	Stream Corridor	Bluff	Wetland
North Jenner Beaches	Hwy. 1, from Russian Gulch to Jenner									•		•			•		•		•			•	
Rivers End	Hwy. 1, Jenner	•	•	•		•				•				•	•		•			•	•	•	
Jenner Visitor Center	Hwy. 1, Jenner		•	•									•	•	•							•	
Vista Points	Hwy. 1, Jenner to Bridgehaven		•																•			•	
Duncans Mills Camping Club	Hwy. 116, Duncans Mills	•	•	•		•	•	•				•		•			•			•	•		
Willow Creek/Pomo Canyon Campgrounds	Willow Creek Rd., E. from Hwy. 1	•	•	•		•		•			•	•							•	•	•	
Casini Ranch Family Campground	Moscow Rd., .75 mi. E. of Hwy. 116, Duncans Mills	•	•	•		•	•	•					•	•	•		•			•	•	

NORTH JENNER BEACHES: The state owns the land from the Vista Trail to the Russian River mouth at Jenner. Several beaches are reached by steep trails down eroding bluffs.

RIVERS END: Privately owned; provides day use, camping, and fishing. Facilities include boat launch and ramp, cabins, restaurant, and bar. For information, call: (707) 865-2484.

JENNER VISITOR CENTER: The Visitor Center has displays on Jenner and the Sonoma coast; other facilities include parking, a small boat launching ramp, picnic tables, and restrooms. The Visitor Center is wheelchair accessible. Call: (707) 875-3483.

VISTA POINTS: Several pull-outs along Hwy. 1 offer views of the Russian River mouth, Penny Island, and the coast.

DUNCANS MILLS CAMPING CLUB: Private, membership-only campground on the Russian River with public day use. Day-use fee per vehicle; no more than ten people per vehicle are permitted. Restrooms and picnic tables available. Call: (707) 865-2573.

WILLOW CREEK/POMO CANYON CAMPGROUNDS: A short walk from parking lots leads to environmental campsites. Willow Creek has 11 sites near the Russian River; the 20 Pomo Canyon sites are in a redwood grove. Campgrounds have toilets, picnic tables, and fire rings, but campers must carry in water and wood. Dogs are not allowed. Two wheelchair-accessible sites at Willow Creek. A trail leads from the Willow Creek Campground to Shell Beach on the coast. The campgrounds are closed during the rainy season. The parking lots are located along Willow Creek Rd., east of Highway 1. Campers must check in at Bodega Dunes Campground. Call: (707) 875-3483.

CASINI RANCH FAMILY CAMPGROUND: Privately owned campground on the Russian River. Public day use allowed when the campground is not full. 160 tent sites, 214 trailer sites with hookups. Fishing licenses, bait, and tackle; boat launch, dock, and boat rentals; playground, picnic areas, and recreation hall. Information: Box 22, Duncans Mills 95430; (707) 865-2255.

Seabirds

As their name implies, seabirds spend most of their time flying, feeding, or resting on the open ocean. Like many shorebirds and waterfowl, they migrate from summer nesting areas in the north to wintering grounds in milder climes.

There are three general categories of seabirds. The "nearshore" or "inshore" birds are those that are typically seen along the coast at beaches, bays, and rocky intertidal areas. Birds such as cormorants, grebes, loons, and scoters usually stay near land, feeding on organisms found in shallow water.

"Offshore" seabirds, such as shearwaters and common murres, usually stay out several miles from land and feed in deep water. These birds usually can be seen only from a boat or when they are nesting on offshore islands such as San Miguel or the Farallones. The third group, the "pelagic" or open ocean species, such as albatrosses, skuas, and Arctic terns, feed and rest in oceanic waters, often several miles from land.

The nesting and breeding activities of seabirds are related to the ocean currents. In the spring, nutrient-rich cooler waters upwell offshore, causing a phytoplankton bloom. This in turn causes a rise in the numbers of fish that provide a source of food for the seabirds.

In contrast to terrestrial birds, seabirds have evolved in an environment that is mainly free from competition. As a result, seabirds have long lifespans, low adult mortality rates, late sexual maturity, and small clutch sizes. Individual birds have been known to live for 20 to 30 years. The Farallon and Channel islands are important breeding areas for seabirds. The entire breeding populations of brown pelicans, black storm-petrels, and Xantus' murrelets nest on the Channel Islands, while large populations of several species of seabirds nest on the Farallones.

The Farallon Islands have long been a significant nesting site. In the 1850s the Farallon Egg Company was formed with the purpose of taking murre eggs from the islands for the booming population of San Francisco. This caused a massive drop in the murre population, and only recently have their numbers been increasing.

Seabirds generally eat fish of all sizes and types, although the anchovy is a favorite. Very few animals prey on the seabirds; however, recent evidence indicates that seabirds are very susceptible to oil spills and pesticides. Populations of many seabird species have begun to decline because of these factors as well as from human disturbance of nests.

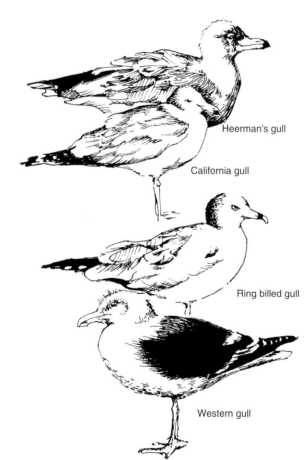

Heerman's gull

California gull

Ring billed gull

Western gull

Pelagic cormorant

Brown pelican

Western grebe

Sonoma County

JENNER TO BODEGA BAY

NAME	LOCATION	Entrance/Parking Fee	Parking	Restrooms	Lifeguard	Campground	Showers	Firepits	Stairs to Beach	Path to Beach	Bike Path	Hiking Trail	Facilities for Disabled	Boating Facilities	Fishing	Equestrian Trail	Sandy Beach	Dunes	Rocky Shore	Upland from Beach	Stream Corridor	Bluff	Wetland
Sonoma Coast State Beaches	Hwy. 1, Russian River to Bodega Head		•	•			•	•	•			•			•		•	•	•		•	•	
Goat Rock	Goat Rock Rd., off Hwy. 1		•	•			•		•			•			•		•	•			•	•	
Wright's Beach	Hwy. 1, 6 mi. N. of Bodega Bay	•	•	•		•		•				•			•		•						
Duncans Landing	Hwy. 1, 5 mi. N. of Bodega Bay		•	•					•				•			•		•			•		
Salmon Creek Beach	Hwy. 1, 2.5 mi. N. of Bodega Bay		•	•				•	•			•			•		•	•			•		•
Bodega Dunes Campground	Hwy. 1, .5 mi. N. of Bodega Bay	•	•	•		•	•	•		•		•	•	•	•	•	•	•			•	•	

SONOMA COAST STATE BEACH: A series of beaches stretching from the Russian River to Bodega Head. Two camping areas are included: Wright's Beach and Bodega Dunes. The park office is at Salmon Creek Lagoon: (707) 875-3483; or call the Russian River district office: (707) 865-2391. A ranger leads walks each Sunday during the summer. An offshore underwater park for divers extends the length of the beaches.

Sonoma Coast State Beach includes Willow Creek, upland from the Russian River, and Penny Island, near the mouth of the Russian River. The other beach units are: Russian Gulch, Goat Rock, Blind Beach, Shell Beach, Wright's Beach, Duncans Landing, Duncans Cove, Gleasons Beach, Portuguese Beach, Schoolhouse Beach, Carmet Beach, Marshall Gulch, Arched Rock Beach, Colemans Beach, Miwok Beach, Salmon Creek Beach, Bodega Dunes Campground, Bodega Head, and Campbell Cove.

All beaches have extremely unsafe currents and waves; even the surf area is unsafe. No lifeguards are provided. Each beach has parking and a path leading to the beach. Many have restrooms. Duncans Cove, Portuguese Beach, Schoolhouse Beach, Carmet Beach, and Colemans Beach all have especially steep trails to the beach. Beaches with other facilities are listed below.

GOAT ROCK: Open grassy peninsula between the Russian River and the ocean. Access to beaches on both the river and the ocean. Picnic tables. Restrooms are wheelchair-accessible; the parking lot is adjacent to the beaches. The 2.6-mile Kortum Trail follows the blufftops from Goat Rock to Shell Beach.

WRIGHT'S BEACH: 30 campsites at beach level. Restrooms are wheelchair accessible. For reservations, call: 1-(800) 444-7275.

DUNCANS LANDING: Used in the 1860s and 1870s for loading timber from Duncans Sawmill on the Russian River to ships anchored in the cove on the south side of the point. There is a small beach at the cove accessible by a steep trail.

SALMON CREEK BEACH: Salmon Creek channel crosses this large, popular, sandy beach. There is a shallow swimming area and waterfowl habitat. A trail leads south to Bodega Head; stay on the designated trail.

BODEGA DUNES CAMPGROUND: 98 campsites, trailer sanitation station, campfire center, picnic tables, horseback riding, and hiking trails. Day-use and camping fees. For information, call: (707) 875-3483. For camping reservations, call: 1-(800) 444-7275.

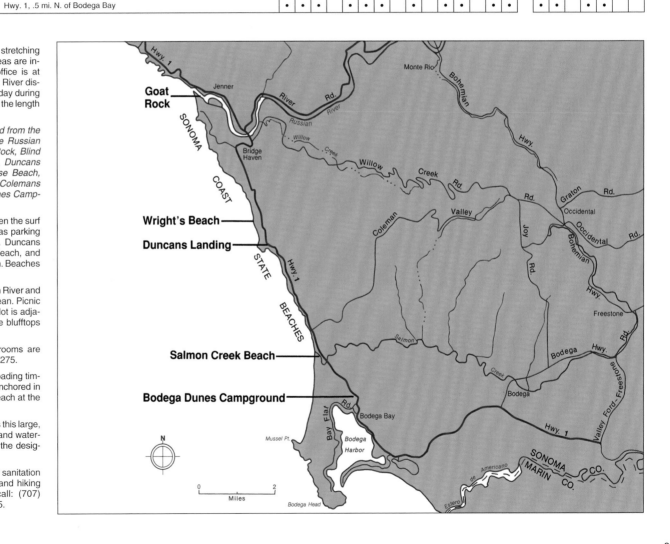

Shorebirds and Waterfowl

Shorebirds and waterfowl are a familiar sight on the coast. Shorebirds are found along the shore or in shallow ponds; they use their narrow, pointed bills to probe intertidal areas for small crustaceans, worms, and other invertebrates. Waterfowl, such as ducks and geese, use open water areas, diving for fish or vegetation.

Almost all shorebirds and waterfowl found in California migrate, either going south in the fall from summer breeding grounds in Canada or Alaska, or going north in the summer from wintering habitat in California, Mexico, or Central Amer-

ica. These birds follow the same routes every year, which are part of the Pacific Flyway. The Pacific Flyway covers the western part of North America and parts of the Arctic in eastern Asia, and is one of four major migration routes across North America.

California's coastal wetlands provide food and resting areas during the fall and winter months for migrating birds. Most waterfowl are not confined to any one area, but move between the wetlands and bays along the coast, and between the coast and more inland areas. These movements are in response to changes in the season and weather, water conditions, and food availability.

Shorebirds are differentiated by their feeding habits. Different species have different bill lengths or shapes and feed in different water levels. Black turnstones, for example, are usually found on rocky shores; snowy plovers and sanderlings on sandy flats; and avocets, western sandpipers, and long-billed dowitchers on mudflats. Willets, in contrast, can feed in a variety of habitats.

Waterfowl are categorized by their feeding habitat; dabbling, or puddle, ducks such as the mallard feed in marshes or nearby grain fields and rest on adjacent open water, while diving ducks such as the canvasback prefer bays, lagoons, or lakes and feed on aquatic vegetation or marine animals. Sea ducks such as scoters seem comfortable in a variety of habitats and, as their name implies, feed primarily on marine invertebrates.

Habitat destruction and disturbance are primary factors in population losses of these birds. For example, the U.S. Fish and Wildlife Service has listed the snowy plover, a small shorebird that nests mostly on sandy beaches, as a threatened species. Much of its nesting habitat has been damaged by development and invasive species and, in other areas, recreational use of the beach disturbs nesting birds. Impacts from recreational activities include direct destruction of eggs and nests, chasing of plovers away from their nests, and interfering with foraging by young birds. In many areas, land managers close portions of public beaches for snowy plover protection during the nesting season, March 1 through September 30.

Willet

Pintail duck

Canvasback duck

Great blue heron

Sanderling

Snowy egret

Avocet

Sonoma County

BODEGA BAY

NAME	LOCATION	Entrance/Parking Fee	Parking	Restrooms	Lifeguard	Campground	Showers	Firepits	Stairs to Beach	Path to Beach	Bike Path	Hiking Trail	Facilities for Disabled	Boating Facilities	Fishing	Equestrian Trail	Sandy Beach	Dunes	Rocky Shore	Upland from Beach	Stream Corridor	Bluff	Wetland
Bodega Bay Harbor	Off Hwy. 1 and Bay Flat Rd., Bodega Bay		•	•										•	•			•					•
Mason's Marina	1820 Westside Rd., Bodega Bay		•	•										•	•			•					
Spud Point Marina	1818 Westside Rd., Bodega Bay		•	•									•	•	•			•					
Westside Regional Park	Off Westside Rd., Bodega Bay	•	•	•		•	•	•		•				•	•		•	•					•
Bodega Marine Laboratory	Off Westside Rd., Bodega Bay		•	•															•	•	•		
Bodega Head	End of Westside Rd., Bodega Bay		•	•						•		•					•	•	•			•	
Campbell Cove	On East Shore Rd., Bodega Bay		•	•						•							•						•
Porto Bodega	1500 Bay Flat Rd., Bodega Bay		•	•		•	•			•				•	•	•		•					
Viewpoint	599 Coast Hwy. 1, Bodega Bay		•	•																			
Bird Walk Trail	Hwy. 1 and N. Harbor Dr., Bodega Bay		•	•								•											•
Pinnacle Gulch Trail	Mockingbird Dr., Bodega Bay	•	•						•			•					•			•	•		
Doran Beach Regional Park	Doran Park Rd., Bodega Bay	•	•	•		•		•		•			•	•	•		•	•					

BODEGA BAY HARBOR: Busiest harbor between San Francisco and Fort Bragg. Numerous parks and privately operated boating facilities. Annual Fishermen's Festival is held in April. Information: (707) 875-3422 or (707) 875-3866.

MASON'S MARINA: Private marina with berths, docks, fuel, haul-out facility, and marine supplies. For information, call: (707) 875-3811.

SPUD POINT MARINA: Intended mainly for commercial fishing boats, the marina also has slips for about 48 recreational boats, boat repair yard, fuel dock, commercial flake-ice machine, hoist, laundromat, hardware store, and chandlery. Restrooms are wheelchair accessible. A breakwater surrounding the marina serves as a public fishing pier. Call: (707) 875-3535.

WESTSIDE REGIONAL PARK: Day-use boating area with launch ramp; 48 campsites; fees are charged. Picnic tables, trailer sanitation facility, and wheelchair-accessible restrooms are available. For camping reservations, call: (707) 565-2267; for information, call: (707) 875-3540.

BODEGA MARINE LABORATORY: University of California research facility, which includes a major aquaculture program. The laboratory is open to the public Fridays 2 PM-4 PM; guided tour includes an overview of marine research laboratory and aquariums. Information: (707) 875-2211.

BODEGA HEAD: Unit of Sonoma Coast State Beaches. The fenced-off area was the site of a proposed nuclear power plant. Hiking trails criss-cross the park; spectacular views of Marin and Sonoma coasts. Winter whale watching. A trail leads from the parking area north to Salmon Creek Beach; stay on the designated trail. For information, call: (707) 865-2391.

CAMPBELL COVE: A Sonoma Coast State Beach. Trail from parking lot to small, sandy cove; boardwalk opposite cove leads to an observation deck overlooking a small lagoon. There is a picnic table and a restroom.

PORTO BODEGA: Facilities include boat dock and launch, 77 berths, 57-space R.V. park with hookups, restaurant, and tackle shop. Information: (707) 875-2354. Charter boats: (707) 875-3344.

VIEWPOINT: Lucas Wharf Restaurant and public fishing pier; accessway on outside deck.

BIRD WALK TRAIL: Paved trail; parking fee. Adjacent salt marsh is home to thousands of birds.

PINNACLE GULCH TRAIL: Take the South Harbour Way exit off Hwy. 1, turn left on Heron Dr. and left on Mockingbird Rd.; the parking lot is on the left at the top of the hill. Fee for parking. A long trail that begins across the street follows a narrow canyon to the sandy beach.

DORAN BEACH REGIONAL PARK: 138 campsites. Facilities include picnic tables, wheelchair-accessible restrooms, trailer sanitation station, boat launch, fish cleaning station, and ocean fishing pier; fees. For camping reservations, call: (707) 565-2267. For information, call: (707) 875-3540. Adjacent U.S. Coast Guard Station is open to the public weekends 1 PM-4 PM.

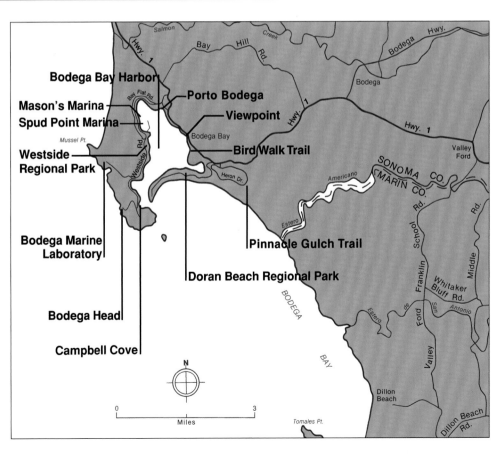

Point Bonita, Golden Gate National Recreation Area

Marin County

North of the Golden Gate, steep headlands form the Marin County coast, with grassy ridges and forested ravines dividing the numerous coastal bluffs. At Bolinas Lagoon the coastline arcs northwestward, creating the crescent-shaped Drakes Bay along the southern shore of the Point Reyes peninsula. The long, straight Point Reyes and McClures beaches, which are frequently buffeted by strong winds and waves, stretch north from the western tip of this peninsula.

The rocky Marin coast, with its frequent fog, has a long history of shipwrecks; over 100 ships are known to have been wrecked here since the first recorded shipwreck in 1595, of the *San Augustin* from Manila. Two lighthouses were built in the late 1800s, one at Point Bonita and one at Point Reyes, to help ships navigate; today, electronic navigation equipment is also used. The Point Reyes Lighthouse, open to the public, is a popular spot to watch migrating gray whales.

A primary force in the formation of the Marin County coast has been the San Andreas Fault, whose movements have created the southwest trending rift zone that contains the Olema Valley, Tomales Bay, and Bolinas Lagoon. This rift zone geologically separates the Point Reyes peninsula from the rest of the county, and roughly outlines the eastern boundary of the 64,000-acre Point Reyes National Seashore, one of only eleven national seashores in the United States.

The sixteen-mile-long Tomales Bay supports a thriving commercial shellfish industry that began in the late 1800s. Giant Pacific oyster seeds were imported from Japan and are the most significant commercial oyster type harvested in California. Drakes Estero, on the Point Reyes peninsula, produces 20% of California's commercial crop. Oyster beds can also be seen on the east side of Tomales Bay near Millerton Point. The southern end of the bay is a wildlife refuge for migrating waterfowl, while silver salmon and steelhead trout spawn in adjacent White House Pool.

The Miwok Indians were the native settlers of Point Reyes; it is estimated that they inhabited more than one hundred villages on the peninsula at the time of the English explorer Francis Drake's supposed visit in 1579. The Miwok thrived on the bay and ocean shores until the Spanish missionaries arrived in 1820. Subsequent Spanish settlement of the coast led to the eventual destruction of the Miwok culture, primarily because of Spanish changes in land tenures and the introduction of European diseases.

The remoteness of Point Reyes and the presence of heavy fog and strong winds have been major factors in preserving the rural character of the area. For over a century the main industries on the peninsula have been dairy and cattle ranching; in 1962, after long debate with developers, dairymen, and concerned citizens, Congress and President John F. Kennedy passed legislation creating the Point Reyes National Seashore, which formally preserved the land while allowing ranching to continue. Today, Point Reyes National Seashore adjoins the Golden Gate National Recreation Area (GGNRA), established in 1972, which extends from Marin County south through San Francisco to San Mateo County.

The grassy Marin Headlands, just across the Golden Gate from San Francisco, have also remained undeveloped; the Army maintained these lands for defense purposes until recently. Now almost all the Army land in the Headlands, including the decaying bunkers, housing barracks, and an old balloon hangar, are part of GGNRA.

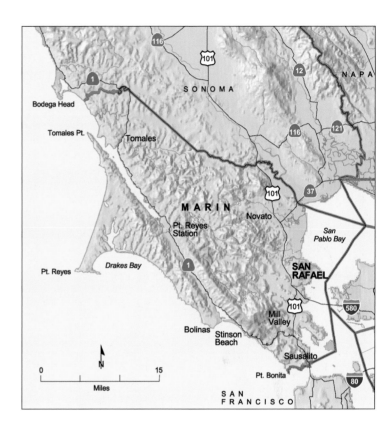

For additional information on the Marin County coast, write or call the West Marin Chamber of Commerce, P.O. Box 1045, Point Reyes Station 94956, (415) 663-9232, or see www.pointreyes.org; or the Marin County Convention and Visitors Bureau, 1013 Larkspur Landing Circle, Larkspur 94939, (415) 499-5000, www.visitmarin.org or www.marin.org. For transit information, call Golden Gate Transit: (415) 923-2000, TDD 257-4554, or see www.goldengate.org; and San Francisco's MUNI: (415) 673-MUNI, or see www.sfmuni.com.

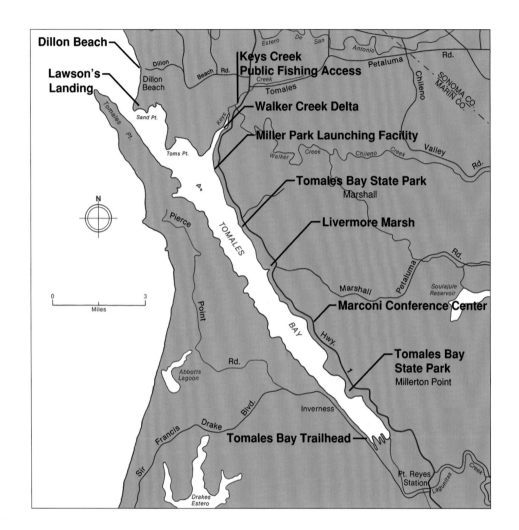

Dillon Beach

Lawson's Landing

Keys Creek Public Fishing Access

Walker Creek Delta

Miller Park Launching Facility

Tomales Bay State Park

Livermore Marsh

Marconi Conference Center

Tomales Bay State Park
Millerton Point

Tomales Bay Trailhead

Dillon
Dillon Beach
Beach Rd.
Tomales
Sand Pt.
Toms Pt.
TOMALES
Pierce
BAY
Point
Rd.
Abbotts Lagoon
Blvd.
Inverness
Francis Drake
Sir
Drakes Estero
Estero De San Antonio
Petaluma
Rd.
SONOMA CO.
MARIN CO.
Chileno
Creek
Walker Creek
Chileno Creek
Valley
Rd.
Marshall
Petaluma
Rd.
Marshall
Soulajule Reservoir
Hwy. 1
Pt. Reyes Station
Lagunitas Creek

N

0   3
Miles

Dillon Beach

Marin County

TOMALES BAY EAST

NAME	LOCATION	Entrance/Parking Fee	Parking	Restrooms	Lifeguard	Campground	Showers	Firepits	Stairs to Beach	Path to Beach	Bike Path	Hiking Trail	Facilities for Disabled	Boating Facilities	Fishing	Equestrian Trail	Sandy Beach	Dunes	Rocky Shore	Upland from Beach	Stream Corridor	Bluff	Wetland
Dillon Beach	Dillon Beach Rd., off Hwy. 1	•	•	•										•	•		•	•					
Lawson's Landing	5 mi. W. of Tomales, off Hwy. 1, S.W. of Dillon Beach	•	•	•		•								•	•		•	•					
Keys Creek Public Fishing Access	Hwy. 1, 1 mi. S.W. of Tomales		•	•											•						•		•
Walker Creek Delta	Delta area 2 mi. S. of Tomales, W. of Hwy. 1																				•		•
Miller Park Launching Facility	Hwy. 1, 3 mi. N. of Marshall at Nick's Cove		•	•										•	•		•		•				
Tomales Bay State Park-Marshall	Hwy. 1, 1.25 mi. N. of Marshall																•	•					
Livermore Marsh	Marshall-Petaluma Rd., Hwy. 1 intersection																		•	•			•
Marconi Conference Center	E. side of Hwy. 1, 1 mi. S. of Marshall at Marconi Cove		•	•									•	•						•			
Tomales Bay State Park-Millerton Point	On Hwy. 1, 4.5 mi. N. of Pt. Reyes Station		•	•				•		•		•	•	•	•		•		•			•	•
Tomales Bay Trailhead	Edge of Tomales Bay Ecological Reserve, 2.5 mi. N. of Pt. Reyes Station		•									•											

DILLON BEACH: Privately owned beach, with crabbing, clamming, and fishing. Day-use parking fee. Picnic tables on the wide, sandy beach.

LAWSON'S LANDING: Take the toll road from Dillon Beach; car day-use and camping fees. Popular area for gaper clamming. Trailer and tent camping on 50-acre meadow; no reservations. Picnic tables; some fire rings. A "clam barge" runs to clam beds, February through July (fee). Sandy boat launch area and fuel dock open 7 AM-5 PM year-round; launch assistance available for fee. The boathouse is open for bait and tackle, fishing licenses, and information from 7 AM-5 PM, February through November. For information, call: (707) 878-2443.

KEYS CREEK PUBLIC FISHING ACCESS: A trail leads from the gravel parking lot down to and along the shore of Keys Creek. Shoreline fishing only. For information, call: (415) 945-1455.

WALKER CREEK DELTA: Part of the 500-acre Cypress Grove area of Audubon Canyon Ranch, a land preservation and education organization sponsored by four Bay Area Audubon Society chapters. Audubon Canyon Ranch lands on Tomales Bay include Livermore, Marshall, Shields, and Olema marshes, Tom's Point, Hog and Duck islands, and other tideland strips along the eastern shoreline. All are sensitive marsh areas and important migratory waterfowl habitats. Access is reserved primarily for educational and scientific purposes and is by appointment only, except at Shields Marsh and at the Bolinas Bay headquarters. For appointment or information, call: (415) 663-8203.

MILLER PARK LAUNCHING FACILITY: Open sunrise to sunset. Pier, jetty, and concrete launch ramp; paved parking.

TOMALES BAY STATE PARK-MARSHALL: Approximately 60 acres of undeveloped land north of the town of Marshall.

LIVERMORE MARSH: Undeveloped Audubon Canyon Ranch land. The marsh is part of the Cypress Grove freshwater marsh restoration area, which also includes Marshall Marsh. Access is by appointment only; call: (415) 383-1644.

MARCONI CONFERENCE CENTER: Originally built as the intended Pacific link of Marconi's global radio communications system, this 62-acre State Historic Park includes 15 historic buildings and land-scaped and wooded grounds on the blufftop above Marconi Cove. The conference center is available for group use by reservation only. For conference center information or reservations, call: (415) 663-9020. Day-use facilities, renovation of major historic buildings, and a summer arts program are planned. For information, call: 1-800-970-6644.

TOMALES BAY STATE PARK-MILLERTON POINT: This 180-acre state park unit includes Alan Sieroty Beach. Day-use area with picnic tables and grills. Restrooms are wheelchair accessible with assistance. Trail to the beach; Tomasini Point, which is undeveloped, is one mile to the north. For more information, call: (415) 669-1140.

TOMALES BAY TRAILHEAD: Trail to the southern tip of Tomales Bay at the edge of the Tomales Bay Ecological Reserve; some sections are steep. Small gravel parking lot.

Dillon Beach

Camping

There are many campgrounds within state and national parks, as well as hundreds of private campgrounds, located near the coast. Campsites can be found in redwood forests, along streams, in grassy fields, or even right on the sand.

A typical campsite in a developed park provides a place to park your car or recreational vehicle, a picnic table, room to pitch a tent, a fire ring or grill, and nearby water; restrooms serve a cluster of campsites. At some campgrounds there may be cupboards at each site, firewood for sale, hot showers, R.V. hookups and sanitation stations, groceries and other concessions. There is a fee charged at all drive-in campgrounds.

There are also walk-in campsites for backpackers and bicyclists often available within large campgrounds, or at parks that are otherwise only for day use. A nominal fee is usually charged. Environmental and enroute campsites are now available in certain state parks throughout the state.

During the summer and on holiday weekends campgrounds fill quickly, often by mid-day. If you do not have reservations, start early! Your stay is often limited to a few nights; in the off-season, however, the limit of stay may be extended. In Northern California where there is persistent summer fog, spring and fall can be the best times to camp.

Most state campgrounds allow pets for a small fee; they must be kept on a leash during the day and in your vehicle at night. Always check the campground policy ahead of time. Permits are required in order to camp and to build fires at some national parks; check at the headquarters of the area.

Most campsites at state park campgrounds can be reserved in advance for a service fee through ReserveAmerica™. Reservations, which are recommended for popular areas, can be made 7 months in advance on the first day of the month beginning at 8:00 AM Pacific Standard Time in one of three ways: online at www.reserveamerica.com; by mail to ReserveAmerica™, P.O. Box 1510, Rancho Cordova, CA 95741-1510; or by phone: 1-800-444-7275 (TDD 1-800-274-7275). See www.parks.ca.gov for more information on different types of camping, how reservations work, and for a list of parks with first-come, first-served campsites.

NAME	LOCATION	Entrance/Parking Fee	Parking	Restrooms	Lifeguard	Campground	Showers	Firepits	Stairs to Beach	Path to Beach	Bike Path	Hiking Trail	Facilities for Disabled	Boating Facilities	Fishing	Equestrian Trail	Sandy Beach	Dunes	Rocky Shore	Upland from Beach	Stream Corridor	Bluff	Wetland
Marshall Beach	Access Rd. off Pierce Point Rd., 5.5 mi. N. of Inverness		•	•						•							•					•	
Tomales Bay State Park	2 mi. N. of Inverness on Pierce Point Rd.	•	•	•		•	•	•		•	•	•	•				•			•	•		
Chicken Ranch Beach	Sir Francis Drake Blvd., .5 mi. N. of Inverness		•	•						•							•			•		•	
Dana Marsh/Path to Beach	Sir Francis Drake Blvd., .25 mi. N. of Inverness									•							•						
Martinelli Park	Sir Francis Drake Blvd., .5 mi. S. of Inverness																•					•	
Tomales Bay State Park-Inverness Ridge	Sir Francis Drake Blvd., 1 mi. S. of Inverness											•					•						
Shields Saltmarsh Study Area	Sir Francis Drake Blvd., 1 mi. S. of Inverness		•									•					•						•
Tomales Bay Ecological Reserve	Southern portion of Tomales Bay		•																				•

MARSHALL BEACH: Part of Point Reyes National Seashore; a steep trail leads to the beach, which is on Tomales Bay. Restrooms are available; parking along road shoulder.

TOMALES BAY STATE PARK: Main parking and picnic areas are at Heart's Desire Beach, off Pierce Point Road. Indian Beach and Pebble Beach are accessible only by trail. Shell Beach can be reached by a 4.1- mile-long trail from Heart's Desire Beach, or by a .5-mile-long path from a parking area at the end of Camino Del Mar. Trails from Shell Beach also lead to Indian and Pebble beaches. Miwok Indian relics and a forest of ancient Bishop pines are within the park. Camping facilities are accessible to hikers and bicyclists only. Restrooms, parking, and picnic tables at Heart's Desire Beach are wheelchair accessible. For information, call: (415) 669-1140.

CHICKEN RANCH BEACH: This small sandy beach on Tomales Bay is reached via a bridge over a creek. There is limited street parking; chemical toilet available.

DANA MARSH / PATH TO BEACH: Dana Marsh is a small restored wetland adjacent to Tomales Bay. There is an unpaved path to a small beach on the bay, accessible from Sir Francis Drake Blvd. at the foot of Rannoch Way. Adjacent private property; do not trespass.

MARTINELLI PARK: Small park with viewing areas adjacent to Tomales Bay.

TOMALES BAY STATE PARK-INVERNESS RIDGE: An undeveloped unit of the park. There are some hiking trails and some picnic tables above the beach among the pines. Check at the main park for more information: (415) 669-1140.

SHIELDS SALTMARSH STUDY AREA: Audubon Canyon Ranch land. Day-use area provides a trail into the marsh, an overlook with a small bench and interpretive displays, and a view of Tomales Bay. Limited parking in roadside pull-out.

TOMALES BAY ECOLOGICAL RESERVE: 500-acre reserve for migrating waterfowl and other wildlife. Many vista points from Hwy. 1 and Sir Francis Drake Blvd. along Tomales Bay. Entering the marsh is not encouraged due to sensitive resources. The land area is closed March 1 to June 30.

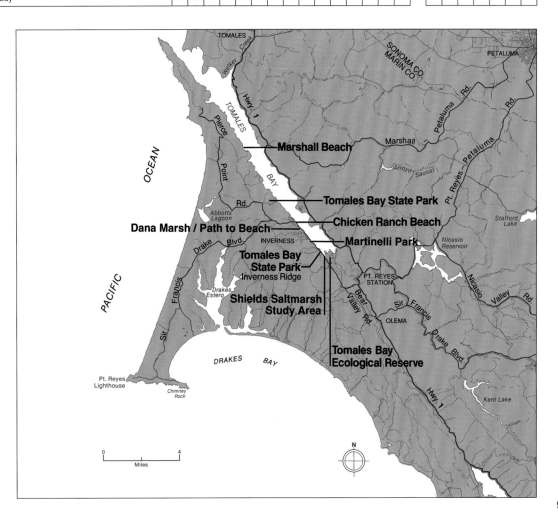

Point Reyes Lighthouse

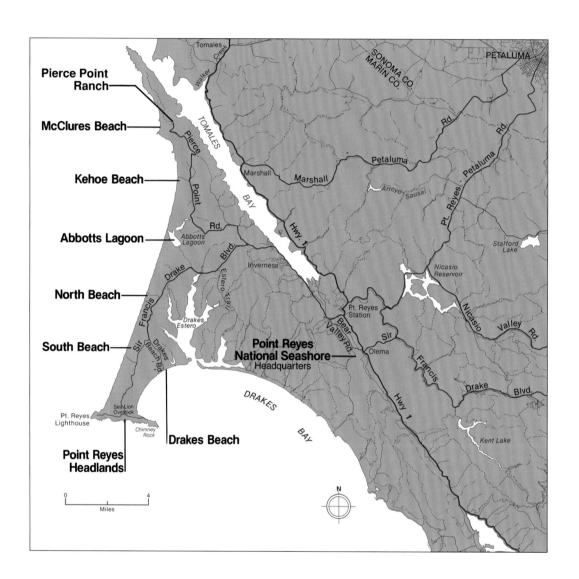

Pierce Point Ranch

McClures Beach

Kehoe Beach

Abbotts Lagoon

North Beach

South Beach

Point Reyes
National Seashore
Headquarters

Drakes Beach

Point Reyes
Headlands

Marin County

POINT REYES WEST

NAME	LOCATION	Entrance/Parking Fee	Parking	Restrooms	Lifeguard	Campground	Showers	Firepits	Stairs to Beach	Path to Beach	Bike Path	Hiking Trail	Facilities for Disabled	Boating Facilities	Fishing	Equestrian Trail	Sandy Beach	Dunes	Rocky Shore	Upland from Beach	Stream Corridor	Bluff	Wetland
Point Reyes National Seashore	Headquarters on Bear Valley Rd., .25 mi. N.W. of Olema	•	•		•		•		•		•		•	•			•		•	•			
Pierce Point Ranch	End of Pierce Point Rd., off Bear Valley Rd.	•					•				•	•								•			
McClures Beach	End of Pierce Point Rd., off Bear Valley Rd.	•	•				•				•						•	•					
Kehoe Beach	Trail off Pierce Point Rd., S. of McClures Beach	•	•								•						•						•
Abbotts Lagoon	Trail from Pierce Point Rd., off Bear Valley Rd.	•	•								•												•
North Beach and South Beach	Off Sir Francis Drake Blvd.	•	•								•						•						
Point Reyes Headlands	End of Sir Francis Drake Blvd.	•	•																•			•	
Drakes Beach	Drakes Beach Rd., off Sir Francis Drake Blvd.	•	•		•	•					•						•				•		

POINT REYES NATIONAL SEASHORE: Headquarters for the seashore is located on Bear Valley Rd.; the visitor center features a bookstore, exhibits, and interpretive displays. Interpretive programs include a Miwok village, a wheelchair-accessible Earthquake Trail, and a Morgan horse farm. Trailhead for over 100 miles of trails, many of which lead to beaches; many species of birds and mammals in the area. Camping is permitted at four walk-in camps by reservation only; Wildcat walk-in camp is a group site. The visitor center is open 9 AM-5 PM weekdays, 8 AM-5 PM weekends. Permits are required for camping, and are available at the visitor center or by phone reservation from 9 AM-noon daily. 36,000 acres of the 71,000-acre National Seashore have been designated the Phillip Burton Wilderness Area. The Wilderness Area comprises a patchwork of tracts interspersed among grazing and other private lands. A network of foot trails leads through the wilderness; there are no structures or pavement. No bicycles or motorized vehicles are permitted. For National Seashore information or camping reservations, call: (415) 464-5100 or see www.nps.gov/pore.

The Point Reyes National Seashore Education Program offers classes and field seminars on environmental education and natural history, a hostel program for seniors, and a youth camp. The Clem Miller Environmental Education Center, an outdoor overnight facility available to groups for educational purposes, features hiking trails and interpretive programs. For information or reservations for field programs or the Clem Miller Center, write: Pt. Reyes National Seashore Association, Pt. Reyes Station 94956, or call: (415) 663-1224.

The following are within the Point Reyes National Seashore: Pierce Point Ranch, McClures Beach, Kehoe Beach, Abbotts Lagoon, North Beach, South Beach, Point Reyes Headlands, and Drakes Beach.

PIERCE POINT RANCH: A wheelchair-accessible interpretive trail passes by the historic buildings of the ranch. A 3.5-mile-long trail leads north from the ranch to Tomales Point, with views of Tomales Bay, Bodega Head, and Point Reyes. An unmarked, unmaintained trail follows the old ranch road to White Gulch on Tomales Bay. No dogs are permitted on Tomales Point.

McCLURES BEACH: There is a parking lot west of the Pierce Point Ranch parking area; a steep trail leads from the parking lot to a sandy beach and tidepools. Strong surf makes swimming extremely unsafe. Granite seastacks and rocky intertidal areas host abundant marine life, including cormorants, endangered California brown pelicans, seals, and giant sea anemones.

KEHOE BEACH: A .5-mile-long trail to the beach runs beside Kehoe Marsh, a migratory waterfowl and year-round bird habitat. Most of the trail is wheelchair negotiable. Wheelchair-accessible restrooms are available; parking along road shoulder. This beach supports snowy plovers, a threatened species, and there may be some restrictions on public access in this area to protect the habitat.

ABBOTTS LAGOON: The lagoon is a migratory waterfowl habitat; canoeing is permitted. Park along road shoulder; two-mile-long trail to the beach. Wheelchair-accessible restroom.

NORTH BEACH AND SOUTH BEACH: A long windy strand with two main access areas; also called 10-Mile, Point Reyes, or Great Beach. Dogs on leash and campfires allowed on the beach. No swimming. Restrooms at North and South beaches are wheelchair accessible. North Beach supports snowy plovers, a threatened species, and there may be some restrictions on public access in this area to protect the habitat.

POINT REYES HEADLANDS: From the parking area near Point Reyes Light, trails lead to two overlooks, Chimney Rock and Sea Lion. A .5-mile-long paved path leads to the Point Reyes Light Visitor Center, view platform, and long stairway to the lighthouse. During whale-watching season (December to April), a shuttle runs between the lighthouse and the Drakes Beach parking area, weekends and holidays only. Special parking and drop-off areas and arrangements for visitors with limited mobility are available. Interpretive walks are led through the lighthouse; Visitor Center open Thursday through Monday, 10 AM-4:30 PM; stairs closed when wind and weather conditions warrant. For information, call: (415) 669-1534.

DRAKES BEACH: A popular broad sandy beach with chalk white cliffs; swimming and beach fires permitted. The Kenneth C. Patrick Visitor Center has interpretive displays and ranger-led walks; open variable hours weekends and summers only. Facilities include snack bar, picnic tables, and wheelchair-accessible restrooms. Near the visitor center is a monument to Sir Francis Drake's landing in 1579. For information, call: (415) 669-1250.

McClures Beach, Point Reyes National Seashore

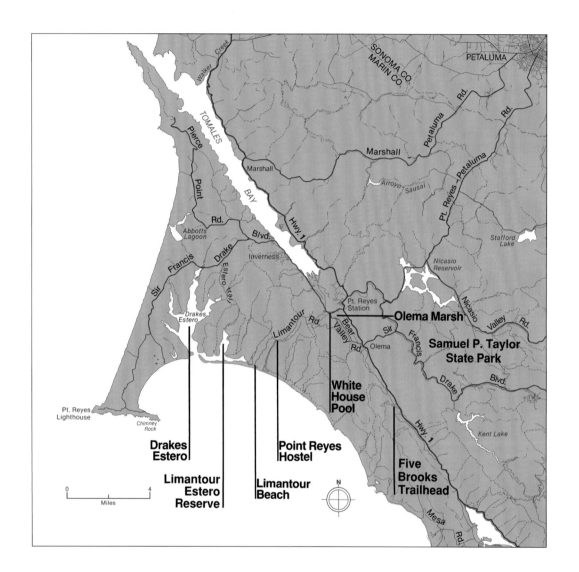

Pt. Reyes
Lighthouse

Chimney
Rock

0 ——— 4
Miles

TOMALES
BAY

Walker Creek

Marshall

Pierce Point

Abbotts Lagoon

Rd.

Blvd.

Inverness

Drake

Estero Trail

Francis

Slr

Drakes Estero

Limantour Rd.

Bear Valley Rd.

Pt. Reyes Station

Olema

Marshall

Arroyo Sausal

Hwy. 1

Pt. Reyes–Petaluma Rd.

Rd.

Rd.

PETALUMA

SONOMA CO.
MARIN CO.

Stafford Lake

Nicasio Reservoir

Nicasio Valley Rd.

Slr

Francis

Drake

Blvd.

Kent Lake

Hwy. 1

Mesa Rd.

N

Olema Marsh

Samuel P. Taylor State Park

White House Pool

Drakes Estero

Limantour Estero Reserve

Point Reyes Hostel

Limantour Beach

Five Brooks Trailhead

Five Brooks Trailhead

White House Pool

Marin County

POINT REYES SOUTH

NAME	LOCATION	Entrance/Parking Fee	Parking	Restrooms	Lifeguard	Campground	Showers	Firepits	Stairs to Beach	Path to Beach	Bike Path	Hiking Trail	Facilities for Disabled	Boating Facilities	Fishing	Equestrian Trail	Sandy Beach	Dunes	Rocky Shore	Upland from Beach	Stream Corridor	Bluff	Wetland
Drakes Estero	Estero Trail off Sir Francis Drake Blvd.		•	•						•		•											•
Limantour Estero Reserve	End of Limantour Rd.		•	•								•											•
Limantour Beach	End of Limantour Rd.		•	•					•		•	•			•		•	•					•
Point Reyes Hostel	Off Limantour Rd., 7 mi. from Bear Valley Rd.	•	•	•			•	•				•	•							•			
White House Pool	Sir Francis Drake Blvd. and Bear Valley Rd. intersection, N.E. corner		•	•							•	•			•						•		
Olema Marsh	Sir Francis Drake Blvd. and Bear Valley Rd. intersection, S.E. corner																						•
Samuel P. Taylor State Park	Sir Francis Drake Blvd., 5.2 mi. E. of Olema	•	•	•		•	•	•			•	•	•		•					•	•		
Five Brooks Trailhead	Hwy. 1, 3 mi. S. of Olema		•	•								•			•					•		•	

The following are within the Point Reyes National Seashore: Drakes Estero, Limantour Estero, Limantour Beach, Point Reyes Hostel, and Five Brooks Trailhead.

DRAKES ESTERO: The largest saltwater lagoon along the Marin Coast. Oysters are available commercially at the north end of the estero; take the dirt road off Sir Francis Drake Blvd. west of Estero Trail. Parking and restrooms are available at Estero trailhead.

LIMANTOUR ESTERO RESERVE: 500-acre marine reserve along the north side of Limantour Spit. Removal of any marine life is prohibited without a scientific collecting permit from the Department of Fish and Game; permits are also available from Point Reyes National Seashore or the Golden Gate National Recreation Area. No permits are required for viewing and observation.

LIMANTOUR BEACH: Long sandy spit with grassy dunes between the ocean and the estero. A wheelchair-accessible trail runs along the lagoon southeast of the main access road (watch for sign). The beach itself can also be reached via the Coastal Trail and the Bay Area Ridge Trail from the south end of the National Seashore. Sculptured Beach, a sandy beach three miles south of Limantour, is accessible only by trail (Coastal Trail from Bear Valley), or along the beach from points north or south at very low tide. For weather, trail, and tide information, call: (415) 464-5100.

POINT REYES HOSTEL: Seven miles from Point Reyes Headquarters at Bear Valley, via Limantour Road. Advance reservations are strongly recommended year-round; available by mail only. 44 beds, including bunkhouse space for groups; wood stoves and full kitchen. A wheelchair ramp leads from the parking lot to the main house. Restrooms are wheelchair accessible. Overnight fees are charged. For reservations and information, write P.O. Box 247, Pt. Reyes Station 94596, or call (415) 663-8811 from 7:30-9:30 AM or 4:30-9:30 PM only.

WHITE HOUSE POOL: Steelhead and salmon fishing stream; season is November-February. Benches and pedestrian bridge provide views of the stream, Tomales Bay, and the ecological reserve; path to bridge and benches are wheelchair accessible. Good bird-watching site. Large paved parking lot is open 4 AM-9 PM.

OLEMA MARSH: Undeveloped Audubon Canyon Ranch land; the 40-acre marsh is reserved for scientific study and educational uses. Visit by appointment only; for information, call: (415) 868-9244.

SAMUEL P. TAYLOR STATE PARK: 2,600 acres of wooded canyons along Lagunitas (or Paper Mill) Creek. The park is the former site of the first paper mill on the Pacific Coast, built in 1856. The park features picnic areas, trails, and campgrounds, including two group sites, 60 developed sites, a large hiker/bicyclist site for groups, 30 enroute sites, and three separate equestrian sites. For camping reservations, call: 1-800-444-7275.

Picnic areas, all campground restrooms, and two campsites are wheelchair accessible; there are four miles of paved trails. Streams in the park are closed to fishing and boating year-round; there is a children's swimming area in the creek. Paved and unpaved cycling paths available. A visitor center features park information and exhibits. For information, call: (415) 488-9897.

FIVE BROOKS TRAILHEAD: A network of trails begins here for the southern part of Point Reyes. Many trails lead over wooded Inverness Ridge to beaches in the Phillip Burton Wilderness Area of the National Seashore. Trails also lead to the Coastal Trail and to walk-in campsites. Check with the Bear Valley Visitor Center for trail conditions; trail maps available at all park visitor centers. Parking and restrooms at the trailhead; main trail is partially wheelchair accessible with assistance. For park and trail information, call: (415) 464-5100. Horse rentals available; call: (415) 663-1570.

Limantour Estero Reserve

California Gray Whale

The gray whales (*Eschrichtius robustus*) are the most frequently seen whales along the California coast. They have a lengthy migration that takes them from their summer feeding grounds off Alaska in the Bering Sea and Arctic Ocean to the winter breeding lagoons along Baja California. During this migration, thousands of Californians head for the coast to witness the passing of 23,000 gray whales.

The gray whale is mostly a bottom-feeder who lives on small amphipods. The whales are 15 feet long at birth, 25 feet at one year, 35 feet at sexual maturity (eight years), and 45 feet at physical maturity (30-40 years). A mature adult weighs about 50 tons, and has a series of ridges along its back instead of a dorsal fin. Gray whales display their 10-foot-wide flukes at the onset of deep dives and their blow usually has a single plume, reaching a height of 10 feet.

In winter, from November through February, the gray whale travels south along the coast, staying only about one-half mile offshore. The whales swim in small groups, with the pregnant females heading south first, followed by the other mature females and males, then the immature young. The whales swim about 20 hours a day at an average speed of four knots. This migration is a three-month, 13,000-mile round trip—one of the longest migrations of any mammal. There is little evidence that the whales feed during the migration period, and their body weight drops about 20 percent.

The lagoons of Baja host the gray whales during winter. Here, the females give birth and nurse their young. Non-pregnant females often assist in the births. Gestation is 13 months, and females produce one calf every two years at most.

After the young are born the whales head for the northern waters. Newly pregnant females are the first to leave, followed by the adult males, then the mothers with calves.

The California gray whale once had a population estimated at 30,000. Whaling off California and Baja during the late 1800s reduced their numbers greatly. With few whales left, whalers no longer found it profitable to hunt the gray whale, and left them alone until the 1920s and 1930s. Hunting was resumed at that time until the whale population was again reduced. In 1938, an international treaty gave the gray whale complete protection. Since then, the population has steadily grown.

Whale watching has become quite an attraction along the coast during the gray whale's migration. The whales can be seen from many points along the coast, such as the Point Reyes Lighthouse, but many enthusiasts head offshore in powerboats. Organized whale-watching expeditions leave from several points along the coast including San Francisco, Monterey, Mission, and San Diego bays, Redondo Beach, and San Pedro.

Further information on whale-watching expeditions can be obtained from:

Oceanic Society Expeditions
Fort Mason Center
San Francisco, CA 94123   (415) 474-3385

or

The American Cetacean Society:
Whale Adventures
P.O. Box 1391
San Pedro, CA 90733   (310) 548-7821

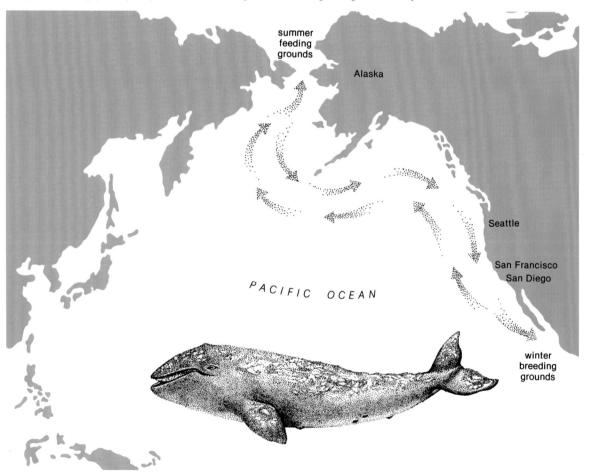

summer feeding grounds

Alaska

Seattle

San Francisco
San Diego

PACIFIC OCEAN

winter breeding grounds

Marin County

BOLINAS

NAME	LOCATION	Entrance/Parking Fee	Parking	Restrooms	Lifeguard	Campground	Showers	Firepits	Stairs to Beach	Path to Beach	Bike Path	Hiking Trail	Facilities for Disabled	Boating Facilities	Fishing	Equestrian Trail	Sandy Beach	Dunes	Rocky Shore	Upland from Beach	Stream Corridor	Bluff	Wetland
Point Reyes Bird Observatory	Mesa Rd., 4.5 mi. N. of Bolinas	•								•									•				
Palomarin Trailhead	End of Mesa Rd., 5 mi. N.W. of Bolinas	•	•							•									•			•	
Agate Beach	End of Elm Rd., Bolinas	•	•						•								•		•				
Duxbury Reef	S.W. perimeter of Bolinas Mesa	•	•							•					•				•				
Bolinas Overlook	End of Overlook Dr., Bolinas	•																		•		•	
Bolinas Beach	Ends of Brighton and Wharf Ave., Bolinas	•							•						•		•		•				•
Bolinas Lagoon Nature Preserve	Between Hwy. 1 and Olema-Bolinas Rd., Bolinas	•																					•
Audubon Canyon Ranch	Hwy. 1, 3 mi. N. of Stinson Beach	•	•									•											•

POINT REYES BIRD OBSERVATORY: One of the few full-time ornithological research stations in the United States. Open every day April 1 through November, Wednesdays and weekends only Thanksgiving through March, weather permitting. Visitors can observe bird banding and 'net runs' every morning. A small visitor center is open early morning to 5 PM; self-guided nature trail at the field station. Special tours for groups can be arranged in advance; a small donation is requested. Natural history trips are also offered year-round. For information about the research station, tours, or trips, call: (415) 868-1221.

PALOMARIN TRAILHEAD: Southernmost trailhead in the Point Reyes National Seashore; reached from Hwy. 1 via Olema-Bolinas Rd. or Horseshoe Hill Rd. to Bolinas, then Mesa Rd. west of Bolinas. A network of trails leads along the bluffs and to lakes, beaches, and walk-in camps. For information on trail conditions, call: (415) 464-5100; maps are available at all park visitor centers.

AGATE BEACH: A path runs to the small beach from a parking lot at the end of Elm Road. Walk at low tide to Duxbury Reef Marine Reserve. There is also a trail from the parking lot to Bolinas Overlook at the end of Overlook Drive.

DUXBURY REEF: Marine reserve below the Bolinas Mesa headlands. Access is from Agate Beach. Permission is required from the Department of Fish and Game to take marine life from the reef or tidepools. Restricted public fishing.

BOLINAS OVERLOOK: Spectacular view from Duxbury Reef south.

BOLINAS BEACH: A gated concrete ramp from the end of Brighton Ave. leads to a sand and pebble beach that is covered at high tide. If gate is locked, call during business hours for access: (415) 868-1224. A small county park with restrooms and tennis courts is near the entrance. The beach extends around the mesa to Bolinas Lagoon; access is from the end of Wharf Avenue. Limited parking at each entrance. Horses are allowed on the beach.

BOLINAS LAGOON NATURE PRESERVE: Viewing of the county-managed preserve from either Hwy. 1 on the eastern side, or Olema-Bolinas Rd. on the western side. Migratory waterfowl habitat; harbor seals hauled-out on Kent and Pickleweed islands can be viewed from shore or from non-motorized boats kept at a distance.

AUDUBON CANYON RANCH: Headquarters for the Audubon Canyon Ranch wildlife sanctuaries and nature preserves. 1,500 acres adjoining Bolinas Lagoon Nature Preserve provides access to heavily wooded canyons from mid-March to mid-July. Facilities include a visitor center with bookstore and exhibit hall, a picnic area, and eight miles of trails. The Henderson overlook, a .5-mile walk up a moderately steep trail, is an observation point for great blue heron and egret nesting sites; the overlook has telescopes, benches, self-guided nature trails, interpretive services, and guided walks. A donation is requested. Audubon Canyon Ranch is open to the public for day use only on weekends and holidays; schools and other groups may visit Tuesday-Friday by appointment. For information, call: (415) 868-9244 or see www.egret.org.

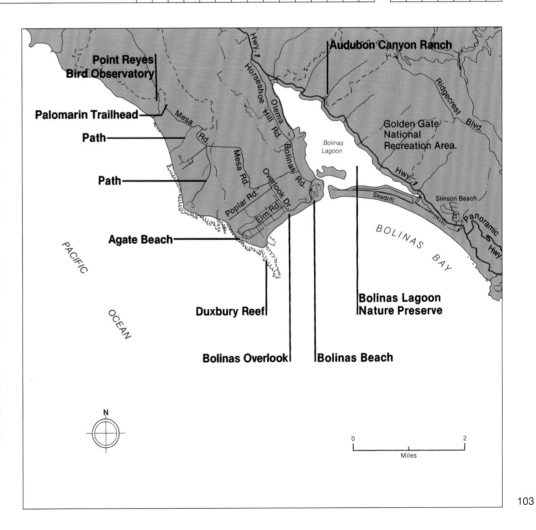

Sharks

Sharks are among the oldest creatures on earth. Sharks first appeared nearly 400 million years ago during the Devonian period of the Paleozoic era, and today there are approximately 350 species of sharks in the world. About 30 species are known to inhabit California's bays and coastal waters; some common species include the sevengill, leopard, soupfin, and blue shark.

Sharks belong to the *Chrondrichthyes* class of fishes, and the *Elasmobranchii* subclass, which includes sharks, rays, and skates. Unlike other fishes, elasmobranchs have no true bones; their skeletal system consists of cartilage.

Sharks vary considerably in size. The devil shark, found in deep water off the coast of Japan, reaches a length of about one foot, while an adult whale shark, found worldwide in warm seas, averages 45 feet long. Some sharks lay eggs; the majority, however, give birth to live pups. Only three species of sharks are known to feed primarily on plankton, while the rest are carnivores. A typical shark's diet consists of fish, mollusks, and crustaceans. Some of the larger sharks, notably the great white, sometimes feed on marine mammals such as seals, sea lions, and sea otters.

One characteristic common to virtually all sharks is a lifetime supply of teeth. The teeth are aligned in rows, with the outermost row being the oldest. When a tooth is lost, a tooth in the next row moves in to replace it. Some sharks can grow a new set of teeth in a little more than a week.

For centuries, sharks have been eaten in the Orient, Europe, Africa, and Mexico; however, the Japanese are probably the world's leaders in using sharks for food. Shark meat contains very little oil, fat, or cholesterol, is low in calories, and high in protein. As well as being used for food, sharks are processed for use in commercial products. For example, shark liver oil is used as a base in cosmetics, and the skeleton is processed into fertilizer and feed.

In recent years, the great white shark has been characterized as a vicious, human-eating beast, largely through highly publicized fictional and factual accounts of shark attacks. Many people have applied this reputation to all sharks. Actually, no shark normally feeds on humans. The vast majority of sharks would rather retreat than challenge an object as large as a human, and shark attacks by the great white or any other species of shark are quite rare.

The first known unprovoked shark attack in California occurred in 1926 in San Francisco Bay; through 2002, 82 known attacks by great white sharks have occurred off the California coast, with only six fatalities resulting from these attacks. According to shark researchers at the California Department of Fish and Game, most unprovoked shark attacks on humans can be classed as a search for food or an investigation of an object thought by a shark to be a food item. Other researchers believe that some shark attacks may also be a shark's response to a perceived threat.

The frequency of shark attacks along the California coast has increased in recent years due to the growing populations of prey, such as sea otters, seals, and sea lions, and because of an increase in the number of ocean swimmers and divers. California attacks have ranged from Imperial Beach in San Diego County to the Klamath River area in Del Norte County, with about one-half of the attacks occurring between Año Nuevo Island and Bodega Bay. The persons attacked most often were divers.

The Department of Fish and Game suggests a few common sense approaches to reduce the already rare chances of being attacked by a shark: Avoid areas of higher attack incidence; do not provoke any shark, regardless of its size; if bumped or harassed by a shark, get out of the water or into a kelp bed as soon as possible (there have been no reported attacks in kelp beds); keep speared fish aboard a float, not tied to your leg; and keep your wetsuit on if you are cut. Always report the time and location of a shark attack to local authorities and to the Department of Fish and Game.

leopard shark
to 6.5 feet

blue shark
to 13 feet

soupfin shark
to 6.5 feet

great white shark
to 20 feet

Marin County

STINSON BEACH / MOUNT TAMALPAIS

NAME	LOCATION	Entrance/Parking Fee	Parking	Restrooms	Lifeguard	Campground	Showers	Firepits	Stairs to Beach	Path to Beach	Bike Path	Hiking Trail	Facilities for Disabled	Boating Facilities	Fishing	Equestrian Trail	Sandy Beach	Dunes	Rocky Shore	Upland from Beach	Stream Corridor	Bluff	Wetland
Seadrift Beach	W. of Hwy. 1, N. of Panoramic Hwy., Stinson Beach																•					•	
Walkway	W. of Hwy. 1, N. of Panoramic Hwy., Stinson Beach	•								•							•					•	
Stinson Beach	W. of Hwy. 1, N. of Panoramic Hwy., Stinson Beach	•	•	•		•	•						•				•		•				
Red Rock Beach	Hwy. 1, 1.5 mi. S. of Stinson Beach	•								•							•		•				
Steep Ravine Beach	Hwy. 1, 2 mi. S. of Stinson Beach	•	•		•					•							•		•				
Slide Ranch	2025 Shoreline Hwy. (Hwy. 1)	•	•							•		•								•			
Muir Woods National Monument	Off Hwy. 1, 3 mi. N. on Muir Woods Rd.	•	•									•	•							•			
Mount Tamalpais State Park	Off Panoramic Hwy.	•	•			•		•				•	•							•			

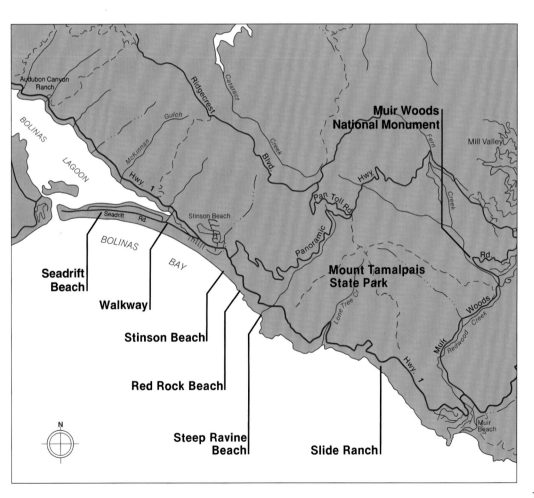

SEADRIFT BEACH: Public beach area is from the water inland to 60 feet from the private seawall. Beach is contiguous with Stinson Beach and provides access to the mouth of Bolinas Lagoon.

WALKWAY: Public accessway from Calle del Arroyo, along Walla Vista to Seadrift and Stinson beaches. Limited road shoulder parking.

The following are within the Golden Gate National Recreation Area (GGNRA), which is administered by the National Park Service: Stinson Beach, Slide Ranch, and Muir Woods National Monument.

STINSON BEACH: Popular broad, sandy beach with lifeguards on duty from May to September. Facilities include a picnic area, snack bar, and wheelchair-accessible restrooms. Open 9 AM-6 PM winter, 9 AM-10 PM summer. A beach wheelchair is available. No pets on the beach. For recorded weather and tide information: (415) 868-1922; for ranger station: (415) 868-0942.

RED ROCK BEACH: Part of Mount Tamalpais State Park. Parking along the highway at the dirt pull-outs; steep trail leads down to the small, popular, clothing-optional beach.

STEEP RAVINE BEACH: Part of Mount Tamalpais State Park. Steep Ravine Environmental Camp consists of ten rustic cabins and six environmental campsites; reservations are required; call: 1-800-444-7275. Campers with reservations may drive to a parking area a quarter-mile down the road leading to the beach. There is a locked gate at the entrance to the road; call State Parks for gate information: (415) 456-1286. Parking for beach day use only is available at several dirt pull-outs along Highway 1.

SLIDE RANCH: Part of the GGNRA. Slide Ranch serves primarily as an environmental education center for groups and families. Environmental education programs are offered by advance arrangement. Call about availability of the ranch for day use. For information or reservations, call: (415) 381-6155.

MUIR WOODS NATIONAL MONUMENT: 554 acres; part of the GGNRA. Muir Woods is the only remaining old-growth redwood forest near San Francisco. Trails connect with the extensive Mount Tamalpais State Park trail system. A paved, wheelchair-accessible path extends for nearly a mile into the woods; there are interpretive signs and exhibits along the path. Facilities include a visitor center with interpretive displays, a bookshop, a snackbar, and wheelchair-accessible restrooms. The monument is open 8 AM-sunset. Call: (415) 388-2595.

MOUNT TAMALPAIS STATE PARK: 6,300-acre park includes Steep Ravine, Red Rock, and other beaches and a popular trail system that connects with trails in Muir Woods National Monument and the Golden Gate National Recreation Area. The three main trails to the coast are the Dipsea, Steep Ravine, and Matt Davis trails. Maps are available at Pan Toll Station, the Marin Municipal Water District office in Corte Madera, and local bookstores. There are 16 walk-in campsites at Pan Toll; a fee is charged; no reservations. Reservations are required for Alice Eastwood group camp; call: 1-800-444-7275. For park information, call: (415) 388-2070.

View of Muir Beach from Overlook

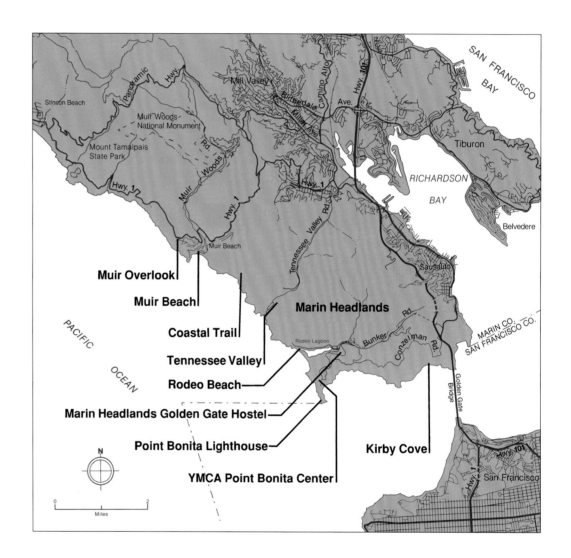

SAN FRANCISCO BAY

Stinson Beach

Mill Valley

Panoramic Hwy

Hwy

Blithedale

Camino Alto

Miller Ave

Ave.

Hwy. 101

Muir Woods
National Monument

Mount Tamalpais
State Park

Muir Woods Rd.

Tiburon

RICHARDSON
BAY

Hwy. 1

Hwy. 1

Hwy. 1

Muir

Tennessee Valley Rd.

Belvedere

Muir Beach

Sausalito

Muir Overlook

Muir Beach

Marin Headlands

Rd.

Coastal Trail

Rodeo Lagoon

Bunker

Conzelman Rd.

MARIN CO.
SAN FRANCISCO CO.

Tennessee Valley

PACIFIC

Rodeo Beach

OCEAN

Golden Gate Bridge

Marin Headlands Golden Gate Hostel

Point Bonita Lighthouse

Kirby Cove

Hwy. 101

N

YMCA Point Bonita Center

Hwy. 1

San Francisco

0   2
Miles

Marin County

MARIN HEADLANDS

NAME	LOCATION	Entrance/Parking Fee	Parking	Restrooms	Lifeguard	Campground	Showers	Firepits	Stairs to Beach	Path to Beach	Bike Path	Hiking Trail	Facilities for Disabled	Boating Facilities	Fishing	Equestrian Trail	Sandy Beach	Dunes	Rocky Shore	Upland from Beach	Stream Corridor	Bluff	Wetland
Muir Overlook	Off Hwy. 1, 1 mi. N. of Muir Beach	•	•										•						•			•	
Muir Beach	Hwy. 1 at Muir Beach Junction	•	•					•		•		•	•		•		•					•	
Coastal Trail	Headlands between Golden Gate Bridge and Muir Beach	•	•									•	•						•	•		•	
Marin Headlands	Off Hwy. 101, N. of Golden Gate Bridge	•	•	•					•	•	•	•	•		•		•		•	•		•	
Tennessee Valley	Tennessee Valley Rd. off Hwy. 1 in Mill Valley	•	•	•						•	•	•	•		•		•			•			
Rodeo Beach	Bunker Rd. off Hwy. 101 (Alexander Avenue exit)	•	•				•	•					•		•		•						
Marin Headlands-Golden Gate Hostel	Off Bunker Rd., near Rodeo Lagoon	•	•			•					•	•	•							•			
Point Bonita Lighthouse	S.W. tip of Marin Headlands end of Conzelman Rd.	•											•							•			
YMCA Point Bonita Center	Off Field Rd., near Conzelman Rd.	•	•										•							•			
Kirby Cove	Conzelman Rd. off Hwy. 101 (Alexander Avenue exit)	•	•	•		•		•				•	•		•		•		•				

The following are within the Golden Gate National Recreation Area (GGNRA): Muir Overlook, Muir Beach, Coastal Trail, Marin Headlands, Tennessee Valley, Rodeo Beach, Golden Gate Hostel, Point Bonita Lighthouse, and Kirby Cove, (415) 331-1540, www.nps.gov/goga.

MUIR OVERLOOK: Spectacular views of the rocky coastline; picnic tables in a sheltered area. Paved parking and wheelchair-accessible restrooms are available.

MUIR BEACH: A popular local beach with fishing, picnic tables, paved parking, and wheelchair-accessible restrooms. Monarch butterflies rest in pine trees near the beach during the winter.

COASTAL TRAIL: The Coastal Trail is part of an extensive network of trails criss-crossing the Marin Headlands. The 35-mile-long trail now links up with the 400-mile-long Bay Area Ridge Trail system. Within the headlands, the trail connects Muir Beach, Tennessee Valley, Rodeo Beach, and East Fort Baker. A new link to the Bay Area Ridge Trail begins from the parking lot at the north end of the Golden Gate Bridge. For maps and information, call the Marin Headlands Visitor Center: (415) 331-1540.

MARIN HEADLANDS: A series of grassy hills and valleys; former military lands, now part of the Golden Gate National Recreation Area. The Headlands include beaches, campgrounds, and spectacular views of the coast and bay areas. The Bay Area Ridge Trail system is planned to link Samuel P. Taylor State Park, Muir Woods, and Mount Tamalpais State Park to the Golden Gate Bridge along several new and existing trails, including the Coastal, Miwok, and Matt Davis trails. For information on the Ridge Trail, call (415) 543-4291. Trailhead parking for all headlands trails is at the northwest end of the Golden Gate Bridge, off Conzelman Road. A wheelchair-accessible visitor center, located in the former Fort Barry chapel, is on Field Rd. off Conzelman Rd. and features books, exhibits, maps, and information; open 9:30 AM-4:30 PM daily. Camping permits are required for camping anywhere in the headlands, and are available at the visitor center. Bicentennial campsite has three sites; each accommodates one-two persons; for information, call (415) 331-1540.

TENNESSEE VALLEY: Gently sloping valley; hiking trails lead north to Muir Beach and south to Rodeo Beach in the headlands area. A two-mile-long trail leads from the end of Tennessee Valley Rd. to the beach at Tennessee Cove. Miwok Livery riding stables offers guided rides by appointment; call (415) 383-8048. Haypress hike-in campsite has five sites, each accommodates up to four persons; for information, call (415) 331-1540.

RODEO BEACH: Sand and pebble beach between the ocean and Rodeo Lagoon, a habitat for several uncommon species of fish and birds, including the endangered California brown pelican. A wheelchair-accessible bridge across the lagoon leads to the edge of the sand. A picnic area and wheelchair-accessible restrooms are available. Trails lead up the bluffs and through Rodeo Valley, connecting to points in the headlands and beyond. Former bunkers house several environmental education programs, including the California Marine Mammal Center and the Marin Headlands Institute. The Marine Mammal Center features a bookstore and is open 10 AM-4 PM daily; educational/interpretive programs by arrangement. For information, call (415) 289-7325. The Headlands Institute offers outdoors education programs, and a science building and coast laboratory that can be toured by prior arrangement, 8:30 AM-5 PM weekdays; call (415) 332-5771.Information center, guided walks: (415) 331-1540.

MARIN HEADLANDS-GOLDEN GATE HOSTEL: 60 beds, bicycle storage room; operated by the American Youth Hostels, Inc. For information, write: Building 941, Fort Barry, Sausalito 94965; or call: (415) 331-2777 or 1-800-909-4776 #62.

POINT BONITA LIGHTHOUSE: Built in 1877, the lighthouse was manually operated until 1980. It can be reached only through a rock tunnel and suspension bridge and is open year-round from 12:30 PM-3:30 PM, Saturday, Sunday, and Monday. Monthly full-moon walks and occasional sunset walks to the lighthouse are available by reservation only. For information and walk schedules, call: (415) 331-1540.

THE YMCA POINT BONITA CENTER: The center offers overnight accommodations, by reservation only, for groups of 20 or more. Trails lead to a small beach also accessible by trail from Battery Alexander. For information, call: (415) 331-9622.

KIRBY COVE: Follow Conzelman Rd. from Hwy. 101 one-half mile to locked gate; walk to the beach at the foot of the dirt road. Picnic facilities and firepits are available. Overnight camping at four sites, each of which can accommodate ten people; for reservations, call: 1-800-365-2267. Wheelchair-accessible facilities include modified picnic tables, a bridge to the beach, and restrooms. Check with the Marin Headlands visitor center for road conditions: (415) 331-1540. For information, call: (415) 388-2070

Seal Rocks

San Francisco County

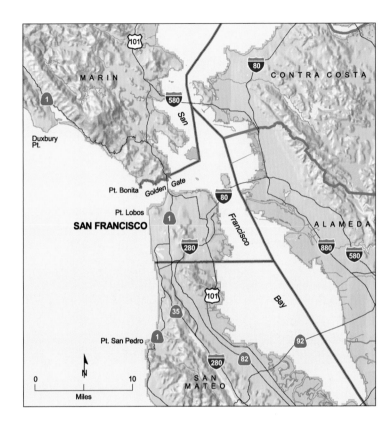

The ocean coastline of San Francisco, from the Golden Gate Bridge to Burton Beach at the San Mateo County line, is almost all public land. The Golden Gate National Recreation Area (GGNRA) encompasses this nearly eight-mile-long coast, stretching from Aquatic Park on San Francisco Bay in the north to the sandy bluffs of Fort Funston in the south, and including coves and cliffs, wooded blufftops, and sandy beaches.

Only one ocean beach in San Francisco, China Beach, is safe for swimming; all others have rip currents and large waves. However, there are other activities that draw visitors to the coast. A trail along the bluffs from the Golden Gate Bridge to the Cliff House offers spectacular views of the Marin Headlands, the Golden Gate, and the San Francisco shore. At Fort Funston, spectators watch hang-gliders soar over the ocean. Even on foggy days, Steller sea lions can be heard barking on the rocks offshore from the Cliff House, San Francisco's first seaside resort, which has been a landmark for over one hundred years.

Ocean Beach, which stretches for about four miles from the Cliff House to Fort Funston, is the largest beach in the Bay Area. It is backed by sand dunes, and a seawall and walkway extend along the northern end. The old Beach Chalet along the Great Highway features WPA murals on the walls. Golden Gate Park, encompassing more than 1,000 landscaped acres near the beach, features many attractions, including museums, gardens, and the Steinhart Aquarium.

The fog that frequently covers the shore kept many early explorers from discovering San Francisco Bay. In 1769, Sergeant José Francisco de Ortega of Portolá's expedition was probably the first European to see San Francisco Bay from the area now called Sweeney Ridge. In 1775, Lieutenant Juan Manuel de Ayala sailed into San Francisco Bay, and a year later Mission Dolores and the Presidio of San Francisco were established.

Under Spanish and Mexican rule, San Francisco was a small port; with the gold strikes in 1848, San Francisco became the main port for travel to the foothills. Its population jumped from 500 residents in 1848 to 40,000 entering the city in 1849. The transcontinental railroad was completed in 1869, creating a land link with the rest of the United States. San Francisco currently has a population of over 700,000.

Despite its rapid population growth, San Francisco has retained much of its shoreline as publicly owned open space. Adolf Sutro, one of the city's more flamboyant citizens, bought the area at Lands End and the decaying Cliff House in the early 1880s. After extensive gardening, he opened the Lands End area as a park. Subsequently, he rebuilt the Cliff House and added the Sutro Baths on the cliffs next to it. These baths, which were housed in a glass building and warmed by the sun, used salt water funneled from the beach below into pools. The pools are now in ruins, although the Park Service has rehabilitated the area so it is now safe for visitors.

Within San Francisco Bay, access to the shore is under the jurisdiction of the San Francisco Bay Conservation and Development Commission (BCDC). For more information, write or call: S.F. Bay Conservation and Development Commission, 50 California Street, San Francisco 94111; (415) 352-3600.

For more information about San Francisco, write or call the Convention and Visitors Bureau, 900 Market Street at Hallidie Plaza, lower level, San Francisco 94102; (415) 283-0177 or see www.sfvisitor.org. For information on transportation within the city, call MUNI (415) 673-MUNI or see www.sfmuni.com. Golden Gate Transit connects San Francisco with Marin and Sonoma counties; call (415) 923-2000, TDD 257-4554 or see www.goldengate.org; SAMTRANS buses travel to southern San Francisco and to San Mateo County; call 1-800-660-4287 or see www.samtrans.org.

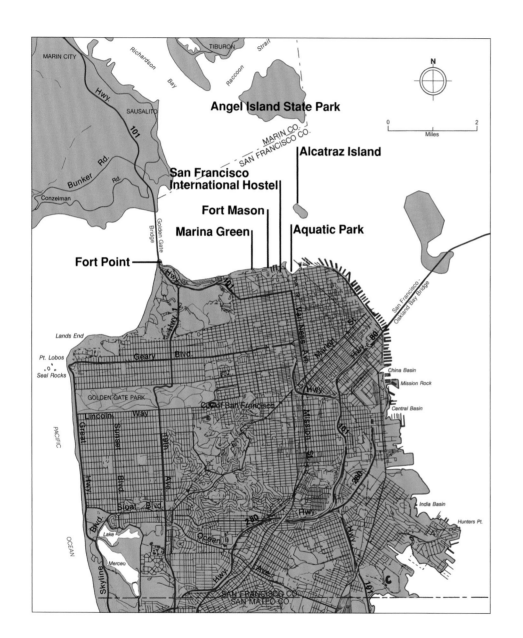

San Francisco Bay

Pier, Fort Mason

San Francisco County

NORTHERN SAN FRANCISCO

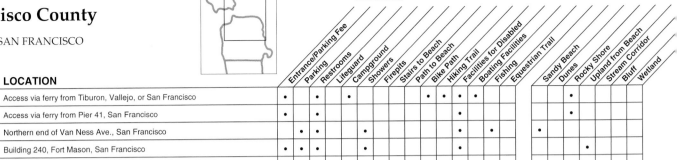

NAME	LOCATION	Entrance/Parking Fee	Parking	Restrooms	Lifeguard	Campground	Showers	Firepits	Stairs to Beach	Path to Beach	Bike Path	Hiking Trail	Facilities for Disabled	Boating Facilities	Fishing	Equestrian Trail	Sandy Beach	Dunes	Rocky Shore	Upland from Beach	Stream Corridor	Bluff	Wetland
Angel Island State Park	Access via ferry from Tiburon, Vallejo, or San Francisco	•		•		•					•	•	•	•					•				
Alcatraz Island	Access via ferry from Pier 41, San Francisco	•		•									•						•				
Aquatic Park	Northern end of Van Ness Ave., San Francisco		•	•			•						•		•		•						
San Francisco International Hostel	Building 240, Fort Mason, San Francisco	•	•	•			•						•							•			
Fort Mason	Marina Blvd. off Buchanan St., San Francisco		•	•									•							•			
Marina Green	Along Marina Blvd., San Francisco		•								•	•					•					•	
Fort Point	Off Lincoln Blvd., beneath the Golden Gate Bridge, San Francisco		•	•								•	•						•	•			

ANGEL ISLAND STATE PARK: 740 acres with picnic areas, environmental campsites, bicycle and hiking trails, and wheelchair-accessible restrooms. For camping reservations, call: 1-800-444-7275. Slips and mooring buoys are available; public transportation is via ferries that leave from San Francisco's Pier 43½, from Tiburon, and from Vallejo. A visitors' center and snack bar are located in Ayala Cove, where the ferries dock at Angel Island. Day-use fee; call (415) 435-1915.

The following are within the Golden Gate National Recreation Area (GGNRA), which is administered by the National Park Service: Alcatraz Island, Fort Mason, Crissy Field, and Fort Point. Headquarters for GGNRA is at Fort Mason; for information, call: (415) 561-4700 or see www.nps.gov/goga.

ALCATRAZ ISLAND: In 1972, "The Rock" became part of the Golden Gate National Recreation Area. Rangers give tours of the former prison that housed maximum security prisoners from 1934 to 1963; facilities are wheelchair accessible. For information, call: (415) 561-4900 or see www.nps.gov/alcatraz. Ferries leave from Pier 41; for schedule, fee, and tour information, call: (415) 773-1188.

AQUATIC PARK: The 1,850-foot Municipal Pier, a favored fishing spot, curves into the bay at Aquatic Park, the site of the National Maritime Museum, the Hyde Street Pier, a sandy beach, bleachers, bocce ball courts, lawns, showers, and wheelchair-accessible restrooms.

The National Maritime Museum contains an extensive collection of model ships and artifacts related to local maritime history. The Streamline Moderne building was built in the 1930s by the Works Progress Administration and was once a public bathhouse. Museum entrance fee; for information, call: (415) 561-6662.

The historic paddlewheel ferryboat *Eureka*, the sail schooner *C.A. Thayer*, and the square-rigged sailing ship *Balclutha* are berthed nearby at the Hyde Street Pier. There are tours and a bookstore; entrance fee. For information, call: (415) 556-6435.

SAN FRANCISCO INTERNATIONAL HOSTEL: Located at Fort Mason; the entrance closest to the hostel is at Bay and Franklin streets. The hostel, with views of San Francisco Bay, can accommodate 170 guests, has kitchen and laundry facilities, and is wheelchair accessible; overnight fee. For reservations and information: Building 240, Fort Mason, San Francisco 94123; (415) 771-7277.

FORT MASON: Once a military installation, Fort Mason now houses special exhibitions, conferences, galleries, museums, music and theater performances, and is headquarters for several environmental organizations. Facilities are wheelchair accessible. Monthly event calendars are available at Fort Mason Center; for information, call: (415) 441-3400 or see www.fortmason.org.

MARINA GREEN: A grassy bayfront recreation area popular for jogging, kite flying, and frisbee tossing; the Green is sited on landfill created for the 1915 Panama-Pacific Exposition. The Golden Gate Promenade, a partially paved bicycle and pedestrian trail, runs along the shore from Marina Green west to Fort Point, passing through Crissy Field, a former military airstrip. Independence Day fireworks, Earth Day celebrations, and other public events are held here. The waters offshore of Crissy Field are a world-class windsurfing spot. The Wave Organ, a sculpture that transmits the sound of bay waters, is located on the tip of land near the St. Francis Yacht Club.

FORT POINT: Built between 1853 and 1861, the brick fort is a National Historic Site. It contains a museum and a bookstore, and has rangers in civil war costumes who lead tours through the fort; the site is partially wheelchair accessible. For information, call: (415) 556-1693 or see www.nps.gov/fopo/.

The Coastal Trail begins at the dirt road above Fort Point and winds along the clifftops, through the Presidio, along Lincoln Blvd. and El Camino del Mar, southwest to Lands End.

Aquatic Park

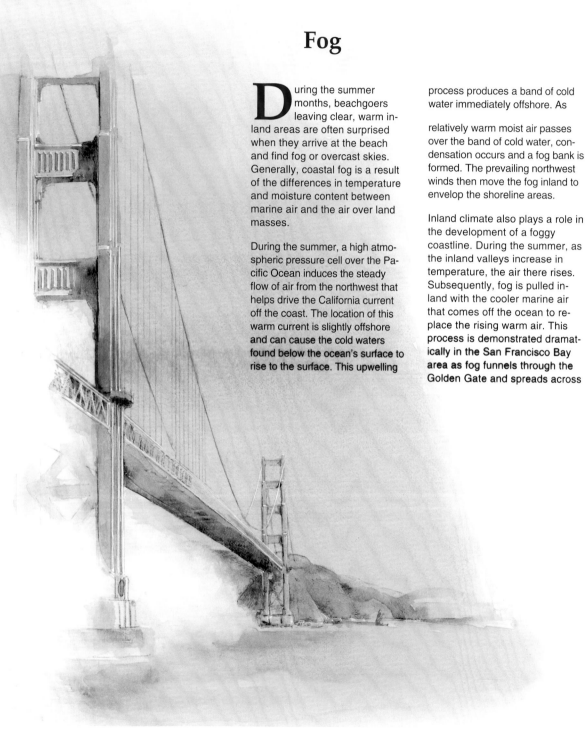

Fog

During the summer months, beachgoers leaving clear, warm inland areas are often surprised when they arrive at the beach and find fog or overcast skies. Generally, coastal fog is a result of the differences in temperature and moisture content between marine air and the air over land masses.

During the summer, a high atmospheric pressure cell over the Pacific Ocean induces the steady flow of air from the northwest that helps drive the California current off the coast. The location of this warm current is slightly offshore and can cause the cold waters found below the ocean's surface to rise to the surface. This upwelling process produces a band of cold water immediately offshore. As relatively warm moist air passes over the band of cold water, condensation occurs and a fog bank is formed. The prevailing northwest winds then move the fog inland to envelop the shoreline areas.

Inland climate also plays a role in the development of a foggy coastline. During the summer, as the inland valleys increase in temperature, the air there rises. Subsequently, fog is pulled inland with the cooler marine air that comes off the ocean to replace the rising warm air. This process is demonstrated dramatically in the San Francisco Bay area as fog funnels through the Golden Gate and spreads across the bay and the surrounding lowland areas.

Usually, as the day progresses, air temperatures above the land masses increase to the point where the fog vaporizes, or recedes away from the coastline ("burns off") by late morning or early afternoon. On some days, however, the fog lifts but does not completely burn off, resulting in partly cloudy or overcast skies. Although the cloud cover may appear to block out most of the sunshine, the sun's ultraviolet light waves (the portion of the spectrum responsible for tanning and sunburn) penetrate through the clouds. Beach users should keep this in mind when visiting the beach on overcast days.

San Francisco County

GOLDEN GATE BRIDGE TO OCEAN BEACH

NAME	LOCATION	Entrance/Parking Fee	Parking	Restrooms	Lifeguard	Campground	Showers	Firepits	Stairs to Beach	Path to Beach	Bike Path	Hiking Trail	Facilities for Disabled	Boating Facilities	Fishing	Equestrian Trail	Sandy Beach	Dunes	Rocky Shore	Upland from Beach	Stream Corridor	Bluff	Wetland
Presidio of San Francisco	Northwest corner of San Francisco	•	•		•		•			•	•	•	•				•		•	•		•	
Coastal Trail	Fort Point to the Cliff House, San Francisco	•	•							•		•							•			•	
Baker Beach	W. of Lincoln Blvd., off 25th Ave., San Francisco	•	•				•			•	•	•			•		•						
China Beach	Seacliff Ave. and El Camino Del Mar, San Francisco	•	•	•		•	•			•			•				•						
Lands End	Pt. Lobos Ave. and 48th Ave., San Francisco	•	•				•			•		•							•	•		•	
Ocean Beach	W. of Great Highway, San Francisco	•	•					•	•				•			•		•	•				

The following are within Golden Gate National Recreation Area (GGNRA), which is administered by the National Park Service: Presidio of San Francisco, Coastal Trail, Baker Beach, China Beach, Lands End, and Ocean Beach. Headquarters for GGNRA is at Fort Mason. For information, call: (415) 561-4700 or see www.nps.gov/goga.

PRESIDIO OF SAN FRANCISCO: A military post under the flags of Spain (1776-1822), Mexico (1822-1848), and the United States (1848-1994), the Presidio is now jointly managed by the National Park Service and the Presidio Trust. The 1,480-acre park contains 11 miles of hiking trails, 14 miles of bicycle routes, historic buildings, coastal defense fortifications, a saltwater marsh, coastal bluffs, and spectacular vistas.

Robb Hill Group Camp has two group sites, each accommodating 30 people; for reservations, call: (415) 561-5444. The William Penn Mott Jr. Visitor Center on Montgomery Street, Building 102, contains photographs and exhibits and is open daily. For information, call: (415) 561-4323. Crissy Field Center on the corner of Mason and Hallek overlooks the marsh and is open Wed.-Sunday. The Exploratorium, a museum of science, art, and human perception exhibits, is located at the Palace of Fine Arts on Marina Boulevard. Open Tuesday-Sunday; for information, call: (415) 561-0360 or see www.exploratorium.edu.

COASTAL TRAIL: Extends from the Golden Gate Bridge to the Cliff House, and connects portions of the GGNRA. The trail begins above Fort Point National Historic Site; take the dirt road above Fort Point that goes under the bridge near the toll plaza. Follow paved path around the west side of the bridge to signed trailhead. The trail follows steep coastal cliffs, and portions of Lincoln Blvd. and El Camino Del Mar. Park at Fort Point, at the vista point on the southeast side of the Golden Gate Bridge, or in the gravel lot off Lincoln Boulevard. Call: (415) 556-8371.

BAKER BEACH: Parking off Lincoln Boulevard. Grills and picnic tables in a cypress tree grove; picnic areas and restrooms are wheelchair accessible. Dangerous surf. Dogs allowed if on leash. Weekend tours of adjacent Battery Chamberlin are available. Call: (415) 556-8371.

CHINA BEACH: Parking off El Camino Del Mar; steep path to the beach, a popular swimming and volleyball spot. Lifeguard April to October. Facilities include a sundeck, picnic area, and restrooms with showers. A concrete service road, open 9 AM-5 PM, leads down to a handicapped parking area and extends out onto the beach for wheelchair access. Call: (415) 831-2210.

LANDS END: Parking off El Camino Del Mar, and at West Fort Miley and the Cliff House. Trail system along the bluffs; a steep trail leads to a small beach. Dogs allowed. Dangerous cliffs off trails. Landslides periodically close certain trails. Picnic area with restrooms at West Fort Miley. Sutro Heights Park at Point Lobos and 48th Ave. is a scenic blufftop area with views of the coast. The Cliff House, San Francisco's first seaside resort, has the Camera Obscura, and food service. The Cliff House Visitor Information Center features historic photographs, and information on seals, whales, and tidepools; call: (415) 239-2366.

OCEAN BEACH: Four-mile strand from the Cliff House to San Francisco Zoo. Stairs lead to the beach from the parking area and pedestrian walkway along the seawall at the northern end; wheelchair access via stairwell #15. Multi-purpose paths (one paved) extend along both sides of Great Highway from Lincoln Way to Sloat Blvd.; horses are allowed on the beach south of Golden Gate Park. Surfing and fishing are popular, but swimming is unsafe due to hazardous rip currents. Wheelchair-accessible restrooms are located at the foot of Balboa St. and east of the Great Highway at Noriega and Taraval streets.

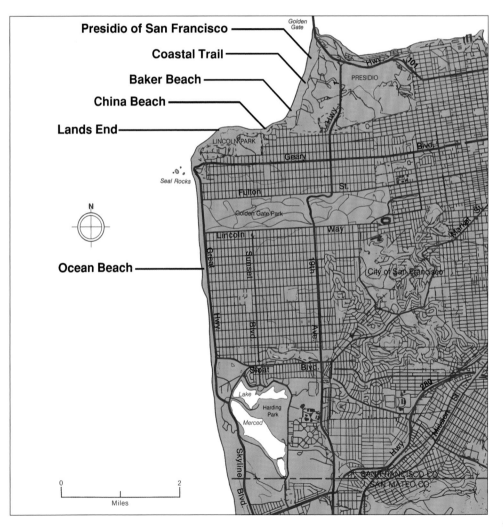

Presidio of San Francisco
Coastal Trail
Baker Beach
China Beach
Lands End
Ocean Beach

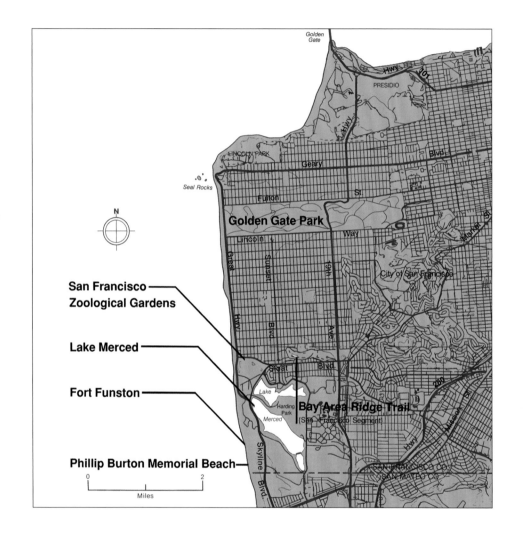

San Francisco Zoological Gardens

Lake Merced

Fort Funston

Phillip Burton Memorial Beach

0 — 2
Miles

Lake Merced

Fort Funston

San Francisco County

GOLDEN GATE PARK TO FORT FUNSTON

NAME	LOCATION	Entrance/Parking Fee	Parking	Restrooms	Lifeguard	Campground	Showers	Firepits	Stairs to Beach	Path to Beach	Bike Path	Hiking Trail	Facilities for Disabled	Boating Facilities	Fishing	Equestrian Trail	Sandy Beach	Dunes	Rocky Shore	Upland from Beach	Stream Corridor	Bluff	Wetland
Golden Gate Park	E. of Great Highway, between Lincoln Way and Fulton St., San Francisco		•	•		•				•	•	•	•	•	•					•			
San Francisco Zoological Gardens	Sloat Blvd. and Skyline Blvd., San Francisco	•	•	•									•							•			
Lake Merced	E. of Skyline Blvd. and Great Highway, San Francisco		•	•						•	•	•	•	•	•					•			•
Fort Funston	W. of Great Highway at Skyline Blvd., San Francisco		•	•					•			•	•		•		•	•				•	
Bay Area Ridge Trail	Fort Funston to Stern Grove, San Francisco		•	•						•	•	•					•	•				•	
Phillip Burton Memorial Beach	S. of Fort Funston, San Francisco								•						•		•	•				•	

GOLDEN GATE PARK: John McLaren converted more than 1,000 acres of sand dunes into an urban park designed by William Hammond Hall, which now includes lakes for boating and fishing, the de Young Museum, California Academy of Sciences (with science and natural history exhibits, Steinhart Aquarium, and Morrison Planetarium), Strybing Arboretum, Conservatory of Flowers, Japanese Tea Garden, rose garden, playgrounds, sports facilities, horse stables, bison paddock, and equestrian trails. Free concerts on weekends. Some streets are closed to motor traffic on Sundays and are used by pedestrians, bikers, and skaters. Windmills near the beach once pumped water for irrigation into the park; one has been restored. Wheelchair-accessible restrooms are available. For information, call: (415) 831-2700.

SAN FRANCISCO ZOOLOGICAL GARDENS: The zoo, located adjacent to Ocean Beach, is open 10 AM-5 PM daily. Special attractions are the Primate Discovery Center, Lion House, Insect Zoo, Gorilla World, Koala Exhibit, Penguin Pool, Storyland, and Children's Zoo. Wheelchair-accessible restrooms are available. For information, call: (415) 753-7080. Nearby is the Recreation Center for the Handicapped, 207 Skyline Blvd., (415) 665- 4100.

LAKE MERCED: Parking off Sunset Blvd. and Harding Park Road. Lake Merced area has a boat launch and rowboat rentals, fishing, benches, paved paths around the lake, par course, restrooms, and food service. Harding Park Golf Course adjoins the lake; call: (415) 664-4690. The historic Broderick-Terry Dueling Site is at the southeast end of the park.

FORT FUNSTON: Parking at the southern end of the park, off Skyline Boulevard. Picnic area and hang-gliding observation deck with scenic views. A steep unmarked trail leads to the beach and sand dunes; Sunset Trail and restrooms are wheelchair accessible. Hang-gliding club facilities; environmental education/day camp. For information, call: (415) 239- 2366.

BAY AREA RIDGE TRAIL (SAN FRANCISCO SEGMENT): The trail extends 3.2 miles from the sand dunes at Fort Funston past Lake Merced to Stern Grove, a steep, tree-lined canyon with a small lake. The trail is mostly paved; restrooms, water, and telephones are available at both ends. Parking at Fort Funston off Skyline Blvd. and in Stern Grove or at Pine Lake Park via Vale Avenue. For information, call: (415) 391- 9300.

PHILLIP BURTON MEMORIAL BEACH: Beach access from Fort Funston in San Francisco County; the beach continues south through the former Thornton State Beach, where road access is now closed due to bluff failure, and to the Daly City shoreline accessible from Mussel Rock City Park.

Viewing Platform, Fort Funston

Pigeon Point Lighthouse

San Mateo County

San Mateo County has 55 miles of Pacific coastline; south of Montara Mountain, the towns of Moss Beach, San Gregorio, and Pescadero still retain the character of small coastside farming and fishing communities, in contrast to the populous suburban areas of Pacifica and Daly City to the north. The south county coast, largely undeveloped and bordering the Santa Cruz and San Gregorio Mountains, has accessible beaches to the neighboring metropolises of San Jose and San Francisco.

Spectacular sea cliffs are characteristic of the county's coastline, rising to steep heights above the shore in the Daly City and Devil's Slide areas. South of Devil's Slide, the rich soils of the marine terraces above the sea cliffs support coastal agriculture; Half Moon Bay is renowned for its fields of autumn pumpkins and Christmas trees and harvests of artichokes and Brussels sprouts.

Although generally too cold and hazardous for swimming, San Mateo's beaches are popular for tidepooling, picnicking, marine mammal watching, and pier and shore fishing. The north coast is noted for summertime striped bass and surfperch. Mussels are also popular, but are quarantined annually, usually between May and October.

Marine mammals can be observed at Año Nuevo State Reserve, where visitors are permitted to view northern elephant seals close-up during their breeding season. Año Nuevo is also noted as one of the first San Mateo County landmarks to be sighted by a Spanish explorer; Don Sebastián Vizcaíno named it Punta del Año Nuevo, New Year's Point, because it was the first landmark he sighted in the new year of 1603. The inhabitants of the region at that time were the Ohlone Indians; their shell middens are still visible at Año Nuevo. The Ohlone culture was eventually destroyed following Spanish settlement in the late 1790s.

Juan Gaspar de Portolá was among the first Spanish explorers to actually visit San Mateo; two plaques commemorate beaches where Portolá's expedition camped in 1769–one at San Gregorio State Beach and the other at the San Pedro Beach Rest Area. It was from San Pedro Beach that Portolá's expedition climbed Sweeney Ridge on November 4, 1769 and first sighted San Francisco Bay.

The rocky shore and heavy fogs of San Mateo's coastline have contributed to numerous shipwrecks throughout the county's history. Pigeon Point owes its name to the wreck of the Boston Clipper ship *Carrier Pigeon* off its rocky point in 1853; the lighthouse was later built in 1872. In 1896, the *Columbia* ran aground off Año Nuevo; Pescadero's residents salvaged the cargo of white paint, used it liberally on the town's buildings, and have since maintained the tradition of painting the houses white.

The steep sea cliffs of Devil's Slide and the former Thornton State Beach were once the planned site of the 1905 Ocean Shore Railroad, intended to take passengers along the coast between San Francisco and Santa Cruz. The railroad was never completed, but remnants of the road cuts are still visible near the base of Devil's Slide. Thornton State Beach to the north is where the San Andreas fault, one of the most active in California, enters the sea to the north. A movement of this fault in 1957 triggered a landslide, destroying the segment of Highway 1 that had previously run along the coast; later landslides destroyed access to the park itself.

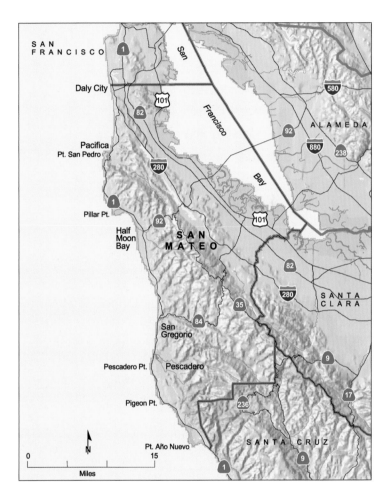

For additional information on San Mateo's coast, contact the Pacifica Visitor Center, 225 Rockaway Beach #1, Pacifica 94044, (650) 355-4122, www.pacificachamber.com; or the Half Moon Bay Chamber of Commerce, 520 Kelly Ave., Half Moon Bay, 94019, (650) 726-5202, www.hmbchamber.com. The San Mateo County Transit District provides transit service linking inland communities to Daly City, Pacifica, and Half Moon Bay and serving Highway 1 both north and south of Half Moon Bay. For transit information, contact: SAMTRANS, 1250 San Carlos Ave., San Carlos, 94070; call: 1-800-660-4287 or see www.samtrans.org.

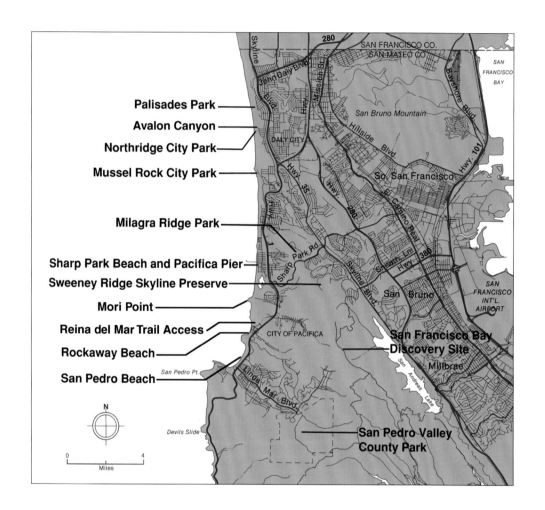

Palisades Park
Avalon Canyon
Northridge City Park
Mussel Rock City Park

Milagra Ridge Park

Sharp Park Beach and Pacifica Pier
Sweeney Ridge Skyline Preserve
Mori Point
Reina del Mar Trail Access
Rockaway Beach
San Pedro Beach

San Francisco Bay
Discovery Site

San Pedro Valley
County Park

Burton Beach

San Mateo County

DALY CITY / PACIFICA

NAME	LOCATION	Entrance/Parking Fee	Parking	Restrooms	Lifeguard	Campground	Showers	Firepits	Stairs to Beach	Path to Beach	Bike Path	Hiking Trail	Facilities for Disabled	Boating Facilities	Fishing	Equestrian Trail	Sandy Beach	Dunes	Rocky Shore	Upland from Beach	Stream Corridor	Bluff	Wetland
Palisades Park	Along Palisades Dr. at Westridge Ave., Daly City		•																	•		•	
Northridge City Park / Avalon Canyon	Along Northridge Dr. at Carmel Ave., Daly City		•																	•		•	
Mussel Rock City Park	Westline Dr., off Skyline Dr., Daly City		•									•			•		•		•	•		•	
Sharp Park Beach and Pacifica Pier	Beach Blvd. and Santa Rosa Ave., Pacifica		•	•									•		•		•						
Milagra Ridge Park	E. of Hwy. 1 on Sharp Park Rd., Pacifica		•								•	•								•			
Mori Point	S. of Sharp Park Beach, Pacifica		•									•								•		•	
Sweeney Ridge Skyline Preserve	Sweeney Ridge, Pacifica		•									•								•			
San Francisco Bay Discovery Site	Sweeney Ridge, Pacifica		•									•								•			
Reina del Mar Trail Access	Hwy. 1 at Reina del Mar Ave., Pacifica		•							•	•						•						
Rockaway Beach	Foot of Rockaway Beach Ave. at San Mario Way, Pacifica		•					•			•	•			•		•		•				
San Pedro Beach	Hwy. 1 between Crespi Dr. and Linda Mar Blvd., Pacifica		•	•			•					•		•	•		•						
San Pedro Valley County Park	E. of Hwy. 1, end of Linda Mar Blvd., Pacifica	•	•	•				•				•	•							•	•		

PALISADES PARK: Distant views of the ocean and Marin headlands; municipal playground.

NORTHRIDGE CITY PARK / AVALON CANYON: Ocean views from Northridge City Park; street parking. Erosion of the bluffs and canyon prevents safe access to the beach; best beach access is from Mussel Rock City Park.

MUSSEL ROCK CITY PARK: Unimproved park at north end of Westline Dr.; steep paths lead from the north end of the parking lot down the bluffs.

SHARP PARK BEACH AND PACIFICA PIER: Shoreline access is along Beach Boulevard. Facilities include a 1,100-foot-long municipal pier, concessions, bait shop, and wheelchair-accessible restrooms; popular salmon fishing location. Open all year. A parking lot offering ocean views is located 1 mile north of the pier area, on Palmetto Ave. south of Avalon Dr. and the R.V. park.

MILAGRA RIDGE PARK: This 232-acre upland area is part of the Golden Gate National Recreation Area, and offers magnificent views of the Pacific Ocean; hiking and bicycling is allowed on old paved roads. Call: (415) 556-8371.

MORI POINT: Spectacular coastal headland, 105 acres, part of the Golden Gate National Recreation Area. Trail starts west of the Moose Lodge, at south end of Bradford Way near Westport St., off Hwy. 1. For information, call: (650) 355-4122.

SWEENEY RIDGE SKYLINE PRESERVE: East of Hwy. 1, .25 mile south of Sharp Park Rd. is the Shelldance Nursery, trailhead, and parking lot for the 1.3-mile-long Mori Ridge Trail, a strenuous climb to the Sweeney Ridge Trail, which leads to the preserve. The 58-acre preserve offers a 360-degree panorama of the Pacific Ocean and San Francisco Bay. There is also a trailhead off Skyline Blvd. from Skyline College Student Parking Lot 3; walk 200 feet south on the road to the maintenance facility, then west to the entrance gate. Call: (415) 556-8371.

SAN FRANCISCO BAY DISCOVERY SITE: In 1769, members of Gaspar de Portolá's expedition first sighted San Francisco Bay; a granite cylinder marks the site, from which there are also outstanding views of the Bay Area. The trailhead gate is at the west end of Sneath Lane off Skyline Blvd.; one mile walk on a paved trail to the summit.

REINA DEL MAR TRAIL ACCESS: Parking area seaward of Hwy. 1 and Reina del Mar Ave.; a paved path offering views of the ocean and headlands winds along Calera Creek to Rockaway Beach.

ROCKAWAY BEACH: Narrow, sandy beach, accessible except at highest tides; popular fishing area. There are two beach parking areas; the northern lot is reached by turning north off Rockaway Beach Blvd. onto Old County Rd., then left on San Marlo Way. A paved bike path leads from the parking lot along the restored Calera Creek wetlands, past the old quarry. The second parking lot is at the south end of Old County Rd., near Highway 1; open dawn to dusk; wheelchair-accessible restrooms. A paved bike path switchbacks south over the ridge to San Pedro Beach. The Pacifica Visitor Center is located at 225 Rockaway Beach Blvd.; for information, call: (650) 355-4122.

SAN PEDRO BEACH: Popular beach for surfing, although often hazardous due to rip currents. Outdoor showers are at the restrooms, next to the parking lot seaward of Hwy. 1; on weekends, additional parking in the park-and-ride lot at Crespi Dr. and Highway 1.

SAN PEDRO VALLEY COUNTY PARK: The park has picnic sites with barbecue pits, a group picnic area, loop hiking trails, and a self-guided nature trail; the visitor center and restrooms are wheelchair accessible. The south and middle forks of San Pedro Creek, spawning areas for migratory steelhead, flow through the park. Day-use fee; for information, call: (650) 355-8289

Mussel Rock

Montara Lighthouse Hostel

Montara State Beach

James Fitzgerald Marine Reserve

NAME	LOCATION	Entrance/Parking Fee	Parking	Restrooms	Lifeguard	Campground	Showers	Firepits	Stairs to Beach	Path to Beach	Bike Path	Hiking Trail	Facilities for Disabled	Boating Facilities	Fishing	Equestrian Trail	Sandy Beach	Dunes	Rocky Shore	Upland from Beach	Stream Corridor	Bluff	Wetland
Gray Whale Cove State Beach	Hwy. 1, .5 mi. S. of Devil's Slide, Montara	•	•	•					•	•							•					•	
Montara State Beach	Hwy. 1 and 2nd St., Montara		•	•					•	•		•			•		•					•	
Montara Lighthouse Hostel	Hwy. 1 and 16th St., Montara	•	•	•			•						•						•	•		•	
James Fitzgerald Marine Reserve	Off California Ave. at N. Lake St. and Nevada Ave., Moss Beach		•	•					•	•		•					•		•	•		•	
West Beach Trail	Princeton Harbor		•	•								•			•		•						•
Coastside Trail	Princeton Harbor to Kelly Ave., Half Moon Bay										•	•										•	

GRAY WHALE COVE STATE BEACH: Accessible by a steep trail and stairway west of Hwy. 1. The parking lot is east of Hwy. 1; dangerous pedestrian crossing to get to the stairway. A grassy picnic area is above the beach along the bluffs. Popular sunbathing beach; clothing optional. Hazardous surf. No dogs or fires permitted. Call: (650) 726-8819.

MONTARA STATE BEACH: Half-mile-long sandy beach. Hazardous surf. Stairs and a parking lot are located at the north end of the state beach. There is also a steep path to the beach opposite the west end of Second St., just south of the Outrigger restaurant; public parking is available during daylight hours until 5 PM at the Outrigger. The stairs and path to the beach seaward of the Outrigger are also for public use. East of Hwy. 1 and north of Martini Creek is the McNee Ranch, an undeveloped part of Montara State Beach that offers a network of trails. Call: (650) 726-8819.

MONTARA LIGHTHOUSE HOSTEL: 45 beds, plus kitchen, laundry, and outdoor bicycle rack, located adjacent to a historic Coast Guard-operated lighthouse. Day-use facilities for groups; daily bicycle rentals available. Check-in hours are 5–9:30 PM daily. Office also open 7:30– 9:30 AM. Overnight fee. Children accompanied by parents pay half price. For reservations: P.O. Box 737, Montara 94037; (650) 728-7177.

JAMES V. FITZGERALD MARINE RESERVE: Three miles long, noted for its diverse marine life. Tidepool walks are occasionally given by rangers at low tides; there is usually a walk given on Saturdays. Groups may make reservations for walks at other times. Popular diving area; it is illegal to remove or disturb marine life or habitat. Tidepool rocks are slippery. Picnic tables are east of the parking lot in a sheltered cypress grove; there are hiking trails along the grassy bluffs to the south. For information and reservations, call: (650) 728-3584.

WEST BEACH TRAIL: From Hwy. 1 take Capistrano Rd. to Prospect Way, jog left to Ocean Blvd., then turn right on West Point Rd. and skirt the marsh to the 16 car parking lot near the base of the hill; wheelchair-accessible restrooms. ½-mile-long trail along Princeton Harbor; good bird watching. Trail ends at a small sandy beach, from where the famous surf break known as Maverick's can be seen.

COASTSIDE TRAIL: Six-mile-long bluff top trail spanning the City of Half Moon Bay. Pedestrian, bicycle, and limited equestrian use. The Coastside trail can be reached from beach parks and from intersecting streets, including Mirada Rd., Kelly Ave., and Poplar Road.

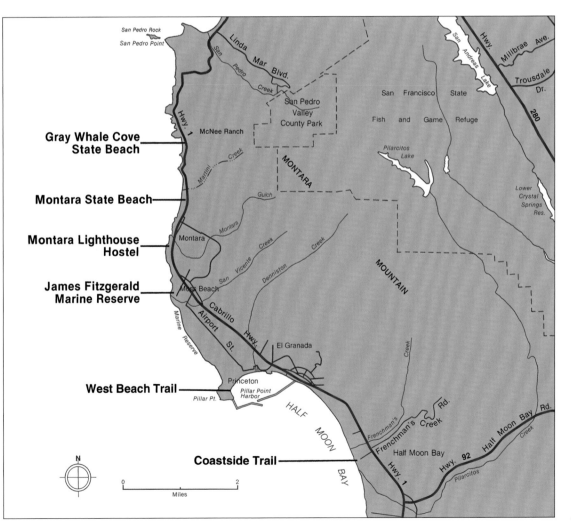

Pillar Point Harbor

Francis Beach

El Granada Beach

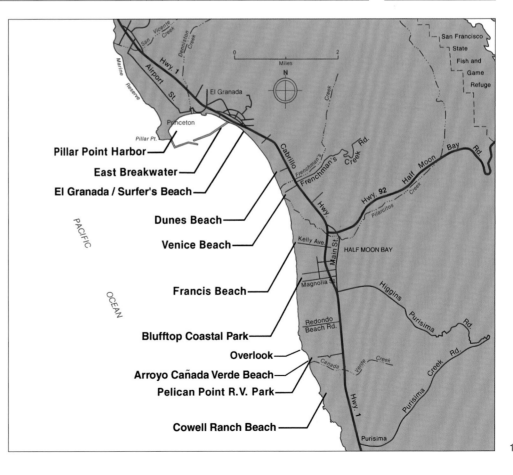

NAME	LOCATION	Entrance/Parking Fee	Parking	Restrooms	Lifeguard	Campground	Showers	Firepits	Stairs to Beach	Path to Beach	Bike Path	Hiking Trail	Facilities for Disabled	Boating Facilities	Fishing	Equestrian Trail	Sandy Beach	Dunes	Rocky Shore	Upland from Beach	Stream Corridor	Bluff	Wetland
Pillar Point Harbor (Johnson Pier)	W. of Hwy. 1 and Capistrano Rd., Princeton		•	•						•				•	•	•		•					
East Breakwater	W. of Hwy. 1, S. of Capistrano Rd., Half Moon Bay		•	•	•										•			•					
El Granada / Surfer's Beach	Between East Breakwater and Mirada Rd., Half Moon Bay		•												•			•					
Dunes Beach	Foot of Young Ave., Half Moon Bay		•	•					•	•						•		•				•	
Venice Beach	Foot of Venice Blvd., Half Moon Bay		•	•					•	•						•		•				•	
Francis Beach	Foot of Kelly Ave., Half Moon Bay	•	•	•		•		•		•	•		•		•	•		•				•	
Blufftop Coastal Park	W. end of Poplar St., Half Moon Bay		•	•					•	•								•				•	
Overlook	Seaward of Ritz-Carlton Hotel, Miramontes Point Rd., Half Moon Bay		•															•					
Arroyo Cañada Verde Beach	Mouth of Arroyo Cañada Verde, Miramontes Point Rd., Half Moon Bay		•	•					•									•			•	•	
Pelican Point R.V. Park	Foot of Miramontes Point Rd., Half Moon Bay	•	•	•		•	•	•							•			•		•	•	•	•
Cowell Ranch Beach	W. of Hwy. 1, S. of Cañada Verde Creek		•	•					•						•			•				•	

PILLAR POINT HARBOR: The only small boat harbor between San Francisco and Santa Cruz. Contains a popular municipal fishing pier. Facilities include a boat ramp (launching fee) and dinghy hoist (2,000 lb. cap.; launching fee), charter boats, picnic area, and food services; 228 commercial and 141 recreational berths, some available for transient use. Open all year, 24 hours a day. Fresh fish available to the public off the boats and in restaurants. For information, call: (650) 726-4382.

EAST BREAKWATER: Popular fishing access. The paved parking lot at the breakwater provides overnight camping for self-contained recreational vehicles; full hookups; fee. Call: (650) 712-9277.

EL GRANADA / SURFER'S BEACH: Wide, sandy beach beneath the seawall along Hwy. 1. Parking available at Pillar Point Harbor.

The following are within the Half Moon Bay State Beach system and are administered by the California State Department of Parks and Recreation: Dunes Beach, Venice Beach, Francis Beach, and Cowell Ranch Beach. For information, contact the ranger station at 95 Kelly Ave., Half Moon Bay 94019; (650) 726-8819.

DUNES BEACH AND VENICE BEACH: Rough dirt road access. Equestrian trail east of parking lot runs from Dunes Beach south to Francis Beach. Horse rentals available at two stables just off Hwy. 1 between Young Ave. and Venice Boulevard.

FRANCIS BEACH: 51 campsites for tent and R.V. camping; reservations recommended; call:1-800-444-7275. There are also enroute campsites available. Overnight and day-use fees. Picnic sites on the bluffs overlooking the ocean; visitor center. Beach and restrooms are wheelchair accessible.

BLUFFTOP COASTAL PARK: Beach access from the end of Poplar street; parking and restrooms are available. Call: (650) 726-8297.

OVERLOOK: Public coastal access parking available in the Ritz-Carlton Hotel garage; see valet for entry. Blufftop trail is located seaward of the hotel; steep bluffs prevent beach access from the hotel property.

ARROYO CAÑADA VERDE BEACH: A paved path leads north to the Ritz-Carlton Hotel and south past a golf course; a stair leads to a cove beach. A 15 car parking lot is west of Miramontes Pt. Rd.; restrooms available.

PELICAN POINT R.V. PARK: 75 recreational vehicle campsites located just upland of the ocean. Laundry facilities, small store, and restrooms. Overnight fee (includes hookups). Tent camping is permitted. Information and reservations: 1001 Miramontes Pt. Rd., Half Moon Bay 94019; (650) 726-9100.

COWELL RANCH BEACH: Located south of Cañada Verde Creek;18-car parking lot adjacent to Highway 1. A dirt trail approximately ½-mile long leads to a blufftop viewing area. Stairs lead to a sandy beach.

Agriculture

Over the last 200 years, California's coastal agriculture has expanded from scattered plots in subsistence farming to include regions of farm land producing diverse varieties of specialty crops. Some of the state's best grazing lands are also located along the coast.

Native Californians were primarily hunter-gatherers and did not engage in agriculture until the establishment of the Spanish missions. Among the earliest coastal farms were those begun in 1812 by the Russians at Fort Ross in Sonoma County to sustain themselves as they conducted their fur seal and sea otter trade. The Spanish settlers to the south had only relatively small gardens, concentrating on raising cattle on their large ranchos. Farming and ranching were major contributors to the growth of early settlements along the coast, and the sea was often important in the commerce of agricultural products.

The stern-wheeler *Vaquero*, for example, made regular trips up and down Elkhorn Slough to take grain and potatoes from Watsonville to Moss Landing and on to San Francisco. Other schooners plied the coastline, loading lumber in the north and cattle in the south.

Since the 1950s, vast areas of productive land have been lost to parcelization, accelerating land costs, and outright conversion to urbanization. In the 1960s alone, one out of twelve acres of coastal cropland was covered by urban expansion. The remaining farmlands, however, are an important resource. Particular combinations of coastal soil and climate create special conditions of high agricultural productivity. The moderating effect of a maritime climate provides timing and yield advantages for national markets, reduces the dangers of large-scale crop loss from freezing, and extends the growing season, enabling farmers to raise and market multiple harvests in a single year.

Certain specialty crops, such as artichokes, avocados, broccoli, Brussels sprouts, celery, strawberries, tomatoes, flowers, and greenhouse products, do especially well in the marine climate. In the event of massive crop failures elsewhere, the California coastal zone's rich soil resources could be converted to grow staple crops such as wheat, oats, and other basic cereals and vegetables.

Grazing lands, which cover numerous acres from Santa Barbara north, are used by cattle and sheep ranchers to produce dairy, meat, wool, and other products. Some coastal river valleys from Monterey County south are used for important agricultural production. The deep alluvial and floodplain soils are highly suited for irrigated crops such as vegetables, citrus fruits, and avocados. Terraces along the southern California coast are also conducive to specialty crop production.

The specific climate and soil conditions along the San Mateo County coast are ideal for a number of valuable crops, most of which are grown in deep, fertile, low-elevation marine terraces and alluvial soils between Pillar Point and Año Nuevo Point. The average daily temperature here varies less than ten degrees from summer to winter, and the growing season averages more than 300 days per year. During the summer months, fog covers much of the county's coastal strip and this foggy environment contributes to high yields of Brussels sprouts, artichokes, and cut flowers.

Mushrooms are also a successful crop along the San Mateo coast because of the climate conditions of high humidity and low evaporation; other crops include broccoli, cauliflower, peas, radishes, chard, and spinach. From late summer to fall, pumpkins become a prominent part of the county's landscape; the Half Moon Bay Pumpkin Festival, held in October, is a popular cultural event that celebrates the harvest of pumpkins and other crops.

Many of California's coastal crop and grazing lands extend virtually to the water's edge. It is very important that beach visitors respect the farmers' and ranchers' efforts to keep their lands in production. Please do not interfere with agricultural operations. Only access trails dedicated to public use should be used to reach the beach; do not walk across cultivated fields to obtain beach access.

NAME	LOCATION	Entrance/Parking Fee	Parking	Restrooms	Lifeguard	Campground	Showers	Firepits	Stairs to Beach	Path to Beach	Bike Path	Hiking Trail	Facilities for Disabled	Boating Facilities	Fishing	Equestrian Trail	Sandy Beach	Dunes	Rocky Shore	Upland from Beach	Stream Corridor	Bluff	Wetland
Martin's Beach	Toll Rd. off Hwy. 1, 6 mi. S. of Half Moon Bay	•	•	•											•		•		•			•	
San Gregorio Private Beach	Off Hwy. 1, 1 mi. N. of San Gregorio Rd.	•	•							•							•					•	
San Gregorio State Beach	Hwy. 1 and San Gregorio Rd.	•	•	•			•			•			•		•		•		•	•	•	•	•
Pomponio State Beach	Hwy. 1 at Pomponio Creek	•	•	•			•			•			•		•		•	•		•		•	
Pescadero State Beach	Along Hwy. 1 from 1 mi. N. of Pescadero Rd. to .75 mi. S. of Pescadero Rd.		•	•					•	•			•				•	•	•		•	•	
Pescadero Marsh Natural Preserve	E. of Hwy. 1, N. of Pescadero Rd.		•										•							•			•

MARTIN'S BEACH: Privately developed fishing cove at the foot of a toll road off Hwy. 1; entrance fee. The beach is open daily all year; the store is open daily during the summer. Popular for surf smelt fishing from April to November. Picnic tables are on the beach. Restrooms are available. Information: Martin's Beach Rd., Half Moon Bay 94019; (650) 712-8020.

SAN GREGORIO PRIVATE BEACH: The beach is located at the end of an unpaved toll road off Hwy. 1; it can be identified by its white gate just north of San Gregorio Road. Entrance fee. Clothing optional. The beach is usually open on weekends, depending on the weather.

SAN GREGORIO STATE BEACH: Memorial beach commemorating Captain Portolá's 1769 Spanish expedition to San Francisco Bay. Large sandy cove directly off Hwy. 1. The San Gregorio estuary and freshwater marsh, important wildlife habitats, extend both east and west of Hwy. 1. Highly eroded bluffs; dangerous rip currents. Facilities include picnic tables and restrooms; day-use fee. For information, call: (650) 726-8820.

POMPONIO STATE BEACH: Facilities include picnic tables overlooking the ocean, day-use cooking grills, and restrooms. Day-use fee. Call: (650) 879-2170.

PESCADERO STATE BEACH: Mile-long beach with main shoreline access points at Pescadero Rd. and at .2 miles and 1 mile north of Pescadero Road. A vista point situated just north of Pescadero Rd. overlooks the Pacific Ocean and the Pescadero Marsh Natural Preserve. Picnic sites are located throughout the beach area. For information, call: (650) 879-2170.

PESCADERO MARSH NATURAL PRESERVE: Undeveloped wildlife sanctuary; the largest coastal marsh between San Francisco Bay and Elkhorn Slough. Sensitive area. Heavy visitor use is discouraged, but informal access for bird watching is permitted along the dirt fill levees off Pescadero Rd. and along the interpretive trails off Hwy. 1 at the north end of the preserve and near Pescadero Creek. The northernmost trail is closed from March 15 to September 1 to prevent hikers from disturbing certain breeding birds. Over 160 species of birds have been sighted in the area; best bird watching is during the late fall and early spring. No hunting or pets allowed. Picnic tables and restrooms are available. For information, call: (650) 879-2170.

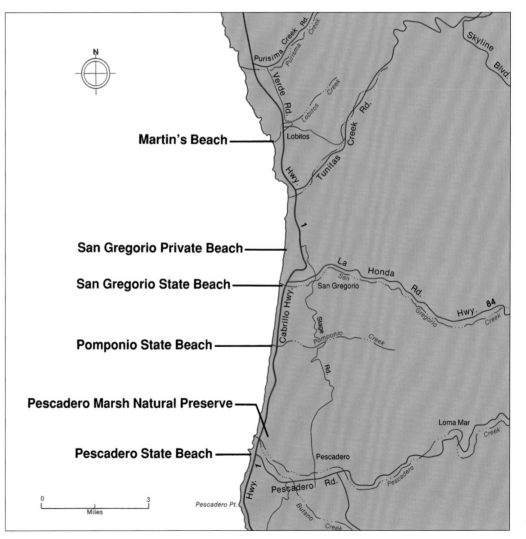

Northern Elephant Seal

The northern elephant seal *Mirounga angustirostris* is the largest of all pinnipeds (seals and sea lions) found in this hemisphere. Adult males reach a length of 14-16 feet with a weight of nearly three tons; they have a distinctive inflatable nasal sac that hangs down almost into their mouths when they rear their heads back. The name elephant seal refers to this proboscis. The females are smaller, reaching 10-12 feet in length and 1,200-2,000 lbs.

Despite their size, both sexes can move swiftly along the sand. Elephant seals come ashore only to breed, give birth, and molt. Their normal diet includes squid, fish, and occasional small sharks, but they do not eat during the three-month breeding season. The elephant seals' only natural enemies are orcas, also known as killer whales, and sharks.

The breeding season begins in late November when the males begin arriving at the breeding site, called a rookery, and establish a dominance hierarchy through brief but violent battles; the bulls at the top of the rank will do most of the breeding. In early December, the females arrive and give birth within six days; pups weigh 75 lbs. at birth and increase to 300-400 lbs. within a month. Mothers nurse their young for 27-29 days.

Mating takes place about 24 days after the birth of the pups and the gestation period lasts eight months; a natural delay causes the fertilized egg to implant in the uterus wall three months later so that the young will be born at the same time the following year, not at sea. By mid-March the adults leave the rookery and head back to sea, leaving the young to follow within a few days. By late March adult females and yearlings begin arriving back on the beaches to molt their fur, a four-to six-week process. Juvenile males begin arriving during May and June and adult males from July to early September. During the fall, young seals born the previous winter haul out on the beaches.

Año Nuevo State Reserve is the site of the world's largest mainland breeding colony of the northern elephant seal.

The reserve encompasses over 4,000 acres of coastal bluffs, sand dunes, and beaches, as well as a 13-acre island one half mile offshore. Although the seals may be seen year round, visits to the reserve during breeding season are extremely popular.

During breeding season, visiting the elephant seal breeding grounds is restricted to participants on guided walks, provided seven days a week from December 15 through March 31. For recorded information, call Año Nuevo State Reserve (650) 879-0227; for walk reservations, call: 1-800-444-4445. Reservations are available beginning in October and weekend spots usually fill immediately. Public transit is available to the reserve in January and February only; for reservations, call SAMTRANS: (650) 508-6441 or 1-800-660-4287. Walks take about 2 hours and cover about 3 miles along blufftops and sand dunes. Visitors should dress warmly and be prepared for cold temperatures, wind, and rain. Walks are given rain or shine.

From April through November, visits are available with permits which are issued at the reserve. From December 1 through 14, the elephant seal breeding area is closed to visitors.

Northern elephant seal, Año Nuevo State Reserve

Northern elephant seal, Año Nuevo State Reserve

San Mateo County

SOUTHERN SAN MATEO COAST

NAME	LOCATION	Entrance/Parking Fee	Parking	Restrooms	Lifeguard	Campground	Showers	Firepits	Stairs to Beach	Path to Beach	Bike Path	Hiking Trail	Facilities for Disabled	Boating Facilities	Fishing	Equestrian Trail	Sandy Beach	Dunes	Rocky Shore	Upland from Beach	Stream Corridor	Bluff	Wetland
Pebble Beach	Hwy. 1, between Hill Rd. and Artichoke Rd.	•	•	•				•		•							•		•			•	
Bean Hollow State Beach	Along Hwy. 1 from Artichoke Rd. to Bean Hollow Rd.		•	•					•								•		•			•	
Overlook	10257 Cabrillo Hwy., Pescadero		•																			•	
Butano State Park	Cloverdale Rd., 5 mi. S. of Pescadero and 3 mi. E. of Hwy. 1	•	•	•		•		•				•	•							•	•		
Pigeon Point Lighthouse Hostel	Pigeon Point Rd., W. of Hwy. 1	•	•	•		•				•				•			•		•	•		•	
Overlooks	Along Hwy. 1 from Pigeon Point Rd. to 4 mi. S. of Pescadero Rd.		•											•					•			•	
Gazos Creek Access	Hwy. 1. N. of Gazos Creek Rd.		•	•					•				•		•		•	•				•	•
Año Nuevo State Reserve	Off Hwy. 1, 27 mi. S. of Half Moon Bay	•	•	•					•			•	•	•	•		•	•	•			•	•

PEBBLE BEACH: Noted for the beach's delicate rock formations called tafoni and for its pebbles; removal of pebbles is illegal. Picnic tables and restrooms are available. A self-guided blufftop nature trail leads south to Bean Hollow State Beach.

BEAN HOLLOW STATE BEACH: Paved parking area and picnic tables; restrooms. Rocky surf and hazardous rip currents. Call: (650) 879-2170.

OVERLOOK: A 10 foot wide path leads from the Highway 1 frontage road to a rocky bluff and scenic overlook. Look for opening in guardrail 8/10 mi. south of Bean Hollow State Beach, then turn north on narrow frontage road; do not block driveway; respect adjacent private property.

BUTANO STATE PARK: 2,200-acre redwood park with 20 miles of hiking trails and a campground; overnight and day-use fees. Special hiker/bicyclist group camping is available. Summertime nature walks and campfire programs. Call: (650) 879-2040.

PIGEON POINT LIGHTHOUSE HOSTEL: Located adjacent to the second tallest lighthouse on the West Coast. Facilities include 52 beds, kitchen, and bike storage. Day-use facilities for non-profit groups available. Daily check-in hours are 4:30 PM-9:30 PM. Office also open 7:30 AM-9:30 AM. Overnight fee. For information and reservations: Pigeon Point Rd., Pescadero 94060; (650) 879-0633.

GAZOS CREEK ACCESS: Paved parking lot; a short wheelchair-accessible path leads to the beach. Provides the only access to Franklin Point to the south, a popular fishing area; stay below the mean high tide line between Gazos Creek and Franklin Point. Restrooms are available. Call: (650) 879-0227.

AÑO NUEVO STATE RESERVE: 4,000 acres; the site of some of the few remaining active dune fields on the California coast, and a protected breeding ground for northern elephant seals. The reserve is also noted for California gray whale sighting in January and March, spring and fall migratory bird watching, and hiking April through November. The reserve's main entrance, parking lot, and office are on New Years Creek Road. The reserve north of Cascade Creek is accessible by three paths, marked with state park signs, off Hwy. 1 at 2.7, 3.2, and 3.3 miles north of New Years Creek Road.

From December 15 to March 31, when the elephant seals are breeding, the reserve south of Cascade Creek is open only to guided walks available by reservation. Also during this time, it is illegal to enter any of the reserve's intertidal areas between Franklin Point and New Years Creek. The guided walks are three miles long and take 2 hours; day-use fees are charged year-round. Open 8 AM-sunset. No dogs allowed. A wheelchair-accessible path and viewing platform is available. For guided walk reservations, call: 1-800-444-4445. Call for recorded ticket information: (650) 879-0227.

Transit: SAMTRANS provides express bus service from the Hillsdale shopping center and Half Moon Bay to the reserve on weekends and holiday Mondays during January and February; includes guided walk at the reserve. For reservations, call: (650) 508-6441 or 1-800-660-4287.

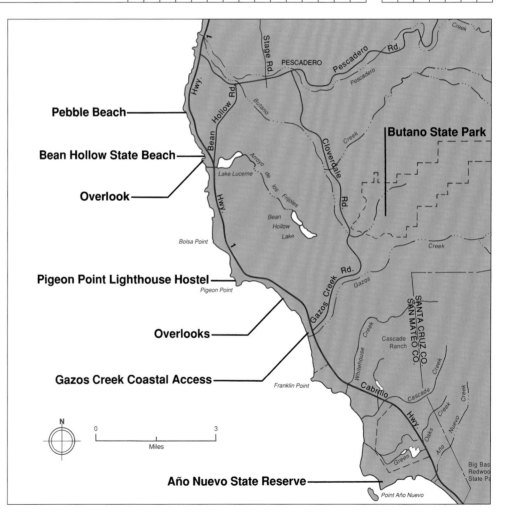

View from West Cliff Drive, Santa Cruz

Santa Cruz County

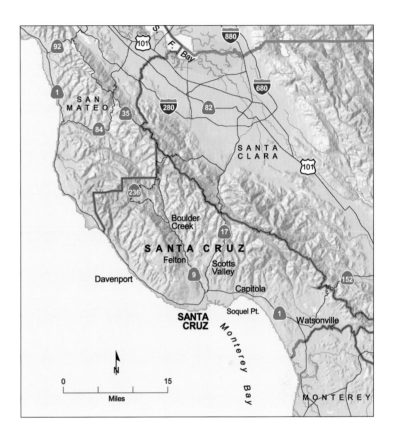

outh of Año Nuevo Point, the varied coastline of Santa Cruz County begins, extending for 42 miles from the north county coastal terraces beneath the Santa Cruz Mountains to the productive farmlands of the Pajaro Valley in the south. Narrow, sandy beaches backed by steep bluffs are characteristic of the coast north of the city of Santa Cruz, in contrast to the wide, sheltered swimming beaches found along the north shore of Monterey Bay.

Agriculture is prevalent along most of the county's coast, providing fields of crops and open space adjacent to the shoreline sea cliffs and Highway 1. The fertile coastal terraces of the north coast produce 30 percent of California's Brussels sprouts, while the more inland Pajaro Valley near Watsonville is noted for crops of lettuce, strawberries, artichokes, and cut flowers.

The rugged Santa Cruz Mountains, which border the coastline north of Monterey Bay, contain beautiful upland forests of coastal redwoods, *Sequoia sempervirens*. These forests have been logged for over 115 years, but because of the establishment in 1902 of the Big Basin Redwoods State Park within the watershed of Waddell Creek, many old-growth redwood groves still remain. A popular hiking trail through redwood groves in Big Basin Redwoods State Park is the scenic 38-mile-long Skyline-to-Sea Trail that begins at Castle Rock State Park near Saratoga Gap and ends at the ocean at Waddell Creek Beach.

Point Santa Cruz, seaward of the city of Santa Cruz, marks the northern limit of Monterey Bay. In 1602, the Spanish explorer Don Sebastián Vizcaíno visited the bay and described it in his ship's log in such enthusiastic terms that the Spanish government sent in Juan Gaspar de Portolá on a coastal land expedition in 1769 to find and settle Monterey Bay for Spain. Although Portolá's expedition in 1769 camped alongside the bay twice, and found San Francisco Bay, it wasn't until a second expedition in 1770 that Portolá recognized it as the same Monterey Bay in which Vizcaíno had originally anchored.

Portolá's expedition also raised the first Catholic cross on the bank of the San Lorenzo River, marking the site of the city of Santa Cruz (Spanish for "Holy Cross"). In 1791 the Santa Cruz Mission was founded by Father Fermín Francisco de Lasuén, and by 1840 a small town had been settled. Today, the city of Santa Cruz is a popular tourist town, with its warm weather, sandy beaches for swimming, surfing, and fishing, and the famous boardwalk amusement park, the last of its kind still open on the California coast. The boardwalk attracts numerous visitors each weekend, and offers food, rides, an arcade, and a mile-long protected sandy beach that stretches in front of it.

Prior to Spanish settlement, the Santa Cruz coast was inhabited by Costanoan Indians, whose diet was dependent upon the abundant shellfish found here. Shell middens left by the Costanoans were enormous; the volume of discarded shells was so great near some abandoned Indian camps that the material was used for road pavement and by farmers as a dietary supplement for their poultry.

For further information on the Santa Cruz County coast, contact the Santa Cruz County Conference and Visitors Council, at 1211 Ocean Street, Santa Cruz 95060, (831) 425-1234 or (800) 833-3494 or see www.santacruzca.org or the Capitola Chamber of Commerce, 716G Capitola Avenue, Capitola 95010, (831) 475-6522. For transit information, write or call the Santa Cruz Metropolitan Transit District (SCMTD), 370 Encinal Street, suite 100, Santa Cruz 95060, (831) 425-8600 or see www.scmtd.com.

Coastal Vegetation

A plant community is a naturally occurring group of different plants inhabiting a common environment. These plants often share certain characteristics that help them adapt to their habitat. Plant communities are not always clearly defined, and their composition changes over time, due to the intentional or accidental introduction of exotic species, or the destruction of native species. Several plant communities are recognizable along the California coast.

The plants found on the sandy beaches and dunes adjacent to the ocean belong to the coastal strand community. Plants in the strand community grow relatively close to the ground as herbs, vines, or low shrubs. Most of these species are adapted to the high amounts of dissolved salts found in the soils bordering the shoreline. Characteristic features of coastal strand are succulent, leathery leaves and stems that retain high amounts of water and resist water loss through evaporation. One of the most common strand plants is ice plant; these low-growing succulents were introduced from South America and South Africa. Other plants in this community include sand verbena, beach primrose, and beach morning-glory.

Saltmarsh and freshwater marsh plant communities are also located near the water's edge, especially where the slope is very shallow, but unlike coastal strand vegetation, they are frequently inundated by the tides or by floods. Because the high amount of water reduces the availability of oxygen, these plants have adaptations to increase the amount of oxygen available to the plant roots.

Salt marsh plants, like coastal strand plants, are adapted to salt stress. Many of these plants, such as pickleweed or glasswort, are succulents, which can hold water to reduce the drying effect of the salt. Other species, such as saltgrass, exude the salt from small glands on their leaves. Generally, salt marshes tend to be characterized by low, shrubby vegetation, often growing in clumps separated by nonvegetated salt flats or mudflats. Freshwater marshes, on the other hand, are typified by fairly tall plants such as cattails or bulrushes.

Above the high water level, coastal sage scrub is the predominant vegetation type along the Southern California coast. An association of woody shrubs ranging in height from one to five feet, it is typified by California sagebrush, white and black sage, California buckwheat, and lemonadeberry. Plants in this community are adapted to arid coastal climates from about sea level to 3,000 feet. The leaves are usually small and leathery, adapted to resist water loss through evaporation.

Similar to the coastal sage scrub, but ranging roughly from Monterey to the Oregon border, is the northern coastal scrub community. Characterized by coyote brush, California blackberry, and monkeyflower, northern coastal scrub is found at elevations below 500 feet on often steep hillsides. The picturesque Monterey cypress, which grows along the blufftops of the central coast, is also found in this community. Most of the species in the northern coastal scrub community, however, rarely grow over six feet in height and are often intermixed with extensive areas of grass.

Chaparral is a common plant community on the hill and mountain sides of coastal California, usually growing at higher elevations than the coastal sage scrub association. This vegetation type is typically found growing in dense, almost impenetrable stands with individual plants from three to ten feet tall. Most of the plants are well-adapted to the harsh climate on these dry slopes; the leaves are often small and covered with a gray fuzz to reflect the sun's rays, and frequently have strong spines. Typical chaparral plants are chamise, manzanita, mountain lilac, toyon, and scrub oak.

Plants that grow along coastal rivers and streams belong to the riparian community. Riparian habitat, because of the relatively high amount of water present, is some of the most valuable habitat for wildlife. The riparian plant community is dominated in the south by trees such as the big-leaf maple, canyon sycamore, willow, and alder. From San Francisco north, the creek bottoms are often dominated by redwoods and attendant species such as madrone and Douglas-fir.

sand verbena
Abronia latifolia

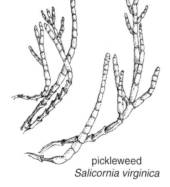

pickleweed
Salicornia virginica

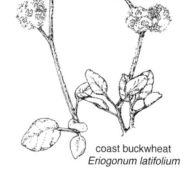

coast buckwheat
Eriogonum latifolium

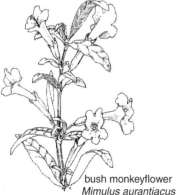

bush monkeyflower
Mimulus aurantiacus

manzanita
Arctostaphylos

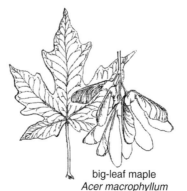

big-leaf maple
Acer macrophyllum

Santa Cruz County

NORTHERN SANTA CRUZ COAST

NAME	LOCATION	Entrance/Parking Fee	Parking	Restrooms	Lifeguard	Campground	Showers	Firepits	Stairs to Beach	Path to Beach	Bike Path	Hiking Trail	Facilities for Disabled	Boating Facilities	Fishing	Equestrian Trail	Sandy Beach	Dunes	Rocky Shore	Upland from Beach	Stream Corridor	Bluff	Wetland
Big Basin Redwoods State Park	14 mi. N. of Santa Cruz on Hwy. 9, 7 mi. N.W. on access rd.	•	•	•		•	•	•			•	•			•					•	•		
Waddell Creek Beach (Unit of Big Basin Redwoods State Park)	Hwy. 1, 1 mi. S. of San Mateo County line	•	•	•			•					•				•	•				•	•	•
Greyhound Rock Fishing Access	Hwy. 1 and Swanton Rd., 4 mi. S. of Año Nuevo		•	•				•		•	•		•			•	•		•			•	
Scott Creek Beach	Hwy. 1, 11 mi. N. of Santa Cruz		•							•						•	•			•		•	
Davenport Landing Beach	Hwy. 1, 10 mi. N. of Santa Cruz	•	•							•	•	•				•	•				•		
Davenport and Davenport Beach	Hwy. 1, 9 mi. N. of Santa Cruz		•							•		•				•	•				•		
North Santa Cruz Beaches	Hwy. 1, between Davenport and Red, White and Blue Beach		•							•						•	•	•		•	•		
Red, White and Blue Beach	5021 Hwy. 1, 5.5 mi. N. of Santa Cruz	•	•	•		•	•	•		•						•	•			•			
Wilder Ranch State Park	Hwy. 1, 2 mi. N. of Santa Cruz	•	•	•			•	•	•			•			•	•		•	•	•	•	•	

BIG BASIN REDWOODS STATE PARK: California's oldest state park was established in 1902. Over 18,000 acres with old growth redwoods, numerous waterfalls and a variety of wildlife. There are 80 miles of hiking trails, an equestrian trail and camp, picnic campsites, 35 walk in tent-cabins, 5 backpacker trail camps. No trailer hookups; trailer sanitation station at Huckleberry Campground. Park Headquarters and the Nature Lodge museum are 9.5 miles west of Boulder Creek on Big Basin Way. Call (831) 338-8860. Rancho Del Oso, the coastal side of Big Basin Redwoods State Park, includes Waddell Marsh in the Theodore Hoover Natural Preserve. A Visitor Center has historical and natural history exhibits and provides guided nature tours. Call: (831) 425-1218.

WADDELL CREEK BEACH: Located across Highway 1 from the Rancho Del Oso entrance, known worldwide as an exceptional windsurfing beach and popular with boogie boarders, fishermen, and naturalists. Parking and toilets are available.

GREYHOUND ROCK FISHING ACCESS: Popular rock fishing spot with a steep but paved pedestrian access to beach, picnic tables, large parking lot, vault toilets and disabled-accessible view platform.

SCOTT CREEK BEACH: A wide sandy beach with fishing and surfing. Popular with birders. Improved shoulder parking, viewing platform and bicycle rack.

DAVENPORT LANDING BEACH: A sandy pocket beach 1000 feet north of the town of Davenport on Davenport Landing Road. Favored for surfing and kite flying. Restroom, pedestrian stairs and handicapped ramp to beach. Bike rack. Limited shoulder parking.

DAVENPORT AND DAVENPORT BEACH: Dirt lot parking across from historic town of Davenport. Bluff top trails known for whale watching January through March. Informal access trail to large sandy pocket beach.

NORTH SANTA CRUZ BEACHES: Access is available to a series of pocket beaches between Davenport and Red, White and Blue Beach. Panther Beach, shoulder parking and trail to beach. Bonny Doon Beach, improved roadside parking; trail to beach. Yellowbank Beach, shoulder parking and trail to beach. Laguna Creek Beach, small parking area on inland side of Highway 1; trail to beach.

RED, WHITE AND BLUE BEACH: Private beach, clothing optional. Look for the red, white and blue mailbox 4.2 miles south of Davenport on the west side of Hwy. 1; turn onto Scaroni Rd. and cross the Southern Pacific railroad track. Day-use and camping fees, must be 21, married, or with parents to enter. Picnic tables available; no cameras or dogs allowed.

WILDER RANCH STATE PARK: The park has 4,505 acres with 34 miles of hiking, biking and equestrian trails. The restored Wilder Ranch complex comprises a California coastal dairy museum including a visitor center and picnic tables. A one-mile long paved multi-use trail extends from the Wilder Ranch complex to Shaffer Lane in the City of Santa Cruz. Both Four Mile and Three Mile beaches are within the park. Fee parking. Call: (831) 423-9703. Visitor Center: 426-0505.

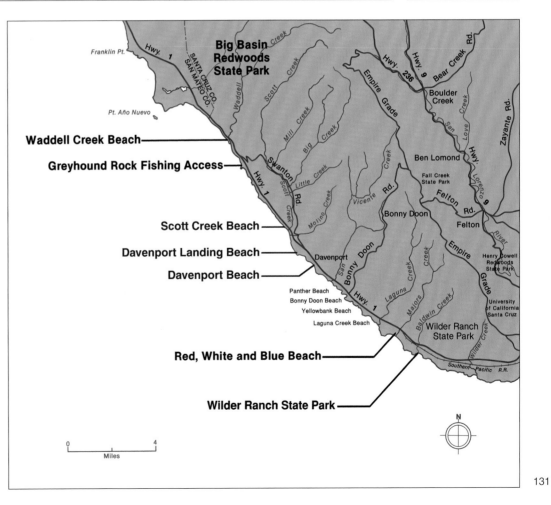

FACILITIES

ENVIRONMENT

Santa Cruz Boardwalk

Mark Abbott Memorial Lighthouse

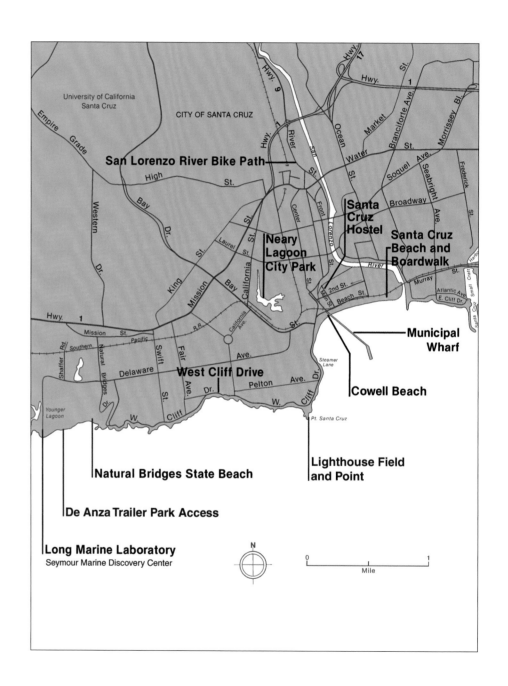

Santa Cruz County

CITY OF SANTA CRUZ

NAME	LOCATION	Entrance/Parking Fee	Parking	Restrooms	Lifeguard	Campground	Showers	Firepits	Stairs to Beach	Path to Beach	Bike Path	Hiking Trail	Facilities for Disabled	Boating Facilities	Fishing	Equestrian Trail	Sandy Beach	Dunes	Rocky Shore	Upland from Beach	Stream Corridor	Bluff	Wetland
Long Marine Laboratory / Seymour Marine Discovery Center	Corner of Delaware Ave. and Shaffer Rd., Santa Cruz	•	•	•								•	•				•	•				•	•
De Anza Trailer Park Access	2395 Delaware Ave., Santa Cruz		•					•										•					
Natural Bridges State Beach	2531 W. Cliff Dr., Santa Cruz	•	•	•			•		•	•	•	•	•		•		•		•	•	•	•	
West Cliff Drive	Runs from Natural Bridges to Cowell Beach, Santa Cruz		•	•							•		•						•			•	
Lighthouse Field and Point	Along W. Cliff Dr. at Point Santa Cruz		•	•						•		•	•	•		•		•		•	•		•
Cowell Beach	W. Cliff Dr. and Bay St., Santa Cruz	•	•	•	•				•				•		•		•	•					
Neary Lagoon City Park	Off California St., N. of Bay Dr., Santa Cruz		•	•								•	•							•			•
Santa Cruz Municipal Wharf	Foot of Washington St. and Beach St., Santa Cruz	•	•	•						•			•	•	•				•				
Santa Cruz Beach and Boardwalk	E. of Municipal Wharf, Santa Cruz	•	•	•	•								•				•						
San Lorenzo River Bike Path	Along San Lorenzo River, Santa Cruz										•	•			•					•	•		
Santa Cruz Hostel	321 Main St., Santa Cruz	•	•	•			•						•									•	

LONG MARINE LAB / SEYMOUR MARINE DISCOVERY CENTER: The only working research lab in California open to the public. Aquarium, lab wide tours, daily programs, marine mammal research tours, children's programs, bookstore. Trails loop around the Discovery Center and the open terrace above the rocky beach. Located at the end of Delaware Avenue. Closed Mondays. Parking. Admission fee. Visitor information (831) 459-3800.

DE ANZA TRAILER PARK ACCESS: DeAnza sits between the Discovery Center site and Natural Bridges State Beach. From the Discovery Center enter gate near bluff edge into DeAnza Trailer Park follow signs along paved way which will lead to a path down to the sandy beach at Natural Bridges. The trailer park trail can also be entered from Delaware Avenue. Daylight hours only.

NATURAL BRIDGES STATE BEACH: 65 acres including eucalyptus groves, dunes, sandy beach and tidepools. Scenic overlook above natural rock bridge formation. Parking and restrooms. Picnic sites inland in eucalyptus grove by creek. Noted Monarch Butterfly Preserve with guided tours during the annual migration from October through February. Wheelchair accessible walkway along the Monarch Trail. Wheelchair-accessible restroom near beach; beach wheelchair available. Admission fee. For information, call: (831) 423-4609.

WEST CLIFF DRIVE: A three-mile long ocean side bicycle/pedestrian path follows West Cliff from Natural Bridges to the Wharf. Numerous scenic overlooks of Monterey Bay with parking bays and benches. Stairs to coves are located at the end of Almar (Mitchell Cove); just upcoast of Lighthouse Point (Its Beach); just downcoast of the Point at the end of Pelton Avenue; and at the end of Monterey Street. Coastal bluffs are highly eroded; tidal rocks are slippery when wet.

LIGHTHOUSE FIELD AND POINT: A 40 acre state park with open uplands and a scenic viewpoint overlooking internationally recognized surfing spot Steamer Lane. Parking, trails, restroom, drinking fountain, picnic tables. The Mark Abbot Memorial Lighthouse houses the Santa Cruz Surfing Museum with the largest collection of surfing books on the West Coast.

COWELL BEACH: At the foot of the Santa Cruz Wharf; has a small overlook plaza, restrooms, fee parking and stairway access to the beach.

Volleyball standards and nets provided during summer months. A plastic ramp for wheelchair access to the beach and water is installed during the summer.

NEARY LAGOON CITY PARK: A nature refuge inland of Cowell Beach. Facilities include a wildlife sanctuary with a floating platform extending into the lagoon for wildlife viewing. No dogs, hunting, or bathing. Paved roadway to parking lot and lagoon. Picnic area, tennis courts, and children's playground.

SANTA CRUZ MUNICIPAL WHARF: A public fishing wharf. Accommodates bicycles, pedestrians and cars. Over a half-mile long. Boat and motor rentals, fishing licenses and equipment. Shops and restaurants, interpretive kiosks. Outdoor tables and benches at end of wharf. Open all year. Vehicle entrance fee. For information on boat rentals, call: (831) 423-1739.

SANTA CRUZ BEACH AND BOARDWALK: Located on a mile long beach, the historic amusement center has rides, arcades, gift and candy shops. A pedestrian path from the amusement center boardwalk crosses the San Lorenzo River via the railroad trestle. The boardwalk promenade has multiple stairways to the wide sandy beach. Swimming. Free admission. Open 7 days a week all summer. Open most other weekends. Operation subject to weather. See www.beachboardwalk.com for current information or call: (831) 426-7433.

SAN LORENZO RIVER BIKE PATH: A bike path runs along both sides of the San Lorenzo River levee from Highway 9 (River Street) to the river mouth at Santa Cruz Beach.

SANTA CRUZ HOSTEL: Located at 321 Main Street between 2nd and 3rd streets on Beach Hill between the Wharf and the Boardwalk. The hostel has 40 beds, showers, lockers, no laundry, street parking. Open 8-10 AM and 5-10 PM. Overnight fee, reservation not necessary; in summer accepts Hostelling International (AYH) members only. The hostel is fully wheelchair accessible. For information, call: (831) 423-8304.

Tidepool, Natural Bridges State Beach

Waves and Surf

Most of the breakers that crash upon California's beaches originate hundreds of miles offshore as a result of winds blowing across the ocean's surface and creating ripples; as the wind continues to blow, ripples turn into progressively larger waves. The size of waves generated by the wind is proportional to wind velocity, wind duration, and fetch (the extent of the open ocean across which the winds blow). A mariner's rule of thumb is that wave height in feet will be approximately half the wind speed, in knots.

Waves that are at a storm center are called "seas." After these waves leave the area of generation, they are called "swells." As swells travel from the generation area, the shorter period waves are filtered out, and the longer period waves radiate out from the source. As the swells continue to travel across the ocean, they tend to become grouped into a series of long waves, followed by a series of shorter waves.

Waves in the open ocean often appear as a confused and constantly changing progression of crests and troughs traveling in numerous directions. These waves can be combinations of locally generated seas and numerous swells that are constantly augmenting and cancelling each other. When waves move into shallow water they begin to "feel" the bottom and start to align parallel to the bottom contours. This alignment, called refraction, begins when waves reach a depth of about half their wave length (the distance between two successive crests). As waves move closer to shore, their length shortens, height increases, speed is decreased, and they break on the land.

Shallow waves will refract over humps and depressions in the sea floor. The part of the wave in shallower water will slow while the portion of the wave in deeper water will move at a higher velocity, causing a bending of the wave. When waves converge or bend toward a central point, they concentrate their energy at this point and can cause shoreline erosion. Divergence of a wave crest will spread the wave energy over a broad section of shoreline, and wave erosion would be less.

Breaking waves can be caused by two different processes. As a wave steepens, it can reach a point when it oversteepens. A steepness greater than about 1 to 13 (height to length) will cause the wave to break. Waves will also break when they reach a water depth that is approximately 5/4 the breaker height. The way the wave breaks depends on the wave and the beach steepness. For example, the classic surfing wave is a plunging breaker where the wave crest falls forward of the lower part of the wave. This type of breaker is most likely to occur with a moderately steep wave and a steep beach. The same wave on a gently sloping beach could become a surging or spilling breaker, where the wave crumbles at the top and spills down its front.

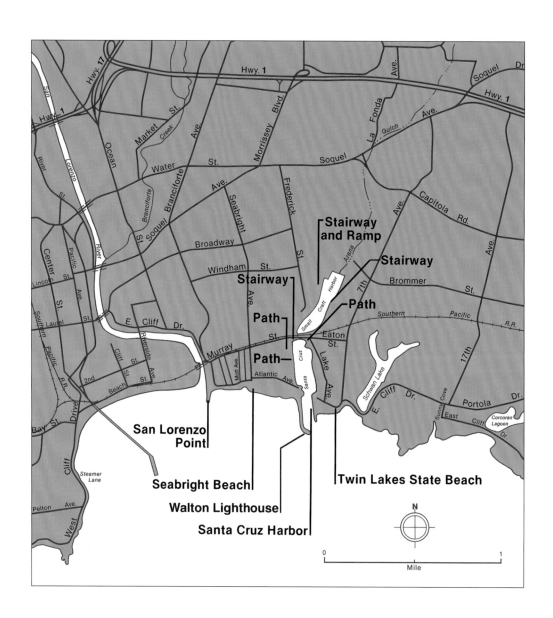

Santa Cruz County

SANTA CRUZ HARBOR AREA

NAME	LOCATION	Entrance/Parking Fee	Parking	Restrooms	Lifeguard	Campground	Showers	Firepits	Stairs to Beach	Path to Beach	Bike Path	Hiking Trail	Facilities for Disabled	Boating Facilities	Fishing	Equestrian Trail	Sandy Beach	Dunes	Rocky Shore	Upland from Beach	Stream Corridor	Bluff	Wetland
San Lorenzo Point	Foot of E. Cliff Dr. off Murray St., Santa Cruz		•																•	•	•		
Seabright Beach (Unit of Twin Lakes State Beach)	E. Cliff Dr., San Lorenzo Point to Seabright Ave., Santa Cruz		•	•	•		•		•				•				•					•	
Walton Lighthouse	West Jetty, Santa Cruz Harbor														•								
Santa Cruz Harbor	Eaton St. at Lake Ave., Santa Cruz	•	•	•	•				•	•			•	•	•		•						
Bike and Footpath	Around Harbor, Santa Cruz		•								•	•											
Stairways to Harbor	W. side of Harbor, S. side of Murray St.; end of Mello Lane, Santa Cruz								•														
Stairway and Ramp to Harbor	Upper Harbor at Frederick Street Park, Santa Cruz								•	•												•	
Path to Harbor	E. side of Harbor and Eaton St., Santa Cruz								•	•													
Path to Harbor	At Santa Cruz Yacht Club, South Harbor, Santa Cruz									•													
Twin Lakes State Beach	7th Ave. at E. Cliff Dr., Santa Cruz		•	•	•			•					•	•	•		•					•	•

SAN LORENZO POINT: San Lorenzo Point is the long, narrow promontory that projects into Monterey Bay, to the east of the San Lorenzo River; part of Twin Lakes State Beach. The point is accessible via E. Cliff Drive and a path that leads to the end of the promontory. Limited parking on E. Cliff Drive.

SEABRIGHT BEACH: Also called Castle Beach. Drop-off zone at E. Cliff Drive at the foot of Mott and Cypress Avenues. Restrooms are wheelchair accessible. Limited on-street parking only. Call: (831) 429-2850. Blufftop promenade from Mott to 4th Ave.; stairway and parking area at the end of Third Avenue. The Santa Cruz Museum nearby on E. Cliff Drive has natural science and history exhibits; open 10 AM-5 PM Tuesday–Sunday. Museum: (831) 420-6115.

WALTON LIGHTHOUSE: Gravel path along west jetty to the lighthouse; fishing is popular.

SANTA CRUZ HARBOR: North and South harbors are crossed by Murray/Eaton St. bridge; there is a pedestrian/bike path around the perimeter of the harbor. Harbor path connects to a City operated loop trail at the northern terminus of the harbor. North Harbor provides 600 boat berths and 12 self-contained RV spaces; call: (831) 475-3279. South Harbor has 400 berths. Ramp is four lanes, concrete, open 24 hours. There is a half-ton fishing hoist at the public launch and a 60-ton travel lift at the boat yard, open 8 AM-5 PM. Hand launch ramp located at southwest side near the bridge. The Joseph G. Townsend Maritime Plaza is on the southeast harbor and provides views of the harbor. Fuel dock open 6:15 AM-4:30 PM; call: (831) 476-2648. Fishing boat rentals and supplies available; restaurants and shops. Open all year. Transient vessels report to the harbor office at the southeast corner of the harbor for berth assignments. For harbor information, call: (831) 475-6161.

STAIRWAYS TO HARBOR: At Murray St.; wooden stairway with rail. Near U.C. Santa Cruz sailing facility. Stairs to harbor at end of Mello Lane.

PATH TO HARBOR: At Eaton St.; paved path serves bicycles and pedestrians.

PATH TO HARBOR: At the Santa Cruz Yacht Club; access via 4th Ave. to harbor.

TWIN LAKES STATE BEACH: Extends on both sides of the harbor; 86 acres. Facilities include volleyball nets and standards and fire rings on the beach; jetty fishing. Schwan Lagoon is a waterfowl refuge. Call: (831) 427-4868.

San Lorenzo Point

Littoral Current

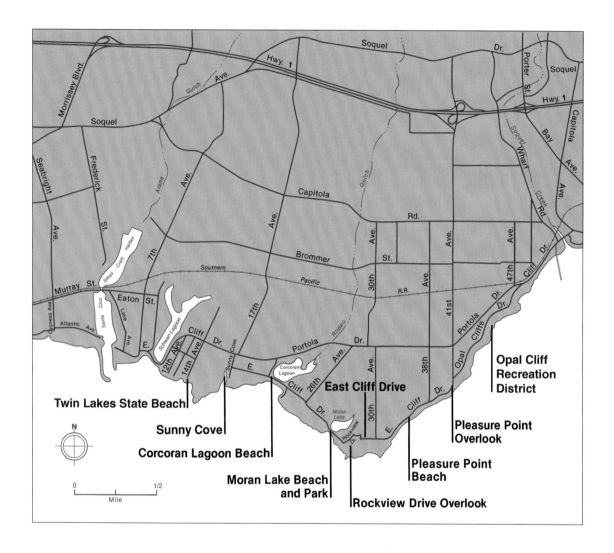

Longshore or littoral transport is a type of sediment movement in the nearshore zone that is caused by waves. As waves approach the shore, they are refracted parallel to the shoreline, but in most cases refraction is incomplete and the waves will strike the land at a slight angle. Consequently, some of the wave energy is transported parallel to the beach, and a weak longshore current is created. As waves approach a sandy shoreline, they often lift bits of sand off the bottom. Once the material is off the bottom, it can be moved by the littoral current.

When an obstacle, such as a headland or jetty, blocks the path of the current, the current often drops its suspended sediment, causing accretion of beach material on the upstream side of the obstruction. Erosion will often occur at the downstream side of the obstacle, where the littoral current returns close to shore. The extent of accretion and erosion will depend on the velocity and duration of the current and the availability of transportable material. The littoral current also will change strength and direction as the orientation of the waves changes, so some sand may make several trips along the same stretch of beach.

The direction of littoral current can often be determined by observing which direction a floating object just offshore moves with respect to a fixed point on shore. Although the force of the littoral current may not be noticeable to people in the water, bathers who find themselves leaving the water up or down coast from where they entered were probably moved there by the littoral current. The predominant long-term direction of littoral current can be determined from the evidence of erosion and accretion along the shore.

Santa Cruz County

LIVE OAK

NAME	LOCATION	Entrance/Parking Fee	Parking	Restrooms	Lifeguard	Campground	Showers	Firepits	Stairs to Beach	Path to Beach	Bike Path	Hiking Trail	Facilities for Disabled	Boating Facilities	Fishing	Equestrian Trail	Sandy Beach	Dunes	Rocky Shore	Upland from Beach	Stream Corridor	Bluff	Wetland
East Cliff Drive	Runs from Twin Lakes to 41st Ave	•									•								•			•	
Twin Lakes State Beach (East of Harbor)	Foot of 14th Ave., Live Oak		•	•			•	•			•		•		•		•					•	•
Sunny Cove	Foot of Sunny Cove, Live Oak							•									•		•			•	
Corcoran Lagoon Beach	E. of 21st Ave. and E. Cliff Dr., Live Oak							•									•				•		•
Moran Lake Beach and Park	2700 block of E. Cliff Dr., Live Oak	•	•	•				•					•		•		•			•		•	•
Rockview Drive Overlook	Foot of Rockview Dr., Pleasure Point	•						•				•					•		•			•	
Pleasure Point Beach	E. Cliff Dr., 34th–36th aves., Pleasure Point	•						•							•		•		•			•	
Pleasure Point Overlook	E. Cliff Dr. at 41st Ave., Pleasure Point	•	•				•	•											•			•	
Opal Cliffs Recreation District	4500 block of Opal Cliffs Dr., Pleasure Point	•						•									•		•			•	

EAST CLIFF DRIVE: Numerous scenic overlooks and parking pull-outs; bike lane. A bike path runs along both sides of 17th Ave. from Soquel Ave. to E. Cliff Drive and connects with the E. Cliff Drive bike lane. The county maintains stairways that provide beach access along E. Cliff Drive at the ends of 3rd, 12th, 13th, 20th, 26th, 36th, 38th, and 41st avenues.

TWIN LAKES STATE BEACH: The wide sandy beach is accessible to pedestrians via the ends of 12th St. and 13th and 14th avenues. Bonita Lagoon, a small wetland, supports some waterfowl.

SUNNY COVE: Privately owned pocket beach that is maintained by the county; the beach is accessible via the foot of Sunny Cove or Johans Beach Drive. Extent of public and private rights is undetermined.

CORCORAN LAGOON BEACH: The beach is seaward of E. Cliff Drive and the lagoon; a path leads from the road to a sandy beach and tidepools. The west side of the beach is accessible via a stairway from 20th Avenue.

MORAN LAKE BEACH AND PARK: The wide sandy beach is south of Moran Lake, a restored lagoon with a nature trail along the northwest shore. The beach is accessible to wheelchairs via a ramp from East Cliff Drive. Parking and wheelchair-accessible restrooms are located at the lake.

ROCKVIEW DRIVE OVERLOOK: Excellent ocean view; popular spot to watch surfers. Short blufftop walkway. Stairs just east of 18 Rockview Dr. and just east of 2970 Pleasure Point Dr. provide access to seasonal sandy beaches and surf areas.

PLEASURE POINT BEACH: Access via stairs at 35th and 38th aves.; pocket beaches are covered at high tide. Popular surfing spot; tidepools.

PLEASURE POINT OVERLOOK: Another popular spot to watch surfers. Hazardous cliffs; stairs are located at the end of 41st Avenue.

OPAL CLIFFS RECREATION DISTRICT: Six parking spaces plus street parking. A trail and stairway lead to pocket beaches, popular with surfers; keys to the locked gate are sold each year at the Freeline Surf Shop at 821 41st Ave.; for information, call: (831) 476-2950.

Pleasure Point Overlook

Tides

The cyclical rise and fall of sea level along the world's shorelines is caused by a long period wave called the tide. Tides are generated by the sun's and moon's gravitational "pull" on the earth. The strength of the gravitational pull between two bodies, such as the earth and the moon, is related to the masses of the two objects and the cube of the distance between them. Although the moon is much smaller than the sun, it has a greater effect on the earth's tides because it is much closer. This gravitational pull creates bulges in the oceans on the sides of the earth closest to (sublunal) and farthest from (antipodal) the moon, with depressions between these points. The bulges follow the position of the moon around the earth causing, at any given location, the vertical rise and fall of sea level, as well as tidal currents, the horizontal flows of water that accompany vertical changes.

There are three types of tides: a diurnal tide with one high and one low tide each day; semi-diurnal with two high and two low tides each day, where the two cycles of high and low water are approximately the same; and a mixed tide, where there are large daily variations in the heights of each high and low water level. Most of the Pacific Coast has mixed tides with two daily high tides (a higher high and a lower high) and two daily low tides (a lower low and a higher low).

In addition to the 24 hour, 50 minute cycle of the tides, there is a tidal cycle that corresponds to the lunar month (the 29.5 days that it takes for the moon to complete one full orbit around the earth). When the moon is at its perigee, or closest point to the earth, its gravitational pull is greatest, and the tidal range or difference between high and low water levels is the greatest. When the moon is at its apogee, farthest from the earth, the difference in tidal range will be least.

The sun has less effect on the tides than does the moon, but it does influence the tides, as do the closer planets. When the sun and moon are in line and are pulling together (called syzygy), such as during the new and full moons, large spring tides occur. When the moon and the sun are at right angles with respect to the earth (called quadrature), as during the first and last quarters, the moon and sun are counteracting each other and smaller, neap tides occur. A combination of a spring tide with the moon at its closest orbit point would create a very high tide. Along the California coast, the highest tides often occur in December, January, or February, normally in the morning. When higher than average tides occur at the same time as a coastal storm, there can be significant damage to coastal areas from wave damage and flooding. The sequence of all possible orientations of the sun, moon, and planets lasts 18.6 years, which is known as a tidal epoch.

Since the configuration of the coast and ocean floor will cause changes in the tidal range, coastal locations separated by only a few miles can have different high and low tides, occurring at different times. Although tides differ from one location to the next, these differences are fairly constant over time; tidal records kept for a few primary reference sites can be used to predict tides at these locations and at other, intermediate sites.

Knowledge of the times of the tides is a useful aid to beach visitors. For example, those interested in tidepool study need to know when low tide occurs so they can view the exposed rock formations that make up the pools. Those visiting pocket beaches (i.e., small cove areas) should also be aware of the tide cycles, in order to avoid being trapped by a rising tide. Most newspapers that serve coastal communities publish tide tables for the local coastal area; it is recommended that these tide tables be consulted before visiting the beach. However, because tide tables provide predictions of tides and cannot take into account the effects of wind and waves, they are not always accurate for actual water levels.

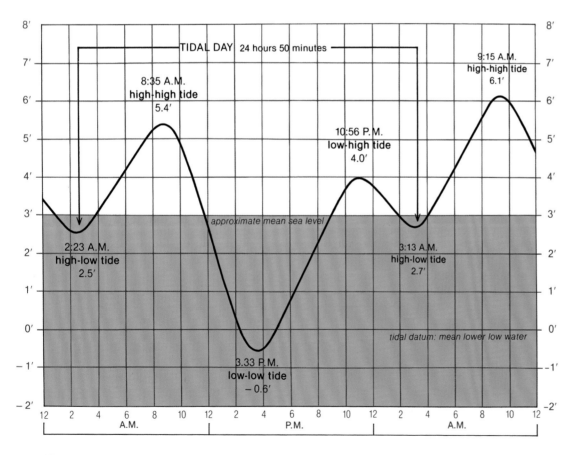

The times and heights of the tides shown here are only for illustrative purposes.

Santa Cruz County

CAPITOLA

NAME	LOCATION	Entrance/Parking Fee	Parking	Restrooms	Lifeguard	Campground	Showers	Firepits	Stairs to Beach	Path to Beach	Bike Path	Hiking Trail	Facilities for Disabled	Boating Facilities	Fishing	Equestrian Trail	Sandy Beach	Dunes	Rocky Shore	Upland from Beach	Stream Corridor	Bluff	Wetland
Cliff Drive	Along Cliff Dr., Capitola		•														•					•	
Hooper Beach	W. of Capitola Fishing Wharf, Capitola		•		•												•			•			
Capitola Fishing Wharf	Foot of Wharf Rd., Capitola			•				•						•	•		•						
Capitola City Beach	S. of Esplanade and Monterey Ave., Capitola		•	•	•												•			•			
Views and Trails	Stairways to Cliff Dr. and trail along Soquel Creek		•																		•	•	
New Brighton State Beach	1500 Park Ave., Capitola	•	•	•	•	•	•	•	•	•	•	•	•		•		•			•		•	

CLIFF DRIVE: A bluff edge sidewalk, scenic overlooks and parking bays on Cliff Drive as it descends to Capitola Village. Just before the first building on the bayside, a stairway winds down the steep bluff to Hooper Beach.

HOOPER BEACH: The section of Capitola Main Beach up coast of Capitola Wharf. Limited small boat storage on beach. Sand ramp at base of Wharf.

CAPITOLA FISHING WHARF: Free public fishing access. Facilities include boat and kayak rentals, fishing classes, rod and net rentals and a restaurant. Public restroom. Daily fishing licenses available at bait and tackle shop For information, call: (831) 462-2208. Wharf open 5AM-12PM. No vehicles allowed on wharf. Stairway at foot of wharf to Capitola City Beach.

CAPITOLA CITY BEACH: The City Beach is a popular swimming area with lifeguards from late May-September. Volleyball standards and nets. The Capitola Village Esplanade which backs the beach offers restaurants, galleries, and shops. A free shuttle operates on summer weekends and holidays from a parking lot at the Capitola/Park Avenue Highway 1 freeway exit to the beach and Village.

VIEWS AND TRAILS: Two stairways lead from El Camino in the Village to Cliff Drive on Depot Hill; from there a pedestrian walkway follows the bluff. A trail runs along the east side of Soquel Creek from Stockton Street Bridge upstream to Blue Gum Avenue.

NEW BRIGHTON STATE BEACH: Four miles south of Santa Cruz on Hwy 1 at Park Avenue exit. One hundred sixty-eight acres of upland and beach: 112 forest campsites, bicycle camping. 12 electrical-only hookups; sanitary station. Restrooms are wheelchair accessible. Shoreline access from beach parking lot to sandy beach via a stairway or a paved trail that is wheelchair accessible with assistance. Stairway access to beach from picnic area and campground. Views of Monterey Bay. Many species of wild birds. A small grunion run occurs spring to late summer. The park is open year round. For information, call: (831) 685-6442.

Capitola City Beach

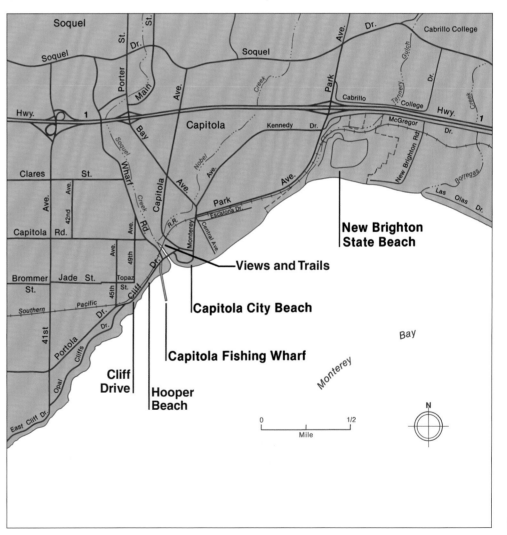

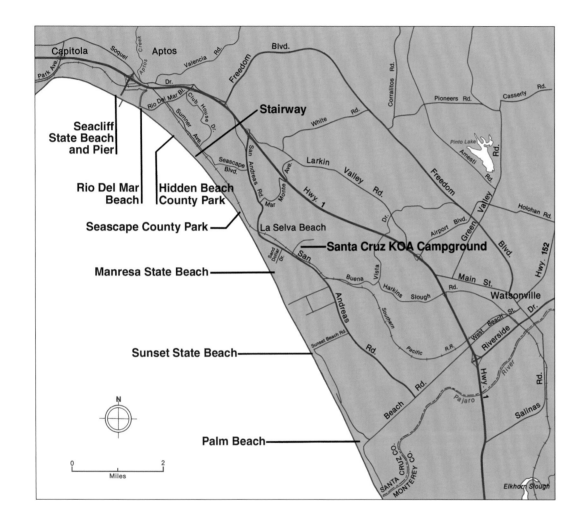

Capitola
Soquel
Aptos Creek
Aptos
Blvd.
Freedom
Blvd.
Valencia Rd.
Park Ave.
Rio Del Mar Bl.
Dr.
Club House Dr.
Sumner Ave.
Stairway
White Rd.
Corralitos Rd.
Pioneers Rd.
Casserly Rd.

Seacliff State Beach and Pier

Seascape Blvd.
San Andreas Rd.
Larkin Valley Rd.
Pinto Lake
Amesti Rd.
Freedom

Rio Del Mar Beach **Hidden Beach County Park**

Monte Mar Ave.
Hwy. 1
Dr.
Green Valley Rd.
Holohan Rd.

Seascape County Park

La Selva Beach
Airport Blvd.
Blvd.
Hwy. 152

Santa Cruz KOA Campground

Sand Dollar Dr.
San
Buena Vista
Harkins Slough Rd.
Main St.
Watsonville

Manresa State Beach

Andreas Rd.
Southern Pacific R.R.
West Beach St.
Riverside Dr.

Sunset State Beach

Sunset Beach Rd.
Hwy. 1
River
Rd.

N

Beach Rd.
Pajaro
Salinas

Palm Beach

0   2
Miles

SANTA CRUZ CO.
MONTEREY CO.
Elkhorn Slough

Seacliff State Beach

Santa Cruz County

CAPITOLA TO THE PAJARO RIVER

NAME	LOCATION	Entrance/Parking Fee	Parking	Restrooms	Lifeguard	Campground	Showers	Firepits	Stairs to Beach	Path to Beach	Bike Path	Hiking Trail	Facilities for Disabled	Boating Facilities	Fishing	Equestrian Trail	Sandy Beach	Dunes	Rocky Shore	Upland from Beach	Stream Corridor	Bluff	Wetland
Seacliff State Beach and Pier	Foot of State Park Dr., Aptos	•	•	•	•	•	•	•	•		•	•			•		•					•	•
Rio Del Mar Beach (Unit of Seacliff State Beach)	End of Rio Del Mar Blvd.		•	•							•				•		•					•	•
Hidden Beach County Park	N. of Manresa State Beach, end of Cliff Dr. and Hidden Beach Way		•	•									•				•						
Stairway	Via Palo Alto off Clubhouse Dr., Seascape								•								•						
Seascape County Park	Sumner Ave. and Seascape Blvd., Seascape		•	•						•			•				•			•		•	
Santa Cruz KOA Campground	1186 San Andreas Rd. at Southern Pacific track, La Selva Beach	•	•	•		•	•	•												•			
Manresa State Beach	W. of San Andreas Rd. at Southern Pacific track, La Selva Beach	•	•	•	•	•	•		•	•					•		•			•		•	
Sunset State Beach	201 Sunset Beach Rd., south Santa Cruz County	•	•	•		•	•	•	•	•					•		•	•			•		
Palm Beach (Unit of Sunset State Beach)	Foot of Beach Rd., south Santa Cruz County		•	•				•		•	•						•	•					

SEACLIFF STATE BEACH AND PIER: Take Hwy. 1 to Aptos-Seacliff exit. The 85-acre beach park has 25 trailer hookup campsites. Thirty overflow self-contained campsites available on a first come, first serve basis. No tent camping. Large picnic area with shaded tables. Visitor center open year round Wednesday through Sunday. The *Palo Alto,* a WWI supply ship, sits grounded in the intertidal zone. Formerly used as an amusement pier with dance floor and later as a fishing pier, the ship is no longer open to the public. For information, call: (831) 685-6442.

RIO DEL MAR BEACH: Southern end of Seacliff State Beach; a pedestrian/bike path leads to Seacliff beach. Parking lot at end of Aptos Beach Drive. Southern restroom wheelchair accessible. No fee.

HIDDEN BEACH COUNTY PARK: Small paved parking lot at end of Cliff Drive. Playground; chemical toilets. Paved trail to beach.

STAIRWAY: Follow Clubhouse Drive past Sumner Ave., left on Via Palo Alto; long stairway to beach between 1074 and 1094 Via Palo Alto.

SEASCAPE COUNTY PARK: Blufftop park developed with parking lot, tot lot, restrooms, loop trail, and sweeping views of Monterey Bay. Entrance is at Sumner Ave. and Seascape Blvd., cross railroad tracks turn left and drive through the Seascape Resort to reach the county park. Resort also provides public parking and a paved path (not wheelchair accessible) to a wide sandy beach.

SANTA CRUZ KOA CAMPGROUND: Take the Larkin Valley exit off Hwy. 1, then take San Andreas Rd. west 3.5 miles to the campground. Private campground with laundry, snack bar, store, paddle-tennis, pool, hot tubs, and playground; provides 238 sites with hookups and 13 walk-in sites. Open all year; camping fee. Pets on leash allowed. Call: (831) 722-0551 or 722-2377.

MANRESA STATE BEACH: Main day-use parking lot, restrooms, and stairway to the beach are located just south of La Selva Beach off San Andreas Road. Another day-use parking lot, stairway, and the Manresa Uplands Campground reached via Sand Dollar Dr. off San Andreas Road. Campground has 64 walk-in sites, restrooms, and hot showers; dirt path to beach. Call: (831) 761-1795. High surf and rip currents create hazardous swimming conditions. For campsite reservations, call: 1-800-444-7275.

SUNSET STATE BEACH: South of Santa Cruz on Hwy. 1 to San Andreas Rd. exit; take San Andreas Rd. to Sunset Beach Road. 218 acres; 90 campsites, no trailer hookups. There is a picnic area on the bluff behind the beach; in March and April poppies and bush lupine bloom in the surrounding meadows. Popular fishing spot. Seven miles of beachfront (including Palm Beach); hazardous swimming. Camping and day-use fees; open all year. Call: (831) 763-7062.

PALM BEACH: South of Santa Cruz on Hwy. 1 to Watsonville 129 exit; take Beach Rd. west. The beach runs to the Pajaro River mouth. Facilities include a picnic area, fitness trail, and wheelchair-accessible restrooms. Open all year; day-use fee.

Rio Del Mar Beach

Big Sur Coast

Monterey County

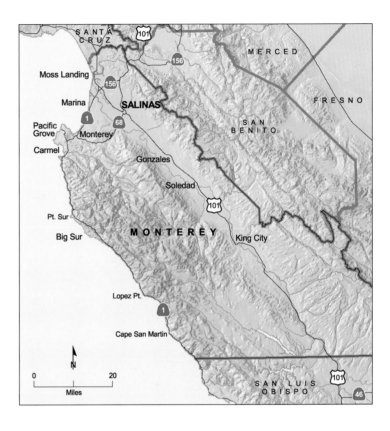

onterey County's coastline is one of the most beautiful in the state, stretching from the flat coastal plain around Monterey Bay in the north, through the steep hills of the Monterey Peninsula, to the magnificent, rugged Big Sur Coast.

Monterey Bay was sighted in 1542 by Juan Rodríguez Cabrillo, and visited again 60 years later by Sebastián Vizcaíno, who named it in honor of his Mexican viceroy, the Count of Monte Rey. In 1770 Gaspar de Portolá and Padre Junípero Serra established the Presidio and the second California mission. The settlement was immediately successful largely because of abundant timber, fertile soil, and mild weather.

Prior to Spanish settlement, the Monterey Bay area was inhabited by the Ohlone, or Costanoan Indians, who were hunter-gatherers, shellfish being their primary food source. During the Spanish occupation the Indians were concentrated in the missions, but following the secularization of the missions by the Mexicans, the Indians were dispersed and eventually disappeared. By 1920, there were only 56 surviving Costanoans.

Moss Landing, the northernmost coastal town in Monterey, was established in the 1860s by Captain Charles Moss, and soon became a bustling harbor and whaling station. Now it is a pleasure and fishing boat harbor, with antique shops, flea markets, and restaurants.

Monterey Peninsula, at the southwest end of Monterey Bay, is the focal point of the county, with its towns of Monterey, Pacific Grove, Pebble Beach, and Carmel. Historically, the city of Monterey was the capital of Alta California under both Spanish and Mexican rule. Today, the "Path of History" meanders through the city and passes many historic buildings and sites. Cannery Row, made famous by John Steinbeck's novel, *Cannery Row*, was the site of flourishing sardine canneries in the 1940s until the sardines suddenly vanished in 1951. Although the sardines later returned, the canning industry did not, and Cannery Row is now a tourist attraction with shops, restaurants, and galleries.

Pacific Grove, on the north side of the peninsula, is noted for its beautiful flowering ice plant, *Mesembryanthemum*, and the millions of Monarch butterflies that winter in the the trees. Seventeen-Mile Drive winds through the forested hills of the Del Monte Forest and along the rocky coast of Pebble Beach. Carmel-by-the-Sea, located at the southwest edge of the peninsula, is a Mediterranean-like village that has become a mecca for both artists and tourists with its shops on the hill, sailboats on the water, and clean, white sandy beach on Carmel Bay.

South of Carmel is Point Lobos State Reserve, a magnificent headland with trails leading through Monterey cypress groves and along the shore, tidepools rich in aquatic life, and abundant marine life such as sea lions and sea otters in the offshore kelp beds.

South of Point Lobos, Highway 1 narrows and winds along the Big Sur Coast between the steep Santa Lucia Mountains and the sparkling Pacific Ocean. There are many pull-outs with spectacular vistas, and several public picnic areas and beaches along Highway 1; Los Padres National Forest, which includes the Ventana Wilderness, begins at the coast and stretches inland for miles, providing numerous hiking trails and campsites. The original inhabitants of the Big Sur Coast were the Esselen Indians, who lived from Point Sur to Lucia; the Salinans, who lived south of Lucia; and the Costanoans, who lived along the coast from the Palo Colorado Canyon to the Big Sur River mouth.

For more information on Monterey County's coast, write or call: Monterey Peninsula Chamber of Commerce, 380 Alvarado (P.O. Box 1770), Monterey 93940, (831) 648-5360 or see www.mpcc.com; Pacific Grove Chamber of Commerce, Forest and Central Avenues (P.O. Box 167B), Pacific Grove 93950, (831) 373-3304 or see www.pacificgrove.org; or Carmel California Visitor and Information Center, San Carlos between 5th and 6th Avenues (P.O. Box 4444), Carmel 93921, (831) 624-2522 or see carmelcalifornia.org.

For transit information, contact Monterey-Salinas Transit: One Ryan Ranch Road, Monterey 93940, (831) 8999-2555 or 424-7695 or see www.mst.org.

Wetlands

Wetlands are areas where the land meets the water in a gradual transition, characterized by wet soils or by plants adapted to a wet environment; a variety of coastal areas are categorized as wetlands, including salt marshes, freshwater or brackish water marshes, shallow-water lagoons, tidal mudflats, salt flats, and fens.

Coastal wetlands are usually created by the flow of sediments into a bay, river mouth, or other shallow area, forming a delta. This delta gradually builds up to an elevation above low tide level; at that point, plants such as cordgrass and other salt marsh species move in. These plants slow the currents and trap more sediments, causing the wetlands to expand further. As more sediment becomes trapped on the delta, upland plant species take over, converting prior wetlands to uplands.

Thus, like most natural landforms, wetlands are continually subject to periodic creation and change.

Up until 500 years ago, wetlands covered over 300,000 acres of California's coastal areas, not including the San Francisco Bay area. Only 70,000 acres of coastal wetlands remain today. Diking and filling of wetlands for development projects account for much of this loss. In addition, soil erosion from hillsides has tripled the rate of sedimentation of wetlands during the last hundred years. If wetland sedimentation continues at this level, many remaining wetlands will be lost as well.

Ecologists have found that coastal wetlands are essential habitats for certain fish, birds, and mammals; in addition, many migratory ducks, geese, and shorebirds depend on wetlands for feeding and nesting. San Diego Bay wetlands, for example, seasonally support more than 180 species of birds, 50 species of mammals, 43 species of fish, and thousands of smaller organisms such as crabs, mussels, and ghost shrimp. Elkhorn Slough in Monterey County contains similar numbers of species, including more than 90,000 gaper clams, one of the species sought by clammers.

Many of the species of birds and mammals listed by the Department of Fish and Game as endangered in California are either directly dependent on or somehow associated with wetlands; these include the salt marsh harvest mouse, the California clapper rail, and the California least tern.

Wetlands also provide a number of direct and indirect benefits to humans. They help stabilize shorelines and absorb flood waters, lessening the need for costly flood control measures; futhermore they purify coastal waters through natural sewage treatment, trapping sediments that would otherwise fill navigation channels. Wetlands are also significant recreation resources and provide opportunities for fishing and bird watching, as well as for nature study and scientific research.

Since wetlands are so valuable from both an economic and biologic standpoint, the California Coastal Act, along with many other federal and state statutes and regulations, mandates governmental regulation in these areas to protect and restore California's wetlands.

Elkhorn Slough, Near Kirby Park Fishing Access

Monterey County

MOSS LANDING / ELKHORN SLOUGH

NAME	LOCATION	Entrance/Parking Fee	Parking	Restrooms	Lifeguard	Campground	Showers	Firepits	Stairs to Beach	Path to Beach	Bike Path	Hiking Trail	Facilities for Disabled	Boating Facilities	Fishing	Equestrian Trail	Sandy Beach	Dunes	Rocky Shore	Upland from Beach	Stream Corridor	Bluff	Wetland
Zmudowski State Beach	Foot of Giberson Rd., north of Moss Landing		•	•						•					•		•	•			•		•
Moss Landing State Beach	Hwy. 1 at Jetty Rd., Moss Landing		•	•											•		•	•			•		•
Elkhorn Slough	E. of Hwy. 1 at Moss Landing	•	•	•								•	•								•		•
Kirby Park	Kirby Rd., W. of Elkhorn Rd., Moss Landing		•	•									•	•	•						•		•
Moss Landing Wildlife Area	Northwest side of Elkhorn Slough		•											•							•		•
Moss Landing Harbor	Sandholdt Rd., W. of Hwy. 1, Moss Landing	•	•	•			•							•	•						•		•
Salinas River State Beach	W. of Hwy.1 at Potrero Rd., Moss Landing		•	•						•			•		•	•		•	•			•		

ZMUDOWSKI STATE BEACH: 177 acres. Lands adjacent to Giberson Road are private; do not trespass. Boardwalk leads through dunes to sandy beach; popular for fishing and clamming. Hazardous riptides. No fires, no dogs. Natural history tours subject to ranger availability. For information, call: (831) 649-2836.

MOSS LANDING STATE BEACH: Also known as Jetty Beach. 55-acre dune and seashore park; a Pacific Flyway stop over. Birdwatching is popular. Offshore fishing, surfing, and wind-surfing are popular activities. Strong tides are hazardous. Chemical toilets available. No dogs or fires. No fee. Parking lot closes ½ hour before sunset. For information, call: (831) 649-2836.

ELKHORN SLOUGH: Tidal slough used extensively for research and education. The endangered California brown pelican is found here; the peregrine falcon and golden eagle are also seen. Harbor seals and sea otters are also present.

Elkhorn Slough National Estuarine Research Reserve encompasses 1,400 acres of wetland and upland habitat. Five miles of trails; a wheelchair-accessible path leads to an overlook. The visitor center has interpretive displays, docent-led walks on weekends, parking, and wheelchair-accessible restrooms. Fee charged; free entrance with California Wildlife Campaign pass, or valid fishing or hunting license. Reserve open 9 AM-5 PM, Wed.-Sun. No fishing, hunting or boating on the Reserve. For information, call: (831) 728-2822 or see www.elkhornslough.org.

KIRBY PARK: A public fishing access at the northeast end of the slough, off Kirby Rd., west of Elkhorn Road. A mile-long paved wheelchair-accessible trail runs north along the slough. Boat ramp with boating floats and kayaks. Portable restrooms and barrier free access to waterfront for fishing. Access is available to Elkhorn Slough Preserve at north end of Kirby Park. Hours: 5 AM-10 PM daily. For information, call: (831) 633-2461.

MOSS LANDING WILDLIFE AREA: Located on the northwest side of Elkhorn Slough, opposite the National Reserve, the wildlife area features roosting grounds for California brown pelicans. There is a trail system and small parking lot. Seasonal closures to protect the snowy plover. For information, call: (831) 649-2870.

MOSS LANDING HARBOR: The T-shaped harbor is considered to be an extremely safe refuge; used for commercial fishing, research vessels, and pleasure boating. Facilities include slips and dry storage, fuel dock, pump-out station, supplies, bait and tackle, restaurants, showers, and wheelchair-accessible restrooms. The North Harbor area has a 2-lane boat ramp; fee charged. The harbor office is at the South Harbor; parking and fishing at the south jetty. For information, call: (831) 633-2461.

SALINAS RIVER STATE BEACH: 246 acres. Boardwalk and steep dune trails lead to the beach. Main parking lot off Molera Road at end of Monterey Dunes Way. Smaller lot at end of Potrero Road off Hwy. 1. Fishing, clamming and hiking. No dogs and no fires allowed. Horses permitted on the beach and on interdune trail only. Information: (831) 649-2836. The 367-acre Salinas River National Wildlife Refuge is located just south of the State Beach; there is a trail and parking. For information, call: (510) 792-0222.

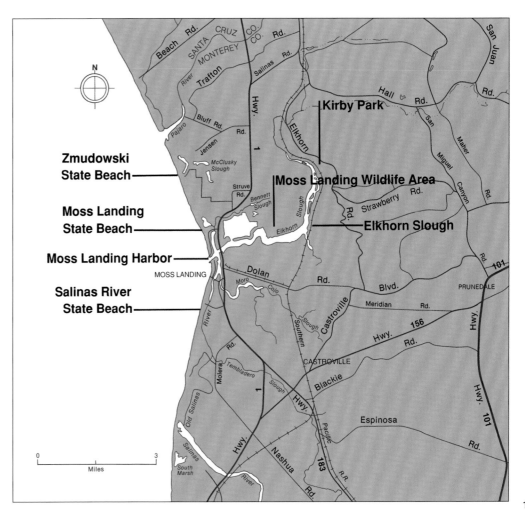

Marina State Beach

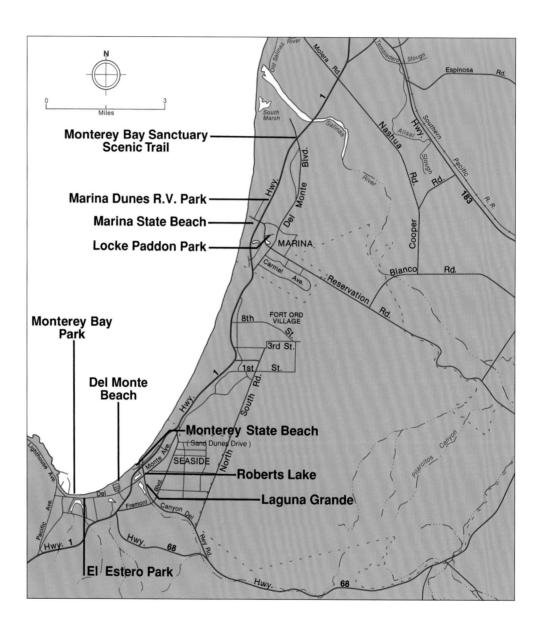

Monterey County

MARINA / SEASIDE / MONTEREY

NAME	LOCATION	Entrance/Parking Fee	Parking	Restrooms	Lifeguard	Campground	Showers	Firepits	Stairs to Beach	Path to Beach	Bike Path	Hiking Trail	Facilities for Disabled	Boating Facilities	Fishing	Equestrian Trail	Sandy Beach	Dunes	Rocky Shore	Upland from Beach	Stream Corridor	Bluff	Wetland
Monterey Bay Sanctuary Scenic Trail	From Castroville to Lover's Point in Pacific Grove		•	•																•			
Marina Dunes R.V. Park	3330 Dunes Dr., Marina	•	•	•			•								•			•		•			
Marina State Beach	Foot of Reservation Rd., Marina		•	•			•	•					•		•		•	•				•	
Locke Paddon Wetland Community Park	Reservation Rd. and Seaside Ave., Marina		•	•							•	•	•							•		•	•
Roberts Lake	Roberts Ave. and Canyon Del Rey, Seaside		•	•													•	•		•			
Laguna Grande	Del Monte Ave. and Canyon Del Rey, Seaside		•							•		•								•			•
Monterey State Beach (off Sand Dunes Drive)	Foot of Sand Dunes Dr., Monterey		•	•									•							•		•	•
Del Monte Beach	End of Beach way off Del Monte Blvd., Monterey		•	•													•					•	
Monterey Bay Park	Upcoast of Municipal Wharf No. 2, Monterey	•	•	•													•			•			
El Estero Park	Del Monte Ave. and Camino El Estero, Monterey		•	•									•	•	•					•			•

MONTEREY BAY SANCTUARY SCENIC TRAIL: The Monterey County section of the Scenic Trail is known as the Monterey Bay Coastal Trail. This 29-mile multi-use trail stretches from Castroville to Lover's Point in Pacific Grove. The path accommodates walkers, joggers, skaters, and cyclists. With limited exceptions the trail is separated from the road. For maps and details see www.mtycounty.com/pgs-parks/bike-path.

MARINA DUNES R.V. PARK: Take the Reservation Road exit off Hwy. 1, go west, then right on Dunes Drive. Private R.V. campground with 65 sites and all hookups. Fee. Pets on leash allowed. Reservations recommended. Beach access is via Marina State Beach to the west. Information: (831) 384-6914.

MARINA STATE BEACH: 170 acres dunes and beach. Main entrance at foot of Reservation Road. Popular fishing beach. Hang-gliding deck with launch ramp at main part of beach. No horses, dogs or fires. Swimming is unsafe due to dangerous surf. The parking lot and restrooms are at the main entrance. Near the parking lot a 2,000 foot long boardwalk leads to the beach. It is wheelchair accessible to an observation platform at the halfway point (sometimes closed due to sand). Pedestrian access is also available to the southern part of the beach from Lake Court in Marina through a gate up a steep trail across the dunes. Information: (831) 384-7695.

The planned Fort Ord Dunes State Park will extend 4 miles along the Monterey Bay shoreline between the cities of Marina and Sand City. Public beach access facilities and coastal dune restoration are expected to be major features of the new park.

LOCKE PADDON WETLAND COMMUNITY PARK: At Reservation and Seaside Roads. A pedestrian dirt path and an asphalt bicycle path run along the perimeter of this wetland area that hosts migratory fowl. Popular for nature walks and bird watching; facilities include picnic tables and wheelchair-accessible restrooms. An interpretive panel provides information on the wetlands. Information: (831) 384-4636.

ROBERTS LAKE: A freshwater lake with beach access, located inland of Highway 1 at the foot of Humboldt Street in Seaside. Used for model boat racing and duck feeding. Parking along Roberts Avenue. Adjoins Monterey Bay Recreational Trail. Information: (831) 899-6825.

LAGUNA GRANDE: Connected to Roberts Lake. A freshwater lake that is a natural preserve for birds. A paved path crosses the lake via a bridge. Grassy picnic areas with barbecue pits and playground with volleyball nets for rent. Wheelchair-accessible restrooms and parking at both west and east sides. Access located off Del Monte Avenue at Virgin and Montecito. Information: (831) 646-3860.

MONTEREY STATE BEACH: (off Sand Dunes Drive) 22 acres of dunes and sandy beach. Roadside parking and wheelchair accessible restrooms. Entrance gate is locked at night. Information: (831) 649-2836.

DEL MONTE BEACH: An 11-acre sandy beach with a small parking area, boardwalk, picnic tables and benches. At the end of Beach Way off Del Monte Boulevard.

MONTEREY BAY PARK: (includes a section of Monterey State Beach) A 19-acre lineal beach and grassy area upcoast of Municipal Wharf No. 2. The surf is hazardous. Benches, restrooms, volleyball courts. Sailboat and kayak launching from beach adjacent to wharf. Dogs are permitted if leashed. Fee parking. Information: (831) 646-3860. Beachgoers can walk the sandy beach from Monterey State Beach to the Municipal Wharf No. 2. The fragile dune ecosystems along this route are under restoration. Do not disturb. Use only designated walkways.

EL ESTERO PARK: El Estero Lake attracts ducks and other migrating birds. No feeding allowed. Facilities include a fishing pier (fishing license required), picnic tables, par course, baseball fields, benches, snack bar and paddle boat rentals. Dennis the Menace Play Area contains a Southern Pacific steam locomotive and other unusual play equipment. Restrooms and play area inside the playground are wheelchair accessible. Open 10 AM-sundown. Information (831) 646-3866.

Monterey State Beach

Monterey Marina, Fisherman's Wharf

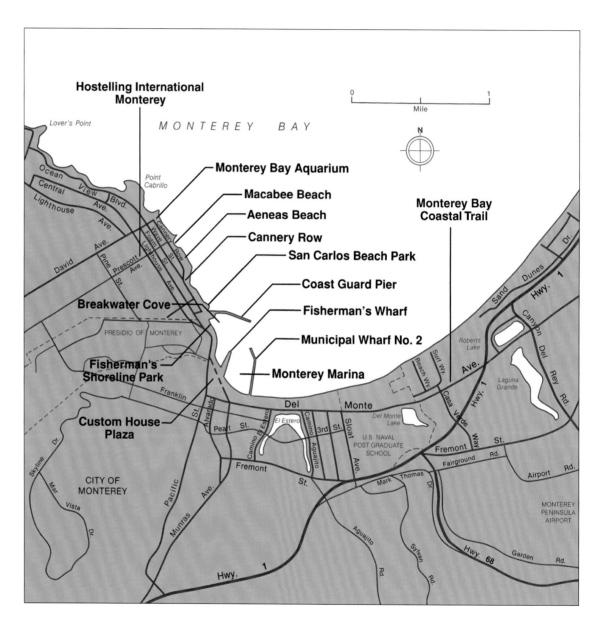

Hostelling International Monterey

MONTEREY BAY

Lover's Point

0   1
Mile

N

Point Cabrillo

Monterey Bay Aquarium

Macabee Beach

Aeneas Beach

Cannery Row

San Carlos Beach Park

Monterey Bay Coastal Trail

Coast Guard Pier

Breakwater Cove

Fisherman's Wharf

Municipal Wharf No. 2

PRESIDIO OF MONTEREY

Fisherman's Shoreline Park

Monterey Marina

Roberts Lake

Franklin

Del   Monte

Del Monte Lake

Custom House Plaza

El Estero

Laguna Grande

Pearl   St.

U.S. NAVAL POST GRADUATE SCHOOL

Fremont   St.

CITY OF MONTEREY

Fremont   St.

Fairground   Rd.

Airport   Rd.

Mark   Thomas   Dr.

MONTEREY PENINSULA AIRPORT

Munras   Ave.

Aguajito   Rd.

Sylvan   Rd.

Hwy.   1

Hwy   68

Garden   Rd.

Monterey County

CITY OF MONTEREY

NAME	LOCATION	Entrance/Parking Fee	Parking	Restrooms	Lifeguard	Campground	Showers	Firepits	Stairs to Beach	Path to Beach	Bike Path	Hiking Trail	Facilities for Disabled	Boating Facilities	Fishing	Equestrian Trail	Sandy Beach	Dunes	Rocky Shore	Upland from Beach	Stream Corridor	Bluff	Wetland
Monterey Bay Coastal Trail	Parallel to waterfront, Monterey										•	•	•						•				
Municipal Wharf No. 2	Wharf No. 2, foot of Figueroa St., Monterey	•	•	•			•						•	•	•								
Monterey Marina	Between Municipal Wharf No. 2 and Fisherman's Wharf, Monterey	•	•	•									•	•									
Fisherman's Wharf	Foot of Olivier St., Monterey	•	•	•										•									
Custom House Plaza	Headquarters at 20 Custom House Plaza, Monterey		•	•								•	•	•								•	
Fisherman's Shoreline Park	Lighthouse Curve at Foam St., Monterey		•									•	•	•								•	
Breakwater Cove Marina	32 Cannery Row, Monterey	•	•											•									
Coast Guard Pier	S.E. end of Wave St., Monterey		•	•					•	•			•	•	•				•				
San Carlos Beach Park	Foot of Reeside Ave., Monterey	•	•	•					•	•	•		•				•		•				
Cannery Row	Cannery Row, Monterey		•						•	•	•	•	•				•		•			•	
Aeneas Beach	400 Cannery Row, Monterey		•							•	•						•		•				
Macabee Beach	Cannery Row, between McClellan and Prescott aves., Monterey		•					•		•	•	•	•	•			•						
Monterey Bay Aquarium	Cannery Row at David Ave., Monterey	•	•	•									•		•		•						
Hostelling International, Monterey	778 Hawthorne St., Monterey	•	•	•			•															•	

MONTEREY BAY COASTAL TRAIL: Runs along the old Southern Pacific Railroad right-of-way the length of the City shoreline for three miles from Seaside to Lover's Point in Pacific Grove. Benches and viewpoints along the wheelchair-accessible trail. Information: (831) 646-3866.

MUNICIPAL WHARF NO. 2: The wharf was built in 1926 and is used by commercial fishermen. Facilities include a restaurant, snack bar, bait and tackle. Fishing from wharf allowed. Wheelchair-accessible restrooms and metered parking. A 3-ton capacity coin operated public boat hoist is available year round. Hoist use requires training. Information: (831) 646-3950.

MONTEREY MARINA: The office is located between Fishermen's Wharf and Wharf No. 2. Facilities include a two-lane concrete public ramp open 24 hours; 412 slips. Visiting boats up to 70 feet accommodated on a first-come, first-served basis. Open anchorage exists. Fishing licenses and bait and tackle available. No fishing in the marina. Fee parking. Restrooms are wheelchair accessible. Information: (831) 646-3950.

FISHERMAN'S WHARF: Features restaurants, shops, galleries and fish markets and the Wharf Theatre. Fishing trips, charters and sightseeing cruises are available. Whale watching cruises are popular and bird watching cruises are also offered. Fee parking. Information: (831) 649-6544.

CUSTOM HOUSE PLAZA: Located at the foot of Fisherman's Wharf, Monterey State Historic Park comprises 10 historic buildings, including the Custom House which was built in 1827, and which displays Monterey's Spanish, Mexican and Native American heritage. The visitor center is located at Stanton Center at 5 Custom House Plaza. Open daily. Picnic areas, gardens and visitor center are wheelchair accessible. Guided tours. Information: (831) 649-7118.

Maritime Museum of Monterey at 5 Custom House Plaza holds the Allen Knight Collection of ship memorabilia and models. It has the first order Fresnel lens used for 90 years at Point Sur Lighthouse; the lens is activated daily for display. Museum gift shop includes ship models. Open daily 10 AM-5 PM. Wheelchair-accessible restrooms. Information: (831) 372-2608.

FISHERMAN'S SHORELINE PARK: Benches and viewpoints along a narrow, 5-acre coast line park along the Monterey Bay Coastal Trail between Fisherman's Wharf and the Coast Guard Breakwater. For park information call (831) 646-3866. On a hill adjacent to the park is the Presidio of Monterey Museum which explores the history of Monterey's military past. Information: (831) 646-3456.

BREAKWATER COVE MARINA: Located at the upcoast foot of the Coast Guard Pier, the marina has 60 slips, a travel lift, boat hoist, boat storage and boat repair facility. Open Mon.-Fri. 8 AM-5 PM. Fuel dock open daily 8 AM-5 PM. Snack bar and marine related sales. Information: (831) 373-7857.

COAST GUARD PIER: There is public access along the rock jetty, often called "The Breakwater", which also serves as a haulout for seals and sea lions. A public parking lot, launch ramp, and wheelchair-accessible restrooms are located the foot of the breakwater. Showers in restroom. Popular fishing and driving area; a stairway provides access to San Carlos Beach.

SAN CARLOS BEACH PARK: At the upcoast end of Cannery Row next to the Coast Guard Pier, San Carlos Park is an important diver entry point. There are landscaping, picnic tables and stairways and a ramp to the beach. Wheelchair-accessible restroom and outdoor shower. A fee parking lot is on the inland side of Cannery Row.

CANNERY ROW: The Row begins near the Coast Guard Pier and parallels the shoreline for half a mile to the Monterey Bay Aquarium. For the first half of the 20th Century a sardine canning industry flourished here. The sardines mysteriously vanished from Monterey Bay in 1951 and Cannery Row now hosts shops, restaurants, galleries, plazas and walkways along the bay. The street was immortalized by John Steinbeck. Offshore are the rich kelp beds that attract divers from around the world. Access to the bay and beach is provided at numerous points along Cannery Row.

AENEAS BEACH: Monterey Plaza Hotel has a large public plaza overlooking the bay and a stairway to Aeneas Beach and its tidepools, a diver entry point. There is also access to a rocky promontory viewpoint via a walkway under the Charthouse Restaurant.

MACABEE BEACH: Between McClelland and Prescott this small sandy beach serves as an entry point for divers. Stairway access from Steinbeck Plaza and at either side of Spindrift Inn. Spindrift Inn provides an outside shower on the bayside of the building. The Cannery Row Welcome Center is located on the Coastal Trail, one block inland from Steinbeck Plaza. Call: (831) 373-1902.

MONTEREY BAY AQUARIUM: Built on the site of the old Hovden Cannery, the Aquarium offers a unique introduction to the marine ecosystems of Monterey Bay. The internationally acclaimed aquarium features numerous habitat tanks, a three-story kelp forest exhibit, sea otters, touch tanks, an open air shorebird aviary and a large outdoor viewing plaza with constructed tidepools.

The Outer Bay Wing features animals adapted to the open sea: ocean sunfish, yellowfin tuna, crystal and comb jellies, and green sea turtles. The million gallon tank is 35 feet deep, 90 feet long and 52 feet from front to back. The Outer Bay galleries also include hands-on exhibits and a learning area for young children. Educational programs, tours, and research programs.

Restaurant and gift shop. Fully wheelchair accessible. The aquarium is a non-profit organization supported by entry fees and member contributions. Open daily except Christmas; regular hours 10 AM-6 PM. Information: (831) 648-4888 or see www.mbayaq.org.

HOSTELLING INTERNATIONAL, MONTEREY: Hostel located in renovated historic carpenter's union hall, open 8-10 AM, 5-10 PM, 11 PM curfew. Free parking; for information, call: (831) 649-0375.

Monarch Butterfly

The Monarch butterfly, *Danaus plexippus*, is one of the best known of the butterflies, noted for its large size, bright colors, and long migrations. The Monarch belongs to a largely tropical family and, because it cannot survive cold winters, migrates south each fall.

Monarch butterflies emerge from their cocoons in late spring and early summer. During the summer months the butterflies are found dispersed throughout the countryside, visiting flowers and feeding on their nectar. In fall, the butterflies assemble in great numbers in trees and bushes and, at some unknown signal, rise up and begin flying southward en masse. The Monarchs return to the same grove of trees in which their ancestors overwintered.

At their wintering grounds they pack themselves tightly together in the trees and remain there in a state of semi-dormancy until spring. The eucalyptus or evergreen pines may be so filled with butterflies that the trees appear orange. When spring arrives, the Monarchs fly northward individually, not as a band.

The life expectancy of a Monarch butterfly is about a year, and not all adults survive the entire journey back north. Some fly part way, lay their eggs and die; their progeny then continue north, even as far as Canada. The butterflies that return south the following fall are the descendants of those that migrated north in the spring.

There are two major populations of Monarchs in the United States. Members of the eastern population migrate as far as 3,000 miles to overwintering spots in central Mexico. The western population returns to wintering grounds along the California coast. Monarchs overwinter in many places along our coast, but the best known spot is Pacific Grove, where thousands of Monarchs may be seen during the winter months in "butterfly trees" throughout the town. Monarchs are protected by law in Pacific Grove, and anyone disturbing them may receive a fine or a jail sentence.

Monarch butterflies are somewhat protected from predation because they feed on milkweed, a poisonous plant that makes the Monarchs themselves inedible to birds. A bird that has tasted one Monarch and become sick learns not to eat other Monarchs. The Monarch's bright, highly visible "warning coloration" serves to remind potential predators of the butterfly's toxicity. The Viceroy butterfly, *Limenitis archippus*, possesses coloration and habits almost identical to those of the Monarch, and enjoys a certain degree of protection from predation because of this "protective mimicry." Potential predators mistake the edible Viceroy for the unpalatable Monarch, and therefore avoid the Viceroy.

The Monarch, like other butterflies, passes through four stages of development, in a process called metamorphosis. The female lays her eggs on the underside of a leaf, usually of a milkweed plant. The striped caterpillar that emerges three to twelve days later immediately starts feeding on the milkweed. After several molts, the full-grown larva finds a sheltered spot and sheds its skin once again, revealing a pale green pupa, or chrysalis. This is a resting stage during which the insect reorganizes physiologically to form a butterfly. After about two weeks the adult Monarch emerges from the chrysalis. The process from egg to adult takes about five weeks.

PACIFIC GROVE

NAME	LOCATION	FACILITIES															ENVIRONMENT						
		Entrance/Parking Fee	Parking	Restrooms	Lifeguard	Campground	Showers	Firepits	Stairs to Beach	Path to Beach	Bike Path	Hiking Trail	Facilities for Disabled	Boating Facilities	Fishing	Equestrian Trail	Sandy Beach	Dunes	Rocky Shore	Upland from Beach	Stream Corridor	Bluff	Wetland
Shoreline Park	Ocean View Blvd., between Pt. Cabrillo and Lover's Pt., Pacific Grove		•						•	•	•						•		•	•		•	
Berwick Park	Ocean View Blvd. and Monterey Ave., Pacific Grove		•						•	•									•	•		•	
Lover's Point	17th St. and Ocean View Blvd., Pacific Grove		•	•			•	•	•	•		•			•		•		•	•		•	
Perkins Park	Along Ocean View Blvd., N.W. of Lover's Point, Pacific Grove		•					•	•	•	•						•		•	•		•	
Point Piños Lighthouse Reservation	End of Lighthouse Ave., W. of Asilomar Ave., Pacific Grove		•									•						•		•		•	
Asilomar State Beach and Conference Grounds	Pico Ave. at Sunset Dr., Pacific Grove		•	•			•			•			•		•		•	•	•	•			

SHORELINE PARK: Follows the shoreline along Ocean View Blvd. from Point Cabrillo to Lover's Point. Paths and benches. Several steep paths allow access to pocket beaches; hazardous surf. Street parking. The Monterey Bay Coastal Trail, a paved pedestrian and bicycle path, runs the length of the park passing through Andy Jacobsen Park, Berwick Park, and Greenwood Park. Information: (831) 648-5730.

BERWICK PARK: Narrow grassy park south of the railroad tracks, with benches and a path; street parking only. Nearby Andy Jacobsen Park, at the foot of 7th St. and Ocean View Blvd., is a small landscaped park on a slope. Greenwood Park, at the foot of 13th St. and Ocean View Blvd., is a grassy park with a foot bridge that crosses a small creek running through the center of the park; view of the bay.

LOVER'S POINT: Grassy blufftop park with a parking lot, benches, paths, picnic tables, and grills, plus a small fishing pier and three small protected sandy beaches accessible by stairways. Restrooms, pier, and beach south of the pier are wheelchair accessible. Popular diving and surfing spot; no dogs allowed. Call: (831) 648-5730.

The Pacific Grove Marine Gardens Fish Refuge is offshore; it is illegal to take any marine or plant life without a permit. The Pacific Grove Museum of Natural History at Forest and Central avenues is free and is open 10 AM-5 PM Tues.-Sun.; exhibits include displays of birds, shells, butterflies, fish, marine mammals, and Indian artifacts. Wheelchair-accessible restrooms are available. Museum: (831) 648-5716.

PERKINS PARK: Blufftop park with benches and a spectacular view of the bay; paths wind through the ice plant. There is a small parking lot at the foot of Beach St. and another at the foot of Asilomar Ave., where there are picnic tables and grills overlooking the shore. Four stairways provide access to small pocket beaches. Undeveloped Esplanade Park is located at Esplanade and Ocean View boulevards.

POINT PIÑOS LIGHTHOUSE RESERVATION: The lighthouse was built in 1855 of 18-inch-thick granite quarried on site; now the oldest continuously operating lighthouse on the west coast. Open to the public Thurs.-Mon. 1 PM-4 PM; small Coast Guard historical museum. The Reservation is a preserve for plants and wildlife, such as the rare Tidestrom's lupine, deer, sea otters, and pelicans. Spectacular coastal views from Point Piños. For information, call: (831) 648 5716.

ASILOMAR STATE BEACH AND CONFERENCE GROUNDS: Thirty six acres of rugged rocky shore, tidepools, sandy beach and dunes. Diving area. Unsafe swimming; hazardous rip currents. Beach wheelchair is available.

Seventy-one acre conference center has conference facilities and 28 lodges (313 rooms) nestled in a restored dune habitat and pine-oak forest. Craftsman style buildings by renowned architect Julia Morgan were designated an historic landmark in 1987. Free guided nature walks and cultural history tours. Conference Center information: (831) 372-8016.

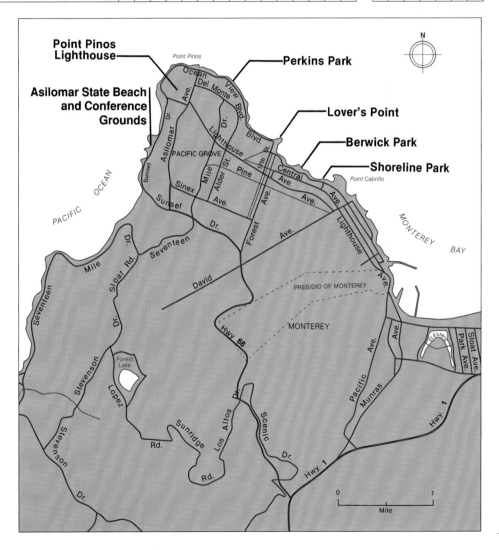

Monterey Pine and Monterey Cypress

Famed for its beauty, the Monterey Peninsula conjures up visions of bluffs meeting the sea. The area's magnificent native trees are an integral part of this scene. Among these, the Monterey cypress and the Monterey pine are perhaps the most well known.

The Monterey pine and cypress are members of the closed cone pine forest plant community–they are evergreen, cone-bearing plants growing on or near the coast from sea level to about 1,000 feet. Both occur in relatively cool climates with much fog, and reach a height of about 100 feet. Although each starts as a symmetrical tree, with age they both become flat-topped and frequently take on a characteristic windblown appearance.

Monterey pine, *Pinus radiata,* is quite easy to distinguish from a cypress by its needle-like leaves that occur in clusters.

There are only four native stands of Monterey pine in the world; the only mainland stands are on Año Nuevo Point in San Mateo County, on the Monterey Peninsula, and in Cambria in San Luis Obispo County. The fourth is located on islands off Baja California. This tree has been planted outside its native range and has, in fact, become the most important cultivated timber tree in the Southern Hemisphere even though its use in California, where it is vulnerable to natural pests and diseases, is severely limited.

The Monterey cypress, *Cupressus macrocarpa,* is restricted to two native stands. Both are on the Monterey Peninsula immediately adjacent to the sea. The larger is at Pebble Beach, the second in Point Lobos State Reserve. Monterey cypress has been extensively planted along the coast as an ornamental and windbreak.

Monterey pine   *Pinus radiata*

Monterey cypress   *Cupressus macrocarpa*

Monterey County

17-MILE DRIVE / CARMEL

NAME	LOCATION	Entrance/Parking Fee	Parking	Restrooms	Lifeguard	Campground	Showers	Firepits	Stairs to Beach	Path to Beach	Bike Path	Hiking Trail	Facilities for Disabled	Boating Facilities	Fishing	Equestrian Trail	Sandy Beach	Dunes	Rocky Shore	Upland from Beach	Stream Corridor	Bluff	Wetland
17-Mile Drive	Between Pacific Grove and Carmel, Del Monte Forest	•	•	•			•	•	•	•		•	•		•	•	•	•	•	•		•	
Spanish Bay Recreational Trail	Sunset Dr. and Asilomar Blvd., Pacific Grove		•								•	•								•			
Spanish Bay Shoreline Pedestrian Trail	Spanish Bay Rd., 17-Mile Drive		•							•		•	•	•			•	•	•		•			
Fanshell Beach	Signal Hill Rd., 17-Mile Drive		•												•		•		•				
Stillwater Cove/Lodge at Pebble Beach	End of Cypress Dr., off Hwy. 1, Del Monte Forest		•						•	•					•		•						
Carmel City Beach	Foot of Ocean Ave., Carmel	•	•					•	•	•			•		•		•	•				•	
Carmel River State Beach	Scenic Rd. at Carmelo St., Carmel	•	•						•	•					•		•		•		•	•	•

17-MILE DRIVE: The famous scenic drive winds along the coast and through pine forests and Monterey cypress groves in the Del Monte Forest, including numerous pull-outs with spectacular views, pocket beaches, picnic areas, equestrian trails, abundant wildlife, and several public golf courses. There are five entrances with toll gates. Shoreline pedestrian and bicycle trails are available; pedestrians and bicyclists may enter free of charge. Call: (831) 624-3811.

The following areas are along 17-Mile Drive: Spanish Bay, Point Joe, Seal and Bird Rocks, Fanshell Beach, Cypress Point Lookout, Crocker Grove, the Lone Cypress, Pescadero Point, and Stillwater Cove.

SPANISH BAY RECREATIONAL TRAIL: Pedestrian/bicycle path runs from Sunset Dr. and Asilomar Blvd. through the native Monterey pine forest behind the Spanish Bay Hotel. One-quarter mile beyond the hotel the 4-foot-wide path merges with 17-Mile Drive. Picnic tables are located along the path near the hotel.

SPANISH BAY SHORELINE PEDESTRIAN TRAIL: Access to the pedestrian boardwalk that crosses the dunes to North and South Moss Beach is from Asilomar State Beach. At Point Joe, a wheelchair-accessible path continues for about one mile to Spyglass Hill Rd. just beyond Seal Rock Creek. This mile-long section offers many scenic vistas and passes by pocket beaches and Seal and Bird Rocks. The trail connects with the Del Monte Forest Equestrian Trail network at Seal Rock Creek. Hikers may continue inland through dunes and forest on the equestrian trails, although there are no signs for wayfinding at the numerous trail junctions. No bicycles or horses are permitted on the pedestrian trail.

FANSHELL BEACH: White sand cove used for picnicking, fishing, and swimming. Sea otters frequent the cove's waters. Closed for seal pupping season in the spring.

STILLWATER COVE / LODGE AT PEBBLE BEACH: A small sandy beach and pier with hoist are located at the southwest end of the private Beach Club. Fishing and diving allowed during daylight hours. Call (831) 625-8507 for boat use fees and launch reservations. Nearby Lodge at Pebble Beach has public parking and trails to beach.

CARMEL CITY BEACH: Fine white sand beach bordered by cypress trees; hazardous surf. 120 parking spaces at the end of Ocean Avenue; crowded on weekends. Along Scenic Road a three-quarter mile long blufftop trail extends from 8th Ave. to Martin Way; there is street parking and eight stairways and a ramp to access the beach. Fires allowed south of 10th Avenue. Volleyball court on back beach. Dog friendly beach; mutt mitts available. Wheelchair-accessible restrooms off Ocean Avenue. Portable toilets downcoast. Information: (831) 624-3543

CARMEL RIVER STATE BEACH: The mile-long beach features a bird sanctuary lagoon (just before the Carmel River empties into the sea) with a wide variety of waterfowl and song birds. No power boats or dogs allowed. Northern beach access, parking and wheelchair-accessible restrooms at south end of Scenic Road. Access stairway at Isabella and Scenic. South of Carmel River off Highway 1 access is from Ribera Road. The southernmost section, off Highway 1, is Monastery Beach, also known as San Jose Creek Beach; popular with scuba divers. The undersea Carmel Bay Ecological Reserve is offshore. Ocean swimming and wading are extremely dangerous. Information: (831) 649-2836.

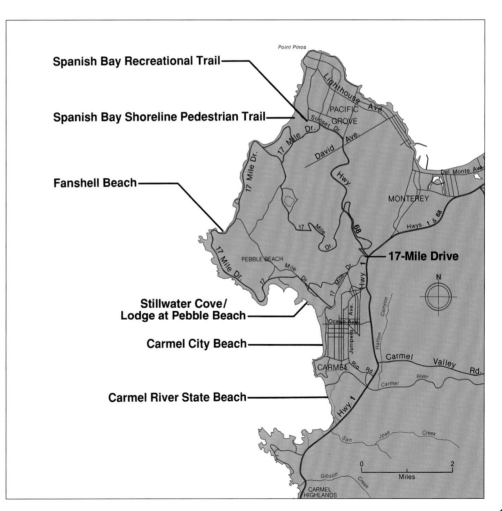

Spanish Bay Recreational Trail
Spanish Bay Shoreline Pedestrian Trail
Fanshell Beach
17-Mile Drive
Stillwater Cove/ Lodge at Pebble Beach
Carmel City Beach
Carmel River State Beach

Bixby Bridge

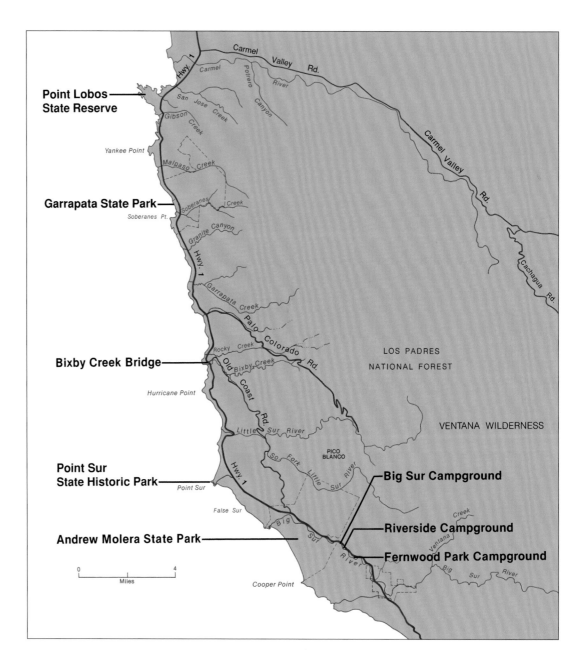

Point Lobos
State Reserve

Garrapata State Park

Bixby Creek Bridge

Point Sur
State Historic Park

Andrew Molera State Park

Big Sur Campground

Riverside Campground

Fernwood Park Campground

Carmel Valley Rd.

Hwy. 1

Carmel

Carmel Valley Rd.

San Jose Creek

Gibson Creek

Potrero Canyon

River

Yankee Point

Malpaso Creek

Soberanes Creek

Soberanes Pt.

Granite Canyon

Cachagua Rd.

Hwy 1

Garrapata Creek

Palo Colorado Rd.

LOS PADRES
NATIONAL FOREST

Rocky Creek

Old Bixby Creek

Hurricane Point

Coast Rd.

Little Sur River

VENTANA WILDERNESS

So. Fork Little Sur River

PICO BLANCO

Hwy 1

Point Sur

False Sur

Big Sur River

Creek

Ventana

River

Cooper Point

Big Sur River

0   4
Miles

154

Monterey County

NORTHERN BIG SUR

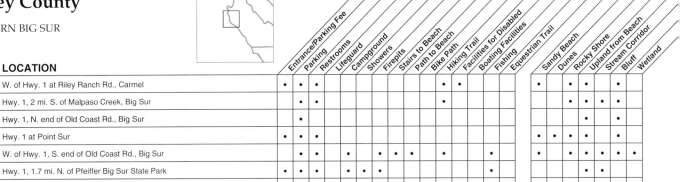

NAME	LOCATION	Entrance/Parking Fee	Parking	Restrooms	Lifeguard	Campground	Showers	Firepits	Stairs to Beach	Path to Beach	Bike Path	Hiking Trail	Facilities for Disabled	Boating Facilities	Fishing	Equestrian Trail	Sandy Beach	Dunes	Rocky Shore	Upland from Beach	Stream Corridor	Bluff	Wetland
Point Lobos State Reserve	W. of Hwy. 1 at Riley Ranch Rd., Carmel	•	•	•						•	•						•		•	•		•	
Garrapata State Park	Hwy. 1, 2 mi. S. of Malpaso Creek, Big Sur		•	•							•								•	•	•	•	
Bixby Creek Bridge	Hwy. 1, N. end of Old Coast Rd., Big Sur		•																•			•	
Point Sur State Historic Park	Hwy. 1 at Point Sur	•	•														•	•	•	•		•	
Andrew Molera State Park	W. of Hwy. 1, S. end of Old Coast Rd., Big Sur		•	•		•		•	•	•	•		•		•		•		•	•	•	•	•
Big Sur Campground	Hwy. 1, 1.7 mi. N. of Pfeiffer Big Sur State Park	•	•	•		•	•	•							•				•	•			
Riverside Campground	Hwy. 1, 1.6 mi. N. of Pfeiffer Big Sur State Park	•	•	•		•	•	•							•				•	•			
Fernwood Park Campground	Hwy. 1, .7 mi. N. of Pfeiffer Big Sur State Park	•	•	•		•	•	•							•				•	•			

POINT LOBOS STATE RESERVE: 1,276 acres of headlands with spectacular views of the coast, sandy coves, beaches and tidepools. Monterey cypress groves and more than 300 plant species and 250 bird and animal species. Numerous sea lions can be seen on the rocks offshore at Sea Lion Point; Bird Island is a sanctuary for thousands of birds. Sea otters can frequently be seen from the bluffs. Interpretive center called Whaler's Cabin contains artifacts from whaling days. Open 11 AM-5 PM. Hiking trails and picnic areas. Swimming in China Cove. Diving access in Whaler's and Bluefish Coves by permit only. No dogs. Admission fee. Open 9 AM-5 PM winter; 9 AM-7 PM summer. Information: (831) 624-4909 or see www.pt-lobos.parks.state.ca.us/

Numerous paved and unpaved highway pull-outs with scenic views are located on Hwy. 1 in Big Sur. These include: Abalone Cove Vista Point at Kasler Point; Notley's Landing just south of Palo Colorado Creek; a turn-out just south of Rocky Creek Bridge; Vista Point, south of Bixby Landing; Hurricane Point, one mile south of Bixby Landing, a paved turn-out just south of Partington Cove; Anderson Landing just south of Julia Pfeiffer Burns State Park; a vista point north of Lucia; and a vista point at Willow Creek Bridge.

GARRAPATA STATE PARK: The park has two miles of beach front including the outstanding coastal headlands at Soberanes Point. The 2,879-acre park offers diverse coastal vegetation with trails leading from ocean beaches into dense redwood groves. Sea lions, harbor seals and sea otters frequent the coastal waters and California gray whales pass close by during their yearly migration. Offshore, the California Sea Otter Game Refuge extends along the entire Big Sur Coast south into San Luis Obispo County. Do not disturb the otters. Hazardous cliffs. Leashed dogs allowed on beach only. Roadside parking; outhouses.

East of Highway 1 are the Soberanes Canyon Trail and the Rocky Ridge Trail which form a four mile loop through the redwoods and across the ridge with sweeping views; approximately 1850 feet of elevation gain. Additional lateral trails. Information: (831) 624-4909.

BIXBY CREEK BRIDGE: A major landmark of the Big Sur Coast, and one of the world's longest concrete arch span bridges, 260 feet high and over 700 feet long, with highway pull-outs on both sides and observation alcoves on the bridge itself; spectacular view.

POINT SUR STATE HISTORIC PARK: 2 ½ hour walking tour of Point Sur Lighthouse every Saturday and Sunday at 10:00 AM. First come, first served. Tour involves ½ mile walk each way. Fees charged. No food, drinks, or dogs. Parking off Highway 1 in front of the light station; no motorhomes or campers. For summer schedule or moonlight tour information, call: (831) 625-4419 or see www.lighthouse-pointsur-ca.org.

ANDREW MOLERA STATE PARK: 4,749 acres of flatland, meadows, mountains, and sandy beach west of Hwy. 1. Day-use areas and walk-in campground .3 mile from the dirt parking lot. Firepits and chemical toilets available; fee charged for camping. Three-night limit. The Big Sur River flows through the park and empties into the ocean, forming a shallow lagoon that is a bird sanctuary; hiking trails lead to the beach adjacent to the lagoon and to the scenic ridgetop to the south where various loops return the hiker to the flood plain. For information, call: (831) 667-2315.

BIG SUR CAMPGROUND: Privately owned, with sites in the redwoods and along the Big Sur River; cabins, tent rentals, water and electric hookups, picnic tables, laundry, store, and playground. Fee charged. Open all year, reservations accepted. Contact: Highway 1, Big Sur 93920; (831) 667-2322.

RIVERSIDE CAMPGROUND: Privately owned, with 46 campsites in the redwoods and along the Big Sur River; facilities include cabins, picnic tables, playground, laundry, and a swinging footbridge over the river. Tent sites and R.V. sites. Fee charged. Information: P.O. Box 3, Big Sur 93920; (831) 667-2414.

FERNWOOD PARK CAMPGROUND: Privately owned, with sites in the redwoods and along the Big Sur River; camping and day use fees. Firewood for sale. Restaurant, bar, motel, and grocery store. For information, call: (831) 667-2422.

Andrew Molera State Park

Pfeiffer-Big Sur State Park

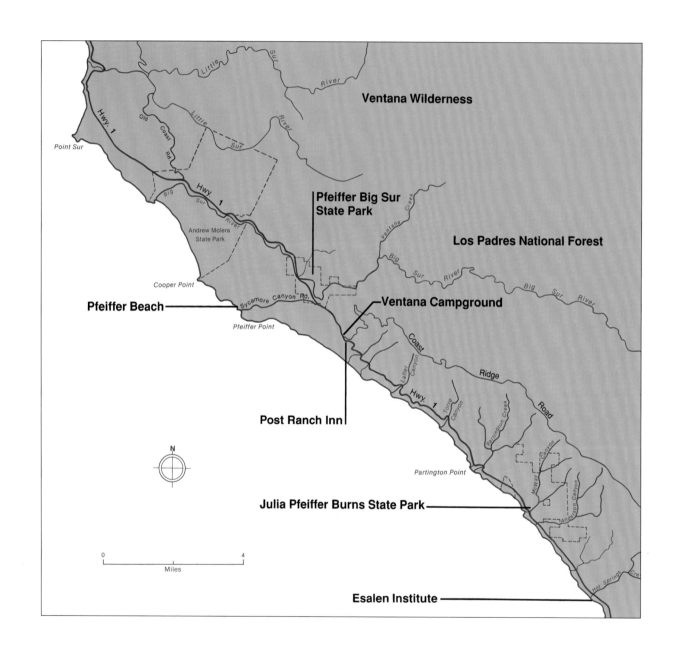

Point Sur

Little *Sur* *River*

Little *Sur* *River*

Hwy. 1

Old Coast Rd

Ventana Wilderness

Pfeiffer Big Sur State Park

Big *Sur* *River*

Hwy 1

Andrew Molera State Park

Ventana Creek

Los Padres National Forest

Cooper Point

Sycamore Canyon Rd.

Big *Sur* *River*

Big Sur River

Pfeiffer Beach

Pfeiffer Point

Ventana Campground

Coast

Ridge

Road

Lafler Canyon

Torre Canyon

Post Ranch Inn

Partington Creek

Hwy. 1

McWay Canyon

Anderson Canyon

Partington Point

N

Julia Pfeiffer Burns State Park

0   4
Miles

Hot Springs Creek

Esalen Institute

Monterey County

BIG SUR

NAME	LOCATION	Entrance/Parking Fee	Parking	Restrooms	Lifeguard	Campground	Showers	Firepits	Stairs to Beach	Path to Beach	Bike Path	Hiking Trail	Facilities for Disabled	Boating Facilities	Fishing	Equestrian Trail	Sandy Beach	Dunes	Rocky Shore	Upland from Beach	Stream Corridor	Bluff	Wetland
Pfeiffer Big Sur State Park	E. of Hwy. 1, 26 mi. S. of Carmel, Big Sur	•	•	•		•	•	•			•	•			•	•				•	•		
Los Padres National Forest/Ventana Wilderness	Big Sur Coast		•	•		•		•		•		•	•		•	•	•		•	•	•	•	•
Pfeiffer Beach	W. of Hwy. 1, end of Sycamore Canyon Rd., Big Sur	•	•	•				•		•	•		•			•	•		•		•	•	•
Ventana Campground	Hwy. 1, 2.4 mi. S. of Pfeiffer Big Sur State Park, Big Sur	•	•	•		•	•	•												•	•		
Post Ranch Inn	W. of Hwy. 1, 2.4 mi. S. of Pfeiffer Big Sur State Park, Big Sur		•									•								•			
Julia Pfeiffer Burns State Park	E. and W. of Hwy. 1, 11 mi. S. of Big Sur State Park, Big Sur	•	•	•		•		•		•		•	•			•	•		•	•	•		
Esalen Institute	W. of Hwy. 1, 14.6 mi. S. of Pfeiffer Big Sur State Park	•	•	•															•		•		

Pfeiffer Beach

PFEIFFER BIG SUR STATE PARK: 821 acres in the redwoods and along the Big Sur River. Facilities include 218 campsites (each with a stove and picnic table), a bicycle camp area, group camp, hiking trails, store, and laundry; no trailer hookups. Some campsites are wheelchair accessible. Swimming in the river; abundant wildlife. Popular spots along the trails include Pfeiffer Falls and the Gorge. Guided walks and campfire programs in the summer.

The park is open all year; for reservations, call: 1-800-444-7275. Seven-day camping limit June 1-Sept. 30. Fee charged for overnight and day use. Information: (831) 667-2315. The Big Sur Lodge has rooms and cabins, swimming pool and sauna, restaurant, and gift shop. Call: 1-800-424-4787 or (831) 667-2171.

LOS PADRES NATIONAL FOREST/VENTANA WILDERNESS: The National Forest, which consists of two sections, comprises almost two million acres and extends into five counties, with 1,750 miles of recreational trails. The smaller section is in Monterey County and includes part of the Big Sur Mountains and the Santa Lucia Mountains. For general information on the National Forest, call: (805) 968-6640.

The 230,000 acre Ventana Wilderness, within the forest, has a system of trails through woodlands and canyons. California campfire permit required during the fire season, approximately May-October. Permits and information from the Monterey District Rangers Office: U.S. Forest Service, 406 S. Mildred St., King City 93930; for information, call: (831) 385-5434 or see www.r5.fs.fed.us/lospadres/

There are three places along the coast where information, maps and permits can be obtained, and where trails lead into the National Forest. Big Sur Station is 1 mile south of the main entrance to Pfeiffer Big Sur State Park. Permits can also be obtained at Bottcher's Gap several miles inland of Highway 1 on Palo Colorado Road which is approximately 10 miles south of Carmel. The Pacific Valley Forest Station is 33 miles south of the Big Sur Valley.

PFEIFFER BEACH: A unit of Los Padres National Forest. A beautiful white sandy beach at the end of sycamore Canyon Rd., which is the second right hand turn off Hwy. 1 south of Big Sur State Park. The turn-off, which is one mile from the State Park entrance, is a sharp right turn downhill. The two mile road to the beach is very narrow and winding. Private property is adjacent; do not trespass.

The beach is at the end of a sandy trail leading from the parking lot through cypress trees, and is surrounded by steep cliffs, sea stacks, and sea caves. Spectacular waves crash through natural arches in the rocks. Sycamore Creek empties onto the beach into a small lagoon. Hazardous surf, gusty winds. Fires are prohibited. Beach open 6 AM-sundown. Wheelchair-accessible restrooms. Fee charged for day use. Information: (831) 385-5434.

VENTANA CAMPGROUND: Privately owned 80 site campground in the redwoods. Fee charged. Reservations recommended; arrive before 6 PM. Restaurant, bar and gift shop up the hill at the renowned Ventana Inn. Information: (831) 667-2712.

POST RANCH INN: Private resort on ridge west of Hwy. 1; entrance opposite Ventana Campground. Loop trail through upland forest open to public on a permit basis; obtain pass at entry kiosk.

JULIA PFEIFFER BURNS STATE PARK: 2,405 acres with wooded hiking trails and picnic tables; the restrooms are wheelchair accessible. A paved footpath leads to a spectacular overlook of McWay Waterfall cascading 80 feet into the ocean near Saddle Rock; another trail leads to a picnic area near McWay Creek. Two environmental campsites available by reservation; call: 1-800-444-7275. Day use and camping fees. Park closes at sunset. Information: (831) 667-2315.

Partington Cove is accessible by a trail on the west side of Hwy. 1, 1.8 miles north of the park entrance; look for an iron gate. The trail leads across Partington Creek via a wooden footbridge and through a 200-foot long tunnel cut into the cliff.

ESALEN INSTITUTE: This retreat was an early center for the "human potential" movement. Classes and workshops are offered on education, religion, philosophy, and the physical and behavioral sciences. Information: (831) 667-3000. Esalen's hot springs are open to the public 1:00 AM –3:00 AM Monday-Saturday mornings; fee charged; reservation required, call: (831) 667-3047.

Southern Sea Otter

The southern sea otter, *Enhydra lutris nereis,* one of the most interesting of California's marine mammals, can be found off the coast between Santa Cruz and Avila Beach in San Luis Obispo County. The sea otter is a member of the weasel family; adult males measure up to 4-1/2 feet long and weigh up to 85 lbs., while females are somewhat shorter and lighter. Otters have dense fur ranging in color from black to dark red, and short front paws used primarily for feeding and grooming; hind feet are webbed and used as flippers for swimming.

Like primates, otters use tools in their daily routines. For example, an otter will dive beneath the water's surface and, with its paws, use a rock to remove shellfish from the sea bottom; once it has surfaced, the otter floats on its back, positions its catch on its chest, and hammers the shell until edible portions are accessible.

The sea otters' diet varies according to environment and the length of time they have stayed in an area. Where kelp beds are present, otters will float in the beds and search for crabs and snails; in coastal areas with sandy beaches, otters will come ashore and forage for crabs and clams. When otters first move into an area, their preferred diet consists primarily of abalone, sea urchins, and crabs if the area is rocky, and clams and crabs in sandy locations. As the population increases and becomes established, abalone and sea urchins become less abundant, and crabs become the otters' chief food source; otters may also feed on other marine species such as squid, mussels, limpets, and sea stars.

To compensate for the lack of blubber that insulates most marine mammals, the average otter consumes food that equals up to 25% of its body weight daily. An otter consumes 2.5 tons of food during a single year; this large quantity of food fuels its metabolism and maintains proper body temperature. Although its fur is not a very efficient insulator in the water, the otter meticulously grooms it to provide maximum warming; part of the grooming process includes rolling vigorously in the water to trap air bubbles in the fur.

The hunters' desire for the otter's thick, attractive fur nearly resulted in the animal's extinction. Although native North Americans hunted the otter from Alaska to southern California for many years, large-scale hunting did not occur until the mid-18th century. Between 1741 and 1911, Russian, American, French, and British fur traders hunted marine mammals off the western coast of the United States and the otter population declined to a near extinct level. In 1911 the Fur Seal Treaty was signed and included provisions to protect the sea otters; subsequently, a number of state and national laws providing for marine mammal protection were enacted. Today, most of the southern sea otters live within the California Sea Otter Game Refuge, a protected habitat area between the Carmel River in Monterey County and Santa Rosa Creek in San Luis Obispo County.

Spotting sea otters in offshore waters can be difficult because otters often remain in kelp beds, and the kelp floats resemble otter heads. One of the best times to observe otters is during feeding times, usually in the early morning and late afternoon. Sea gulls hovering above kelp beds are a good indicator of an otter's presence, as gulls often wait above the beds to feed on scraps left by otters. With binoculars one can observe the otters feeding and grooming themselves and their young; at times it is even possible to hear otters pounding on shells with rocks and pups crying out for their mothers.

Sea otters are protected by state and federal law; it is illegal to take or even to temporarily possess a sea otter. Any person finding a dead, sick, or wounded otter or an apparently abandoned otter pup should not touch the animal but should immediately notify the California Department of Fish and Game at (805) 772-1135 or the Marine Mammal Center at (415) 289-7325.

Monterey County

SOUTHERN BIG SUR

NAME	LOCATION	Entrance/Parking Fee	Parking	Restrooms	Lifeguard	Campground	Showers	Firepits	Stairs to Beach	Path to Beach	Bike Path	Hiking Trail	Facilities for Disabled	Boating Facilities	Fishing	Equestrian Trail	Sandy Beach	Dunes	Rocky Shore	Upland from Beach	Stream Corridor	Bluff	Wetland
Landels-Hill Big Creek Reserve	Hwy. 1, 7 mi. S. of Julia Pfeiffer Burns State Park		•																	•	•		
Limekiln State Park	Off Hwy. 1, 2 mi. S. of Lucia	•	•	•		•	•	•		•		•			•		•			•	•	•	
Kirk Creek Campground	W. of Hwy. 1, 4 mi. S. of Lucia	•	•	•		•		•		•	•	•		•		•		•	•	•	•	
Mill Creek Picnic Ground	W. of Hwy. 1, 5 mi. S. of Lucia		•	•			•			•		•		•		•			•	•	•	
Sand Dollar Picnic Area and Beach	W. of Hwy. 1, 11 mi. S. of Lucia		•	•			•			•	•	•				•		•			•	
Plaskett Creek Campground	E. of Hwy. 1, 11.7 mi. S. of Lucia	•	•	•		•		•										•	•			
Jade Cove	W. of Hwy. 1, 12.4 mi. S. of Lucia								•			•			•		•		•		•	
Willow Creek Picnic Ground	W. of Hwy. 1, 14 mi. S. of Lucia		•	•			•					•				•			•		•	•

LANDELS-HILL BIG CREEK RESERVE: The mountainous 3,911-acre reserve is part of the University of California Natural Reserve System and is used for teaching and research. The reserve supports a variety of wild and endangered plants and animals. The Reserve holds an annual openhouse. Information: (831) 667-2543.

LIMEKILN STATE PARK: 33 campsites in the redwoods and on the beach; wheelchair accessible restrooms; showers. Firewood for sale. Fishing; trails to waterfalls and historic limekilns. Information: (831) 667-2403. For reservations, call: 1-800-444-7275.

The following areas are units of the Los Padres National Forest: Pfeiffer Beach, Kirk Creek Campground, Mill Creek Picnic Ground, Sand Dollar Picnic Area and Beach, Plaskett Creek Campground, Jade Cove, and Willow Creek Picnic Ground.

KIRK CREEK CAMPGROUND: 33 campsites on the bluffs above a sandy beach; magnificent view of the coast. At the south end of the campground a steep trail leads to the beach; diving area. Fee for use of campsites. Wheelchair-accessible restrooms. Information: (831) 385-5434.

MILL CREEK PICNIC GROUND: Two picnic tables overlook the ocean; a steep path leads to the rocky shore, which is a diving area. Pets permitted on leash. The chemical toilets are wheelchair accessible. Hang gliders who launch from Plaskett Ridge use the blufftop as a landing site. All hang gliders must register with the Pacific Valley Ranger Station. Information: (805) 927-4111.

SAND DOLLAR PICNIC AREA AND BEACH: Picnic tables among the cypress trees; trails lead across the field and down to a crescent-shaped sandy beach. Pets must be leashed. Restrooms are wheelchair accessible. Popular hang-glide landing area. Fee charged. For information, call: (831) 385-5434.

PLASKETT CREEK CAMPGROUND: 44 sites with picnic tables and grills at the base of the Santa Lucia Mountains. Fee charged for use of campsites. Group sites for up to 50 persons are available by reservation; for information, call: 1-800-283-2267.

JADE COVE: Look for the Los Padres National Forest sign on Hwy. 1. Shoulder parking only; a steep trail leads down the bluff to several rocky coves. Diving area. Named after the nephrite jade found here. For information, call: (831) 385-5434.

WILLOW CREEK PICNIC GROUND: Day-use area along the rocky shore near where Willow Creek flows into the ocean; view of Plaskett Rock offshore. Information: (831) 385-5434

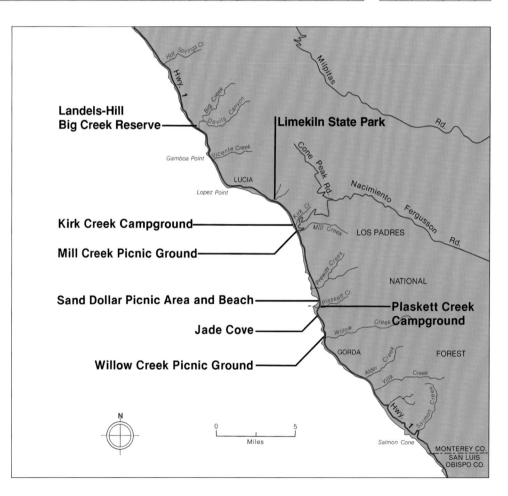

Morro Rock

San Luis Obispo County

anging from the rugged sea cliffs south of Big Sur to the extensive sand dune fields of Pismo Beach, San Luis Obispo's coast is a favorite among visitors. Ninety-six miles long, the county's coast includes numerous wide, sandy beaches, sheltered bays, and many vista points offering scenic views of the Pacific Ocean.

San Luis Obispo County's topography and natural features are a result of the intense geologic activity that has taken place here. Complex folding and uplifting of the earth's crust five million years ago produced the Santa Lucia Mountain Range bordering the shoreline along the north coast, and the San Luis Mountains between Pismo Beach and Morro Bay. This folding formed the tilted strata at the base of the San Luis Mountains, which has since been eroded by the ocean into spectacular tidepool areas accessible from Montaña de Oro State Park and the Shell Beach area.

Morro Rock and the eight other volcanic peaks between Morro Bay and the city of San Luis Obispo were formed by volcanic activity over 20 million years ago. The Rock marks the narrow and once treacherous entrance to Morro Bay, sighted originally by the Spanish explorer Juan Rodríguez Cabrillo in 1542. Cabrillo never landed here, and it wasn't until Juan Gaspar de Portolá's expedition in 1769 that Europeans first explored the land and met the Chumash Indians, the native inhabitants of the region. Chumash villages and camps have been found at Los Osos Creek and Montaña de Oro and Morro Bay State Parks.

Today, Morro Bay is a heavily used fishing port and one of the largest bay wildlife habitats on California's coast. At low tide, 1,400 acres of mudflats are exposed, providing a vast feeding area for over 250 species of birds; the adjacent Morro Bay State Park is noted as one of the best sites for bay bird-watching.

Popular for clam digging and sand dune recreation, Pismo State Beach along the south county coast contains more than 2,000 acres of beach and windswept sand dunes. The beach and dunes are formed here because the effects of longshore drift are interrupted by Point San Luis to the north; longshore drift is a coastal process that, along this part of the coast, would ordinarily carry sediments south. Consequently, sands are deposited here instead of being transported away, and widespread sand dune fields have resulted.

In contrast to the Pismo dunes and beach in the south are the grassy coastal terraces and steep sea cliffs in the north county. Above these terraces overlooking the Pacific Ocean, William R. Hearst, in collaboration with the renowned architect Julia Morgan, built his enormous private castle from 1920 until his death in 1951. In 1958 the Hearst family donated the castle and immediate grounds to the State Department of Parks and Recreation; it has since become one of the most popular visitor attractions in California.

For more information on the San Luis Obispo County coast contact the San Luis Obispo County Visitors & Conference Bureau, 1037 Mill Street, San Luis Obispo 93401, (805) 541-8000 or see www.sanluisobispocounty.com; Cayucos Chamber of Commerce, P.O. Box 141, Cayucos 93430, (805) 995-1200 or 1-800-563-1878; and the Pismo Beach Chamber of Commerce, 581 Dolliver Street, Pismo Beach 93449, (805) 773-4382 or 1-800-443-7778.

For transit information, call: Central Coast Area Transit (CCAT), (805) 541-2228; or call: South County Area Transit (SCAT), Mon.-Fri., (805) 773-5400 or see www.slorta.org/. Buses for all systems can be boarded at regular stops or flagged at any corner or safe location. All systems have some wheelchair access; call for information. Dial-a-ride transit service is available: CCAT (Los Osos/Baywood Park areas, closed Sat., Sun., and holidays), (805) 528-7433; Runabout, service for the entire county (closed Sun. and Christmas), (805) 541-2544; and Dial-a-ride Morro Bay (closed Sun. and holidays), (805) 772-2744.

Hearst Castle

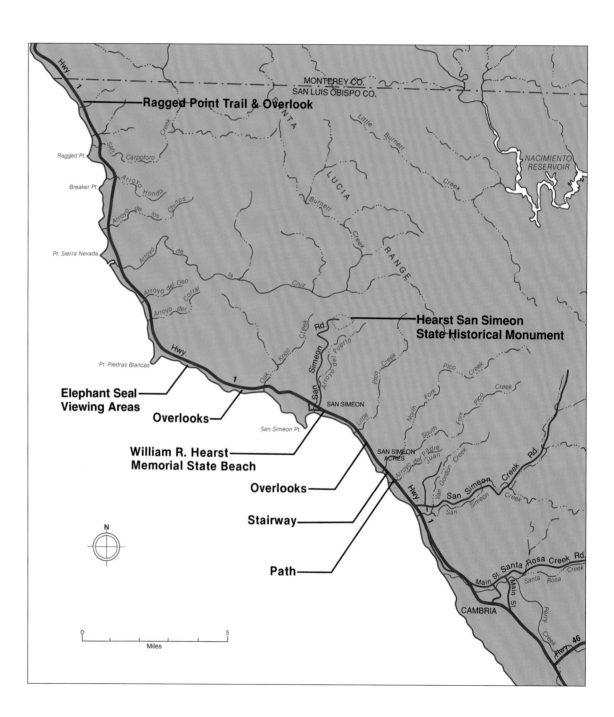

MONTEREY CO.
SAN LUIS OBISPO CO.

Ragged Point Trail & Overlook

Ragged Pt.

Breaker Pt.

Pt. Sierra Nevada

NACIMIENTO RESERVOIR

Hearst San Simeon State Historical Monument

Pt. Piedras Blancas

Elephant Seal Viewing Areas

Overlooks

SAN SIMEON

San Simeon Pt.

William R. Hearst Memorial State Beach

SAN SIMEON ACRES

Overlooks

Stairway

Path

CAMBRIA

Main St. Santa Rosa Creek Rd.

46

N

0 ————————— 5
Miles

San Luis Obispo County

SAN SIMEON

NAME	LOCATION	Entrance/Parking Fee	Parking	Restrooms	Lifeguard	Campground	Showers	Firepits	Stairs to Beach	Path to Beach	Bike Path	Hiking Trail	Facilities for Disabled	Boating Facilities	Fishing	Equestrian Trail	Sandy Beach	Dunes	Rocky Shore	Upland from Beach	Stream Corridor	Bluff	Wetland
Ragged Point Trail and Overlook	W. of Hwy. 1, 15 mi. N. of San Simeon		•	•					•								•		•	•	•	•	
Elephant Seal Viewing Areas	4.5 mi. N. of Hearst Castle		•																•		•		
Overlooks	2.9 and 3.5 mi. N. of San Simeon Rd., along Hwy. 1		•																•		•		
Hearst San Simeon State Historical Monument	E. of Hwy. 1 on San Simeon Rd., San Simeon	•	•	•									•							•			
tWilliam R. Hearst Memorial State Beach	W. of Hwy. 1 on San Simeon Rd., San Simeon		•	•			•		•						•		•					•	
Overlooks	3, 1, and 2 mi. S. of San Simeon Rd. along Hwy. 1		•																•	•			
Stairway to Beach	W. end of Pico Ave., San Simeon Acres		•				•										•		•	•		•	
Path to Beach	Corner of Cliff Dr. and San Simeon Ave., San Simeon Acres		•						•								•		•	•	•		

William R. Hearst Memorial State Beach

RAGGED POINT TRAIL AND OVERLOOK: Seaward of the Ragged Point Inn, located north of the Ragged Point peninsula, is a grassy area and overlook on a high blufftop terrace above the ocean. Scenic views of the Big Sur Coast to the north. A steep switchbacked trail leads past a waterfall to the small sandy beach and rocky shore. Taking of abalone not permitted. The trail and overlook are privately managed.

ELEPHANT SEAL VIEWING AREAS: Approximately 4.5 miles north of Hearst Castle, two parking areas provide excellent viewing of a growing colony of northern elephant seals. A short blufftop trail connects the two areas. Interpretive signs and volunteer docents provide information and instruction on safe viewing. Books and videos are available for purchase at the Friends of the Elephant Seal office in the Cavalier Plaza, San Simeon. For information, call: (805) 924-1628.

OVERLOOKS: Two parking areas on Hwy. 1 north of San Simeon provide views of the ocean.

HEARST SAN SIMEON STATE HISTORICAL MONUMENT: Formerly William R. Hearst's private estate, now open to the public as a 123-acre State Monument located within the 235,000 acres of private Hearst Ranch holdings. Tours of the monument leave from approximately 8:20 AM-3:00 PM daily except on Thanksgiving, Christmas, and New Year's days; there are extended hours during the summer. Constructed between 1920 and 1950, Hearst's architecturally renowned mansion and grounds still contain portions of his famous art collection and unique herds of exotic animals; the estate provides numerous scenic views of San Simeon Bay and the San Luis Obispo coast.

Three different tours leave by shuttle bus daily from the visitor parking lot just off Highway 1. Each tour takes 2 hours and involves considerable walking; tour 1 (150 steps) is recommended as easier than tours 2 and 3 (300 steps each). A fourth tour runs from April through early fall and emphasizes the formal gardens, wine cellar, Neptune Pool dressing rooms, and the main guest house. An evening tour that features a Living History of the 1930s takes 2½ hours and runs March-May and September-December.

Smoking is restricted. A wheelchair vehicle bus is available and requires reservations 10 days in advance; call: (805) 927-2020. Entrance fee; reservations for tours are highly recommended. For reservations, call: 1-800-444-4445. For information write Public Affairs, Hearst-San Simeon State Historical Monument, 750 Hearst Castle Road, San Simeon 93452; or call: (805) 927-2020.

WILLIAM R. HEARST MEMORIAL STATE BEACH: Excellent swimming area, protected from heavy surf and wind by scenic San Simeon Point. 1,000-foot-long pier, fishing equipment rentals, and boat charter service are located at the west end of the main parking lot. Picnic tables are in the eucalyptus grove just north of the pier and in the grassy area near the park entrance. Additional parking lot adjacent to the beach is north of the eucalyptus grove. Hours 8 AM (earlier in summer) to sunset. Entrance fee; for information, call: (805) 927-2020.

OVERLOOKS: Three parking lots along Hwy. 1 between San Simeon and San Simeon Acres overlook the ocean south of San Simeon Bay.

STAIRWAY TO THE BEACH: At the end of the Pico Avenue cul-de-sac, just south of Pico Creek, is an overlook with benches and a short stairway leading to a sandy beach; trail leads inland, under highway bridge. Parking on Pico Avenue.

PATH TO BEACH: A public path along the north bank of Arroyo del Padre Juan Creek leads to a sandy cove beach. The path is located directly south of the Cavalier Inn, west of Highway 1 and Hearst Avenue. On-street parking.

Native Americans of the Coast

At one time as many as 300,000 Native Americans inhabited what is now California. Of the more than 50 different native groups that occupied California, 16 lived, hunted, and fished along the shoreline and on off-shore islands.

Most California Native Americans were hunter-gatherers; that is, they used naturally occurring plant and animal life for food, clothing, and tools, rather than planting crops and maintaining livestock. Along the coast, many Native American groups re-lied heavily on the sea for sustenance. For example, the Chumash, who inhabited the coast from San Luis Obispo to Los Angeles counties, plied the coastal waters in planked ca-noes, called tomols, and hunted sea otters, fished for alba-core, and harvested shellfish such as clams, abalone, and mussels; the Chumash also used tomols to travel to offshore islands and trade with other Native American groups.

The societal structure and specific lifestyle of coastal Native American groups varied from place to place. In general, the members of a particular tribe inhabited small villages; in some cases their villages were governed by a larger, centrally located one. Villages usually consisted of a number of domi-ciles, a ceremonial structure or area, storehouses, and a burial ground; buildings were often constructed of wooden poles woven with grasses. Many tribes had distinct hunting, collecting, and fishing areas.

Trade with inland tribes was also a major characteristic of the coastal Native Americans in California. Artifacts made by vari-ous coastal tribes include baskets, wood trays and boxes (of-ten inlaid with shells or coral), water pots and cookware, and obsidian projectile points; some tribes used shells for dishes and pieces of jewelry.

The San Luis area was occupied by the Obispeño group of the Chumash tribe. The Chumash in general had one of the most highly developed societal structures of all California Na-tive Americans; they evolved from a simple hunter-gatherer culture to a society based on diverse subsistence activities and an extensive trade network.

The Chumash's ability to utilize ocean resources such as sea otters, fish, and shellfish contributed to their affluence relative to other Native American groups in California. The abundance of food resources and the favorable coastal climate allowed the Obispeño Chumash to enjoy a peaceful existence.

Native Americans in California lived according to their own lifestyles and customs until the arrival of the Spanish mission-aries. Native American attitudes toward missionization varied. The Ipai tribe, from the San Diego area, was very hostile and rebellious toward the Spanish missionaries; the Chumash, however, typically were friendly and, for example, even taught the Spanish how to use asphaltum to make watertight roofs and bowls. The establishment of the missions resulted in the introduction of European diseases and a displacement of Na-tive American culture; Euro-American influences virtually ex-terminated the Native American population along much of coastal California by the early 1900s. Today, place names such as Malibu, Hueneme, Pismo, Sinkyone, and Talawa re-main as evidence of Native American existence along the California coast.

San Luis Obispo County

SAN SIMEON / CAMBRIA

NAME	LOCATION	Entrance/Parking Fee	Parking	Restrooms	Lifeguard	Campground	Showers	Firepits	Stairs to Beach	Path to Beach	Bike Path	Hiking Trail	Facilities for Disabled	Boating Facilities	Fishing	Equestrian Trail	Sandy Beach	Dunes	Rocky Shore	Upland from Beach	Stream Corridor	Bluff	Wetland
San Simeon State Beach	Between San Simeon Creek and Santa Rosa Creek	•	•	•		•				•		•	•				•			•	•		
San Simeon Creek Access	W. of Hwy. 1 at San Simeon Creek		•							•							•				•		
Moonstone Beach	Moonstone Beach Dr. off Hwy. 1, Cambria		•									•							•	•		•	
Leffingwell Landing	Moonstone Beach Dr., .25 mi. S. of intersection with Hwy. 1, Cambria		•	•					•	•		•	•	•	•				•			•	
Santa Rosa Creek Access	W. of Hwy. 1, near S. end of Moonstone Beach Dr., Cambria		•							•		•					•				•		•
Shamel County Park	Windsor Blvd. and Nottingham Dr., Cambria		•	•				•		•							•						
Overlooks	W. of Nottingham Dr. at Plymouth and Lancaster streets., Cambria		•																•			•	
East West Ranch	W. of Hwy. 1, at seaward end of Windsor Blvd., Cambria		•									•							•	•		•	
Cambria Hostel/Bridge Street Inn	4314 Bridge St., Cambria	•	•	•			•												•				
Sherwood Drive Access Points	Sherwood Dr., between Wedgewood and Lampton sts., Cambria Pines Manor		•							•							•		•			•	

The following are administered by the California Department of Parks and Recreation as part of San Simeon State Beach: San Simeon Creek Access, Moonstone Beach Drive Vista Point, Leffingwell Landing, and Santa Rosa Creek Access. Call: (805) 927-2035.

SAN SIMEON STATE BEACH: The state beach runs from San Simeon Creek to Santa Rosa Creek, east of Hwy. 1 off San Simeon Creek Rd., and includes 205 campsites, some with wheelchair access; a hiker/bicyclist group site; a trailer sanitation station; and an overflow trailer camp area south of San Simeon Creek.

A short path to the beach is located at the southwest corner of the campground underneath the Hwy. 1 San Simeon Creek overpass. The creek area is accessible by short trails from the lower campground; there is a foot bridge at the east end of the creek. Fee; reservations recommended during the summer, call:1-800-444-7275. Call: (805) 927-2035.

SAN SIMEON CREEK ACCESS: Adjacent to the south bank of San Simeon Creek. The parking area off Hwy. 1 provides access to San Simeon State Beach.

MOONSTONE BEACH: Vista point at northern end overlooks the Moonstone Beach area of San Simeon State Beach; named for the moonstone agates occasionally seen here. Beach access is from San Simeon Creek to the north. Highly eroded bluffs; boardwalk along the edge south of the parking area.

LEFFINGWELL LANDING: Facilities include picnic tables north of the parking lot in a sheltered cypress grove, and a ramp for car-top boat launching. Hiking trails along blufftops to the north connect to the Moonstone Beach Drive Vista Point. Noted as a good spot for watching sea otters, frequently seen resting in the rocky tidepool areas offshore and to the south. Tidepool and offshore areas are part of the California Sea Otter Game Refuge; do not disturb the otters.

SANTA ROSA CREEK ACCESS: Parking area and benches are adjacent to the ocean; Santa Rosa Creek marshland is south of the parking lot. Hiking trails along blufftops to the north overlook the rocky shore and tidepool areas below.

SHAMEL COUNTY PARK: Parking and access to the beach is along the northern edge of the park; the park contains a playground area, a grassy playing field, and a swimming pool that is open in summer.

EAST WEST RANCH: One mile-long blufftop trail provides spectacular views of the rocky shoreline. Otters can be spotted in the kelp beds.Easy walk, dogs allowed, no restroom or camping facilities.

CAMBRIA HOSTEL/BRIDGE STREET INN: Located one mile from Moonstone Beach in a colonial style house; 15 beds. Open 5-9 PM; for information, call: (805) 927-7653.

SHERWOOD DRIVE ACCESS POINTS: Access down low bluffs to the sandy beach and rocky shore is along Sherwood Dr. at the ends of Wedgewood, Castle, Harvey and Lampton Streets. There is also a park at the end of Lampton Street. Take Ardath Dr. off Hwy. 1; Drake St. off Ardath Dr. leads to Sherwood Drive.

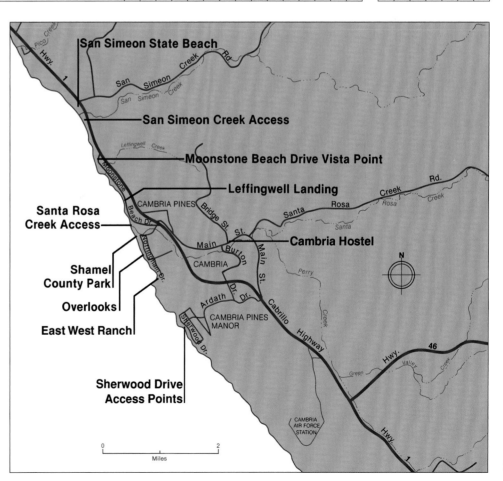

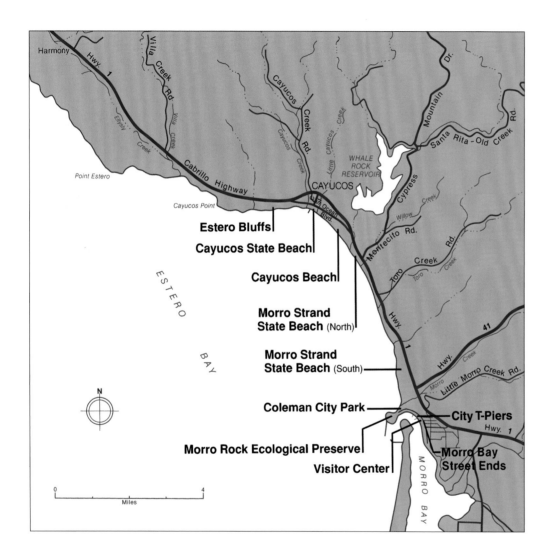

Estero Bluffs

Cayucos State Beach

Cayucos Beach

**Morro Strand
State Beach** (North)

**Morro Strand
State Beach** (South)

Coleman City Park

City T-Piers

Morro Rock Ecological Preserve

**Morro Bay
Street Ends**

Visitor Center

Morro Strand State Beach South

San Luis Obispo County

CAYUCOS / NORTH MORRO BAY

NAME	LOCATION	Entrance/Parking Fee	Parking	Restrooms	Lifeguard	Campground	Showers	Firepits	Stairs to Beach	Path to Beach	Bike Path	Hiking Trail	Facilities for Disabled	Boating Facilities	Fishing	Equestrian Trail	Sandy Beach	Dunes	Rocky Shore	Upland from Beach	Stream Corridor	Bluff	Wetland
Estero Bluffs	Hwy. 1, N.W. of Cayucos		•										•				•				•		
Cayucos State Beach	W. of North Ocean Dr. between Cayucos Rd. and E. St., Cayucos		•	•	•		•	•					•		•		•					•	
Cayucos Beach	Along Pacific Ave. between 1st and 22nd Sts., Cayucos		•						•								•		•			•	
Morro Strand State Beach (North)	Along Studio Dr. between 24th St. and Cody Ave., Cayucos		•	•					•	•							•					•	•
Morro Strand State Beach (South)	W. of Hwy. 1 between Yerba Buena Ave. and Atascadero Rd., Morro Bay	•	•	•		•		•		•							•	•				•	
Coleman City Park	W. of Coleman Dr. and the Embarcadero, Morro Bay		•	•				•									•	•					
Morro Rock Ecological Preserve	W. end of Coleman Dr., Morro Bay		•	•											•		•		•				
City T-Piers	W. end of the Embarcadero, seaward of the power plant, Morro Bay		•	•									•	•	•								
Morro Bay Estuary Visitor Information Center	601 Embarcadero St., suite 11, Morro Bay		•	•																•			
Morro Bay Street Ends	Along the Embarcadero, between Beach St. and Anchor St., Morro Bay		•												•								•

ESTERO BLUFFS: One of California's newest state parks, the Estero Bluffs just north of Cayucos provides 4 miles of blufftop trails through rolling grasslands, with access to beaches and creeks. Seasonal closures at the north end of the park near Villa Creek are necessary to protect endangered snowy plovers. Dogs must be kept on leash while on the trail, and are not allowed on beaches. Management of this property is not finalized. Please respect signs and seasonal closures.

CAYUCOS STATE BEACH: Facilities include a popular fishing pier that is lit at night and is wheelchair accessible; group barbecue and picnic facilities are in the patio area adjacent to the Veterans Memorial building. A beach wheelchair is available. Parking is along Ocean Front Rd. and in the lot adjacent to the patio area. To reserve the group facilities or buildings, call: County General Services Department, (805) 781-5200.

CAYUCOS BEACH: 18 stairways along Studio Dr. and Pacific Ave., between 1st and 22nd streets, lead to the sandy beach below; all are marked "Public Beachwalk." Private property adjoins both sides; do not trespass.

MORRO STRAND STATE BEACH (NORTH): From 24th St. to the south end of Studio Drive. Paved parking lot, restrooms, and picnic tables are at the end of 24th St., with an additional unpaved parking area at the north end of Studio Drive. Stairs and paths to the beach are along Studio Dr. 200 feet north of Juanita Ave., and at the ends of Coronado, Mayer, Mannix, and Cody avenues. Call: (805) 772-7434.

MORRO STRAND STATE BEACH (SOUTH): Formerly called Atascadero State Beach. 1.7 miles long; extends from the creek at the north end of Beachcomber Dr. to the mouth of Morro Creek south of Atascadero Road. 104-site campground is at the west end of Yerba Buena Dr., with pedestrian access at the end of Hatteras St. and through the tunnel at the end of Orcas Street. No fee for beach day use. Clam digging not permitted between Azure St. and Morro Rock. Facilities include dressing rooms and outdoor showers. Additional beach access is at the end of Easter St. and at the end of Atascadero Rd., as well as through the "Cloisters Parcel," which is now state property, located west of Sandalwood Ave. and between Sienha and Azure Streets. For information, call: (805) 772-7434.

COLEMAN CITY PARK: Contains playground equipment, picnic tables, fire rings, and restrooms. Access to the dunes area and Morro Strand directly north. South of the Embarcadero is a small protected sandy beach within Morro Bay.

MORRO ROCK ECOLOGICAL PRESERVE: 576-foot high Morro Rock is the northernmost of the volcanic peaks running from San Luis Obispo City to Morro Bay. Protected state preserve for the nesting of the endangered peregrine falcon; entry or climbing is prohibited. However, fishing, hiking, and parking around the base of the rock are permitted; a large parking area and restrooms are northeast of the rock. The breakwater on the southwest side can be hazardous during heavy surf. Small protected beach on north side of the rock with limited parking.

CITY T-PIERS: The North T-Pier is used primarily for fishing and commercial boat docking. Public parking west of the Embarcadero; wheelchair-accessible restrooms, restaurant, bait and tackle shop, fish-loading facilities, and a public shower are located on or close to the pier. U.S. Coast Guard boats berth at the North T-Pier and are usually open for tours; call: (805) 772-1293 for appointment. For Morro Bay boating information, contact the Harbormaster office at the base of the pier: (805) 772-6254.

MORRO BAY ESTUARY VISITOR INFORMATION CENTER: Includes exhibits about the estuary and its watershed, birds of the estuary, eelgrass, and steelhead trout. Located on the waterfront, upstairs in the Marina Square building. Open 10 a.m. daily, closes at 5 p.m. in winter, later in summer. Call: (805) 772-3834.

MORRO BAY STREET ENDS: All street ends along the Embarcadero between Beach and Anchor Streets lead to the bay and provide public access for viewing and fishing; limited on-street parking. There are shops and restaurants along the Embarcadero.

Morro Bay, City T-Pier

Coast north of Point San Luis

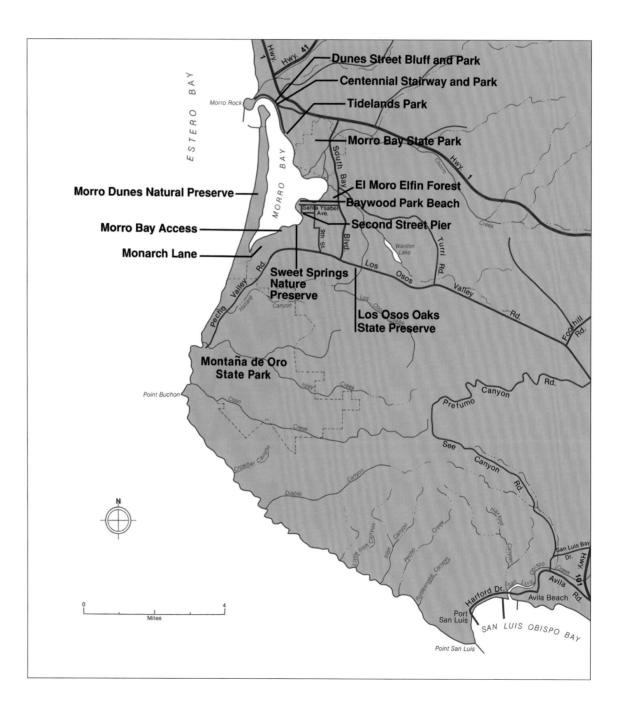

ESTERO BAY

Dunes Street Bluff and Park

Centennial Stairway and Park

Morro Rock

Tidelands Park

Morro Bay State Park

MORRO BAY

El Moro Elfin Forest

Morro Dunes Natural Preserve

Baywood Park Beach

Santa Ysabel Ave.

Second Street Pier

Morro Bay Access

Warden Lake

Monarch Lane

Sweet Springs Nature Preserve

Los Osos Oaks State Preserve

Montaña de Oro State Park

Point Buchon

Hwy. 41

Hwy. 1

South Bay Blvd.

Turri Rd.

Los Osos Valley Rd.

Foothill Rd.

Pecho Valley Rd.

Hazard Canyon

9th St.

Los Osos Creek

Islay Creek

Canyon

Crowbar Canyon

Coon Creek

Diablo Canyon

Little Irish Canyon

Irish Canyon

Pecho Creek

Rattlesnake Canyon

Hartog Canyon

Prefumo Canyon

See Canyon Rd.

Canyon Rd.

San Luis Bay Dr.

Obispo Creek

Avila Rd.

Hwy. 101

Harford Dr.

San Luis Creek

Avila Beach

Port San Luis

SAN LUIS OBISPO BAY

Point San Luis

N

0   4
Miles

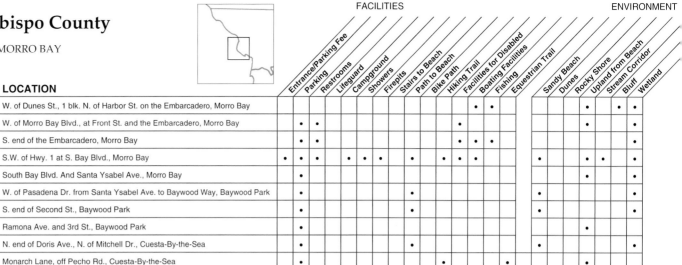

NAME	LOCATION	Entrance/Parking Fee	Parking	Restrooms	Lifeguard	Campground	Showers	Firepits	Stairs to Beach	Path to Beach	Bike Path	Hiking Trail	Facilities for Disabled	Boating Facilities	Fishing	Equestrian Trail	Sandy Beach	Dunes	Rocky Shore	Upland from Beach	Stream Corridor	Bluff	Wetland
Dunes Street Bluff and Park	W. of Dunes St., 1 blk. N. of Harbor St. on the Embarcadero, Morro Bay													•	•					•		•	•
Centennial Stairway and Park	W. of Morro Bay Blvd., at Front St. and the Embarcadero, Morro Bay	•	•										•							•		•	
Tidelands Park	S. end of the Embarcadero, Morro Bay	•	•										•	•	•								•
Morro Bay State Park	S.W. of Hwy. 1 at S. Bay Blvd., Morro Bay	•	•	•		•	•	•		•		•	•	•	•		•			•		•	•
El Moro Elfin Forest	South Bay Blvd. And Santa Ysabel Ave., Morro Bay	•																		•			
Baywood Park Beach	W. of Pasadena Dr. from Santa Ysabel Ave. to Baywood Way, Baywood Park	•								•							•						•
Second Street Pier	S. end of Second St., Baywood Park	•								•							•						
Sweet Springs Nature Preserve	Ramona Ave. and 3rd St., Baywood Park	•															•						
Morro Bay Access	N. end of Doris Ave., N. of Mitchell Dr., Cuesta-By-the-Sea	•															•						•
Monarch Lane	Monarch Lane, off Pecho Rd., Cuesta-By-the-Sea	•										•				•	•						
Morro Dunes Natural Preserve	W. of Morro Bay											•			•		•	•					
Los Osos Oaks State Preserve	S. of Los Osos Valley Rd. at Palamino Dr., Los Osos	•										•								•			
Montaña de Oro State Park	S. end of Pecho Valley Rd.	•	•			•		•		•		•	•	•	•	•	•		•	•		•	

DUNES STREET BLUFF AND PARK: There is a vista point with a public bench at the end of Dunes St. above the Embarcadero. West of the Embarcadero is a small grassy park with benches and tables adjacent to Morro Bay. There is also a small boat dock with slips and a fishing and viewing deck.

CENTENNIAL STAIRWAY AND PARK: A stairway leads from the west end of Morro Bay Blvd. down to a small park that contains benches, a shuffleboard court, and a giant chessboard. Wheelchair-accessible restrooms and parking are west of the Embarcadero at the end of Front Street. For information, call: (805)772-6200.

TIDELANDS PARK: Two-lane boat ramp with two docks is adjacent to a small picnic area with wheelchair-accessible restrooms at the south end of the parking lot. No mooring of boats is allowed at the ramps or buoys. Fish-cleaning station near the ramps; parking spaces for boat trailers. For information, call: (805) 772-6200.

MORRO BAY STATE PARK: Park entrances are one mile south of Hwy. 1 on South Bay Blvd. and at the south end of Main St. in Morro Bay. Most of the park overlooks Morro Bay and the southern mudflats at the mouth of Los Osos Creek; best vista points are at the peak of Black Mountain and from the Museum of Natural History at White Point. The Natural History Museum has exhibits of wildlife, ecology, and Native American history of Morro Bay; open daily 10 AM-5 PM; entrance fee. The museum is wheelchair accessible. Museum: (805) 772-2694.

The campground contains 135 units with tables and stoves; 20 sites have electric and water hookups. Laundry tubs available. Overnight and day-use fees. Two group camps are available at the campground for 30 or 50 persons; reservations required. Some campsites are wheelchair accessible. For camping reservations, call: 1-800 444-7275. Enroute campsites are also available for self-contained R.V.'s. An 18-hole public golf course is accessible from Black Mountain or Golf Course roads. Park facilities include a small car-top boat launch ramp and a path to the beach north of the museum, and a dock with berths and restrooms to the south. Protected heron rookery is south of the park's west entrance; visitors may observe from the road but may not enter or disturb the herons. Great blue herons nest in the eucalyptus treetops from January to August. For park information, call: (805) 772-7434.

EL MORO ELFIN FOREST: Preserves a forest of diminutive oaks and rare coastal dune scrub habitat. The 90-acre area supports approximately 200 plant species, 110 kinds of birds, and 13 species of reptiles and amphibians. There is a boardwalk with benches and interpretive signs. For information, call: (805) 528-0392.

BAYWOOD PARK BEACH: A public path leads from the small parking area on Pasadena Dr. to a sandy beach with picnic tables and benches. The beach provides access to Morro Bay mudflats at low tides.

SECOND STREET PIER: 50-foot-long T-pier, primarily used as a vista point. Benches are in a cypress grove to the east. A county street right-of-way at the south end of First St. also provides bay access.

SWEET SPRINGS NATURE PRESERVE: Managed by Central Coast Audubon Society, includes a ¼-mile long boardwalk, freshwater pond and abundant bird life including shorebirds, ducks, Brant, scaup, wigeon, and snowy egret.

MORRO BAY ACCESS: A 40-foot-wide dirt accessway leads to Morro Bay; popular bird-watching area. Limited on-street parking. The extent of public and private rights is undetermined and is subject to further investigation.

MONARCH LANE: Accessway to Montaña de Oro and the Bay is found at Monarch Lane off of Pecho Road in Los Osos. Pedestrian and equestrian trail through eucalyptus and oak that provide habitat for migrating monarch butterflies in season.

MORRO DUNES NATURAL PRESERVE: Undeveloped 3-mile-long sand spit; popular for bird watching and clamming. Best access is by private boat or by rental boat available at the foot of Pacific Ave. off the Embarcadero. Dirt road access for four-wheel-drive vehicles begins off Pecho Valley Rd., 1/2-mile south of the junction with Los Osos Road.

LOS OSOS OAKS STATE PRESERVE: Two miles of marked trails lead through undisturbed groves of scenic old-growth coast live oak; the preserve contains vista points overlooking the Los Osos Valley and the Santa Lucia Mountains. Fragile environment; do not leave main trails. Poison oak is prevalent. Dogs are prohibited.

MONTAÑA DE ORO STATE PARK: 8,400 acres and 3 miles of coastline. 50 campsites with tables and stoves are east of the ranger's office above the Spooners Cove beach and day-use parking area; overnight fee. Day-use restrooms are wheelchair accessible. There are also four environmental campsites. For camping reservations, call: 1-800-444-7275. The park contains 50 miles of trails, some of which are open to equestrian use. Hazard Canyon Trail is 1.5 miles south of the park entrance on the west side of Pecho Valley Rd.; the ¼-mile-long path leads through eucalyptus groves used by Monarch butterflies for nesting October through March. Several trails to blufftop overlooks begin along the road between the park office and Coon Creek to the south. Special equestrian camp area is ¼ mile south of the park entrance on Pecho Valley Rd., and contains two camps, each with a 25-horse and 12-vehicle limit. Camping fee. Dogs prohibited on trails. Equestrian camps available by reservation only: Morro Bay State Park, Lower State Park Rd., Morro Bay 93442. Call: (805) 528-0513.

Clamming

The Pismo clam is the most sought-after clam along the central California coast. Adapted to an environment of well-oxygenated surf, Pismo clams burrow no more than six inches deep and are usually found in areas of one to three feet of water at low tide.

To dig for Pismo clams, use a clam fork with 8- to 10-inch tines. Attach a "C" type clamming gauge with a 4-½ inch span to the handle and use this gauge to measure every clam removed. Clams less than 4-½ inches across their greatest width (or 5 inches north of the Monterey-San Luis Obispo county line) must be reburied on their edge, with the points of their shells toward the ocean and the small dark buttons near the hinges pointing upward.

To locate clams, work in a line parallel to the edge of the ocean, probing with a clam fork about every two inches until a clam is found. Always face the breaking surf in order to anticipate large waves; rip currents are extremely dangerous. Never clam in water deeper than waist high, or deeper than knee high if wearing waders; large waves can fill them and pull you down. Clam bags should be easily detachable so that they do not become dangerous weights.

Pismo clams may be taken anywhere along the coast except in the following preserves in San Luis Obispo County: along Morro Strand State Beach between Azure Street and Morro Rock; along Montaña de Oro State Park between the southern tip of Morro Bay and Hazard Canyon; along Pismo State Beach between the Grand Avenue vehicle ramp and .3 mile north (Pismo Invertebrate Reserve); and from the mouth of Oso Flaco Creek to the county line at the Santa Maria River. Preserve areas are

subject to change; contact California State Parks for the most up-to-date information: (805) 549-3312.

Many other species of clams grow in the mud and gravel of bays, estuaries, and river mouths. These clams live eight inches to three feet below the surface, depending on the species and age of the clam. Along mudflats at low tide, clam diggers with shovels, clam forks, or trowels are frequently seen digging clams out of their burrows.

Clams may be taken only between one half hour before sunrise to one half hour after sunset. Licenses are required. For bag limits, seasons, and size restrictions, refer to the current California Marine Sport Fishing Regulations, available at sporting goods stores or from the California Department of Fish and Game.

San Luis Obispo County

AVILA BEACH / NORTHERN PISMO BEACH

FACILITIES — ENVIRONMENT

NAME	LOCATION	Entrance/Parking Fee	Parking	Restrooms	Lifeguard	Campground	Showers	Firepits	Stairs to Beach	Path to Beach	Bike Path	Hiking Trail	Facilities for Disabled	Boating Facilities	Fishing	Equestrian Trail	Sandy Beach	Dunes	Rocky Shore	Upland from Beach	Stream Corridor	Bluff	Wetland
Pecho Coast Trail	Harford Dr., Avila Beach											•								•			
Port San Luis Pier and Beach	W. end of Harford Dr., Avila Beach	•	•	•	•		•		•				•	•	•		•						
Avila State Beach	S. of Front St., between Harford Dr. and San Rafael St., Avila Beach		•	•	•		•			•			•		•		•						
South Palisades City Park	W. of Shell Beach Rd., Searidge Ct. to Silver Shoals Dr., Pismo Beach	•	•																			•	
Stairs to Beach	2757 Shell Beach Rd., and 2651 and 2555 Price St., Pismo Beach	•					•		•								•		•	•			
Spyglass City Park	S.W. intersection of Spyglass Dr. and Solano Rd., Pismo Beach	•									•		•						•			•	
Memory Park	Along seaward edge of Seacliff Dr., Pismo Beach	•																	•	•		•	
Vista Points	End of El Portal Dr., and at Naomi Ave. and Seacliff Dr., Pismo Beach	•																	•	•		•	
Ocean City Park	Ocean Blvd. between Vista Del Mar and Capistrano aves., Pismo Beach	•							•										•			•	
Margo Dodd City Park	Ocean Blvd. from Windward Ave. to just S. of Cliff Ave., Pismo Beach	•	•						•										•	•		•	
Dinosaur Caves Park	Price St. and Shell Beach Rd., Pismo Beach	•											•									•	

PECHO COAST TRAIL: The 3.7-mile-long trail above Port San Luis passes Pt. San Luis Lighthouse, and proceeds up the coastal terrace to Rattlesnake Canyon; open by reservation only; docents lead tours. For information, call: (805) 541-8735.

PORT SAN LUIS PIER AND BEACH: The 1,320-foot-long pier is lit at night. Boating facilities include boat hoists, fuel dock, trailer boat parking, and limited visitor moorings. Launching and storage fees. Sandy beach at the foot of the seawall along Harford Dr. between Port San Luis and the Union Oil Pier; two stairways lead to the beach. Disabled persons are allowed to drive onto the pier. For information: Box 249, Pier 3, Avila Beach 93424.

AVILA STATE BEACH: Playground equipment is located along the beach north of the public fishing pier; restrooms, outdoor showers, and fish-cleaning facilities at the pier. Lifeguards on duty spring through summer. A motorized beach wheelchair is available.

SOUTH PALISADES CITY PARK: Blufftop park, lawn, trails and ocean views.

STAIRS TO BEACH: At Cliffs Hotel, public parking, blufftop path, and stairs lead to a sandy beach. At Shore Cliff Lodge and at Shelter Cove Lodge, stairs lead to a rocky pocket beach.

SPYGLASS CITY PARK: Grassy bluffs at the west end of the parking lot; dirt pathway leads to rocky shore and tidepools. Bluffs are highly eroded. Playground equipment, wheelchair-accessible picnic tables, and bike racks. Additional access to bluffs from pathway at the northwest corner of Seacliff Drive. Private property on either side of path; do not trespass.

MEMORY PARK: The park consists of a grassy blufftop overlook with benches for viewing.

VISTA POINTS: Blufftop path at end of El Portal Drive. At the ends of Seacliff Dr. and Park Place, public walkway leads to blufftop viewing pavilion with benches. The concrete walkway above the gravel path and the property on both sides of the accessway are private; do not trespass.

OCEAN CITY PARK: Grassy park on the bluffs overlooking the ocean; facilities include benches, bike racks, picnic tables, and a short concrete walkway. Stairways leading to the sandy beach and tidepools below are located at the end of Vista Del Mar Ave. between Cuyama and Morro avenues, and at the ends of Morro and Palomar avenues.

MARGO DODD CITY PARK: Grassy blufftop park with a gazebo, picnic tables, and benches; dirt parking areas are west of Seaview Ave. and south of Cliff Avenue. Stairway to rocky beach at the end of Pier Avenue. Bluffs are highly eroded.

DINOSAUR CAVES PARK: There are no dinosaurs, and no cave access, but this 10-acre blufftop park at Price St. and Shell Beach Rd. offers picnic tables, ocean views, and a short walking trail.

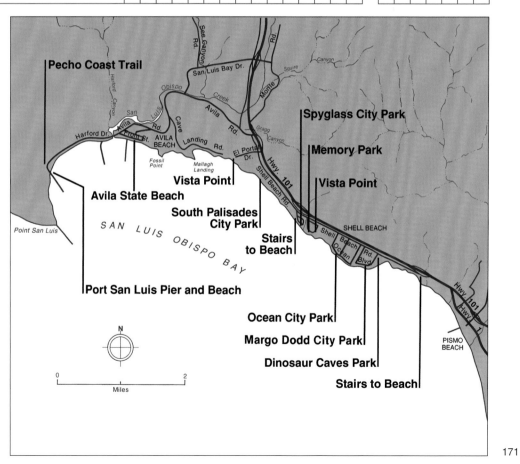

Clams

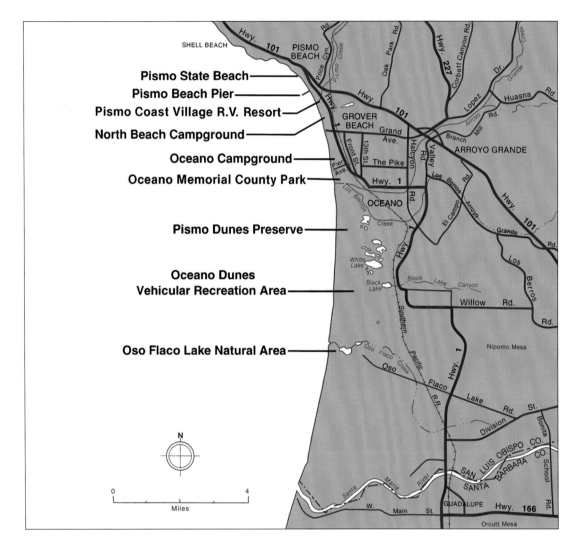

Pismo State Beach
Pismo Beach Pier
Pismo Coast Village R.V. Resort
North Beach Campground
Oceano Campground
Oceano Memorial County Park
Pismo Dunes Preserve
Oceano Dunes Vehicular Recreation Area
Oso Flaco Lake Natural Area

A variety of clams are found along the California coast; bent-nosed, geoduck, gaper, soft-shell, and Washington clams live in the mudflats of bays and lagoons that contain quiet waters with little oxygen content. Chione species and jackknife clams prefer the same environment, but their range is limited to south of Point Conception. Littleneck clams, or rock cockles, live in the gravel areas of bays.

Razor clams, which are much sought after by clammers, are also the fastest burrowers and are found at gently sloping beaches with moderate surf, most commonly along the Del Norte County and Humboldt County coasts. The Pismo clam is probably the most popular, and is found on beaches between Half Moon Bay (in San Mateo County) and Baja California, with the greatest population at Pismo State Beach.

Clams are filter feeders. To feed, they extend a double-tubed siphon to the surface of the sand and draw water and food in one tube while expelling waste through the second tube. Clams can often be located by searching for exposed siphon tubes. One clam, the large gaper, expels from its tube, at fairly regular intervals, jets of water two or three feet in the air, sometimes hitting those who attempt to disturb it.

Clams are potentially prolific reproducers. For example, a single, sexually mature Pismo clam will spawn about 15 million eggs. These eggs hatch into free-swimming larvae that must settle into the sand and attach themselves to grains of sand where they develop into young clams. During this free-swimming period a majority of the larvae succumb to predation and unfavorable environmental conditions.

Those clams that manage to establish themselves in the sand are subject to predation by shorebirds, gulls, surf fish, and people, who are the greatest predators; the California Department of Fish and Game reported that 150,000 people once removed 75,000 pounds of clams during a single weekend at Pismo State Beach. In fact, the Pismo clam, once so abundant it was commercially harvested at Southern California beaches by the wagonload, is now usually found only at low tide by clam diggers who are willing to wade out into sometimes hazardous surf.

common littleneck clam
1.5"-2"

gaper clam
6"-8"

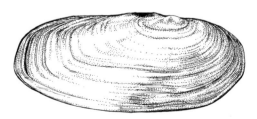

razor clam
4"-6"

Pismo clam
4.5"-5"

San Luis Obispo County

PISMO BEACH

NAME	LOCATION	Entrance/Parking Fee	Parking	Restrooms	Lifeguard	Campground	Showers	Firepits	Stairs to Beach	Path to Beach	Bike Path	Hiking Trail	Facilities for Disabled	Boating Facilities	Fishing	Equestrian Trail	Sandy Beach	Dunes	Rocky Shore	Upland from Beach	Stream Corridor	Bluff	Wetland
Pismo State Beach	Runs from Wilmar Ave. to the county line		•	•	•				•	•					•		•	•					
Pismo Beach Pier	End of Pomeroy and Hinds aves., Pismo Beach		•	•	•				•	•				•	•		•						
Pismo Coast Village R.V. Resort	1615 S. Dolliver St. (Hwy. 1), Pismo Beach	•	•	•		•	•	•									•						
Pismo State Beach North Beach Campground	S. Dolliver St. (Hwy. 1), S. of Addie St., Pismo Beach	•	•	•		•		•		•							•	•					•
Pismo State Beach Oceano Campground	Roosevelt Dr., E. of Pier Ave., Oceano City	•	•	•		•	•	•		•		•	•				•	•				•	•
Oceano Memorial County Park	Along Pier Ave. and Mendel Dr., Oceano City	•	•	•		•	•	•												•			•
Pismo Dunes Preserve	S. of Arroyo Grande Cre														•			•		•			
Oceano Dunes Vehicular Recreation Area	W. of Hwy. 1, S. of Oceano City	•	•	•													•	•					
Oso Flaco Lake Natural Area	W. end of Oso Flaco Lake Rd.		•	•											•			•					•

Pismo State Beach runs from the foot of Wilmar St. to the county line at the Santa Maria River. The area north of Pismo Creek is administered by the city of Pismo Beach; the area to the south is administered by the California Department of Parks and Recreation. For information, call: (805) 489-1869 or 489-2684 (recorded message).

PISMO STATE BEACH: The following street ends have paths and/or stairways leading to Pismo State Beach: Wilmar St., Wadsworth Ave., Main St., Stimson Ave., Ocean View Ave., Park Ave., Addie St., La Sage Dr., Grand Ave., Pier Ave., McCarthy St., Juanita St., Gray St., Surf St., York St., and Utah Street. A stairway at the end of Wadsworth Ave. leads to eight volleyball courts.

PISMO BEACH PIER: A 1,250-foot-long pier in the middle of Pismo Beach, lit at night, with wheelchair access. Stairs to a wide, sandy beach are at the north end of the parking lot. Access road available for wheelchairs at the left of the parking lot; beach wheelchair available. Concession stand, bait sales, and fishing equipment rentals; shops and restaurants nearby. Surfing allowed south of the pier only.

PISMO COAST VILLAGE R.V. RESORT: 400 R.V. sites with full hookups, located adjacent to Pismo State Beach. Laundry, store, and pool. Information and reservations: 165 S. Dolliver St., Pismo Beach 93449. For information, call: (805) 773-1811.

PISMO STATE BEACH NORTH BEACH CAMPGROUND: 300 yards from the ocean, separated by sand dunes and eucalyptus groves. Contains 103 campsites with stoves and tables. Meadow Creek is along the south boundary of the campground; a trail to the beach that extends along the creek begins off Dolliver St. and passes through a Monarch butterfly preserve. Trailer sanitation station is near the entry kiosk. Call: (805) 489-1869 or 489-2684. For camping reservations, call: 1-800-444-7275. La Sage Golf Course, which is open to the public, is south of the campground off Grand Avenue.

PISMO STATE BEACH OCEANO CAMPGROUND: Adjoins the Oceano Lagoon to the east and Pismo Beach to the west; short hike to the ocean. 82 campsites with stoves and tables; 42 have trailer hookups; 1 with wheelchair access; restrooms with showers. Hiking trail around the lagoon begins along the eastern side of the campground. Primitive campsites are along the beach beginning ¾ mile south of Arroyo Grande Creek. For information, call: (805) 489-1869 or 489-2684. For camping reservations, call: 1-800-444-7275.

OCEANO MEMORIAL COUNTY PARK: Overnight camping area is southwest of Mendel Dr. and Pier Ave.; 64 campsites, some with trailer hookups. Day-use grassy park is west of Norswing Dr. and north of Mendel Dr.; contains picnic sites, playground, and the Oceano Lagoon.

PISMO DUNES PRESERVE: The area of the dunes adjacent to the ocean, from Arroyo Grande Creek to approximately 1½ miles south and ¾ mile inland, is restricted from vehicular use. Unique, undeveloped areas of large dunes; popular with hikers.

OCEANO DUNES VEHICULAR RECREATION AREA: Within Nipomo dunes area are approximately 3,600 acres of dunes and hard, sandy beach of which 1500 acres are open to vehicular use. Vehicle access to the beach at low tides is from the ramps at the ends of Grand and Pier avenues. Street-legal vehicles are permitted between Grand Ave. and milepost 2. Off-highway vehicles allowed south of milepost 2 for approximately 3 miles, and in designated dune areas. Fee for day use; reservations required for camping, call: 1-800-444-7275. This beach supports snowy plovers, a threatened species, and there may be some restrictions on public access in this area to protect the habitat.

O.H.V. use must conform to California vehicle codes. Drivers are responsible for knowing all rules. For information, call: (805) 473-7220.

OSO FLACO LAKE NATURAL AREA: 800-acre protected area within the Oceano Dunes Vehicular Recreation Area. The entrance kiosk is at the west end of Oso Flaco Lake Rd., 3 miles west of Hwy. 1; 75-acre lake and marshland area located within the Nipomo Dunes. There is a 1.1 mile long boardwalk from the parking lot, crossing the lake and extending to the beach. Trail and restrooms are wheelchair accessible. The area also provides access to the Guadalupe-Nipomo National Wildlife Refuge to the south.

Jalama Beach County Park

Santa Barbara County

South of the San Luis Obispo County line, the vast sand dunes of Guadalupe give way to the undeveloped Casmalia and Solomon Hills, which border the rugged coastline between Point Sal and Point Conception in Santa Barbara County. The steep hillsides here abut the shoreline, creating narrow, sandy beaches that are secluded because of their distance from Highway 101.

At Point Conception, the shoreline curves abruptly east, providing an unusually long stretch of coast that faces south and is sheltered from waves by the northern Channel Islands, located offshore 20 to 30 miles south. The sandy beaches here, known for good swimming, are situated beneath the bluffs of the narrow coastal terrace that extends from Point Conception to the south county line. Bordering this terrace to the north are the scenic mountains of the Santa Ynez range, characterized by steep-walled canyons and sharp peaks that vary in elevation from 1,500 to 4,000 feet.

Between the south county shoreline and the Channel Islands is the Santa Barbara Channel, noted as a productive fishery resource. The Chumash Indians, who settled in large villages along the Santa Barbara and Ventura coasts, caught such great quantities of swordfish, marlin, tuna, sardines, and other fish in the channel that some early Spanish explorers were prompted to believe that the fishing industry alone could support all future settlers.

Beneath the Santa Ynez Mountains and adjacent to the Santa Barbara Channel is the city of Santa Barbara. The city, with its numerous Spanish-style buildings, was once the site of one of the larger coastal Chumash villages. It was here that Captain Jose Francisco Ortega, accompanied by Father Junípero Serra, founded a military presidio in 1782 to protect the Spanish-claimed coast from Russian explorers. In 1786, the Spanish established the Santa Barbara Mission for the purpose of converting the neighboring Indians to Christianity; this had devastating effects on the Chumash culture.

After secularization of the Mission in 1834, officers of the presidio were given large and very profitable land grant ranches along the Santa Barbara coast. Among these was Rancho Lompoc, which, after the ranch's subdivision in 1874, eventually became the city of Lompoc, now noted as one of the largest commercial flower-growing areas on the west coast. Directly north of Lompoc is the vast Vandenberg Air Force Base, established by the military in 1949 for coastal defense. Point Sal State Beach is on the northern boundary of the Vandenberg base, and is one of the most secluded beaches in the county.

The Union Pacific Railroad line runs along three quarters of the county's coast directly above the shoreline, from the Ventura County boundary to six miles south of Point Sal, before heading inland. Amtrak's Coast Starlight passenger train runs this route daily, providing one of the most scenic railroad trips available on the California coast.

Santa Barbara County includes four of the offshore islands within Channel Islands National Park. Described as America's Galapagos, the isolated islands support many unique species of flora and fauna not found on the mainland. Most of the islands are open to public access.

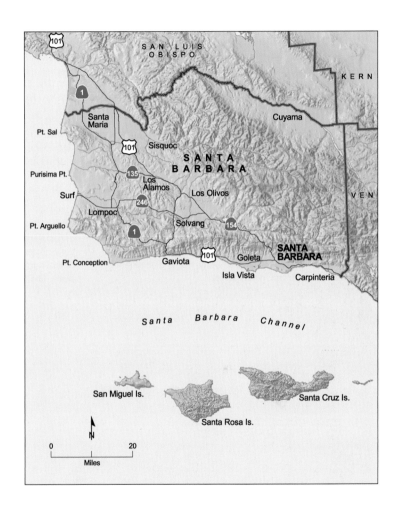

For more information on Santa Barbara County's coast, contact the Santa Barbara Region Chamber of Commerce, 924 Anacapa Street, Suite 1, Santa Barbara 93101, (805) 965-3023 or see www.sbchamber, or the Carpinteria Chamber of Commerce, 5285 Carpinteria Avenue, Carpinteria 93014, (805) 684-5479 or see www.carpcofc.com. For transit information, write or call the Santa Barbara Metropolitan Transit District (MTD), 550 Olive Street, Santa Barbara 93101, (805) 683-3702 or see www.sbmtd.gov.

Sand Dunes

Sand dunes are a prominent landscape feature found at numerous locations along the California coast. Sand dunes are likely to develop in areas where an abundant sand supply restricted to a limited portion of coastline is combined with strong onshore winds, a gently sloping beach, and large differences between high and low tide levels.

Typical dune formation occurs when onshore winds transport loose, dry sand across wide stretches of low-lying beach areas. The sand accumulates around vegetation, driftwood, or other elements that inhibit air flow. Eventually, the stockpiled sand forms continuous ridges perpendicular to the prevailing winds.

Vast areas of the California Coast used to be dune systems. However, these areas often were razed for development and only small dune systems now remain. San Francisco, Monterey, Los Angeles Airport, El Segundo, and Huntington Beach all contain former sand dunes.

Examples of areas in California with large still existing coastal dune fields include Pelican Bay, north of Crescent City; the Humboldt Bay area; Manchester State Beach, north of Point Arena; Inglenook Fen/MacKerricher State Beach, north of Fort Bragg; Ocean Beach, San Francisco; Monterey Bay; and Pismo, Nipomo, and Guadalupe dunes, west of Santa Maria.

Coastal dune systems naturally protect low-lying inland areas from ocean storms. The dunes provide a large buffer of sand that absorbs wave energy and prevents storm waves from eroding areas farther inland. Portions of a dune field may be obliterated by a single winter storm. However, during calmer periods, the dune system will rebuild.

Although sand dunes can provide valuable protection from the erosive power of ocean waves, the dune field itself is a very delicate system. A healthy, stable dune system consists of a series of dune ridges stabilized by salt-tolerant grasses and other vegetation such as hottentot fig, sand verbena, and salt bush. The vegetative cover can be removed by natural causes such as storms or, as is often the case, as a result of paths made by vehicles and people.

A loss of this protective cover can destroy dune stability and cause the eventual loss of a dune system. As dunes lose stability, sand migrates inland with the prevailing wind and covers highways, houses, and inland vegetation, creating an expensive and hazardous maintenance problem. In addition, sand blown inland is no longer available for natural beach replenishment.

Visitors are encouraged to enjoy the beauty of the sand dunes, but are asked to respect the fragile nature of the system; stay on marked trails, avoid trampling dune vegetation, and watch for warning or use information signs.

Evening primrose *Oenothera cheiranthifolia*   Hottentot fig *Mesembryanthemum edule*   Beach morning glory *Convolvulus soldanella*   Salt bush *Atriplex semibaccata*

Santa Barbara County

NORTHERN SANTA BARBARA COAST

NAME	LOCATION	Entrance/Parking Fee	Parking	Restrooms	Lifeguard	Campground	Showers	Firepits	Stairs to Beach	Path to Beach	Bike Path	Hiking Trail	Facilities for Disabled	Boating Facilities	Fishing	Equestrian Trail	Sandy Beach	Dunes	Rocky Shore	Upland from Beach	Stream Corridor	Bluff	Wetland
The Dunes Center	1055 Guadalupe St., Guadalupe		•	•						•	•		•						•				
Rancho Guadalupe Dunes Preserve (Guadalupe Entrance)	W. end of Main St., Guadalupe		•	•													•	•		•		•	
Point Sal State Beach	End of Brown and Point Sal roads, W. of Guadalupe		•							•					•		•	•	•			•	
Vandenberg Air Force Base Fishing Access	From Purisima Point to 8 mi. S. of the Point		•							•					•		•			•		•	
Ocean Beach County Park	Ocean Park Rd. (Hwy. 246), 10 mi. W. of Hwy. 1, Surf		•	•				•		•			•		•		•	•	•	•	•	•	•
Surf Railroad Depot	W. end of Hwy. 246, Surf		•	•						•					•		•	•					
Vandenberg Air Force Base Beach Access	From 1.5 mi. N. to 3.5 mi. S. of Ocean Beach County Park		•												•		•			•			
Jalama Beach County Park	Jalama Beach Rd., 20 mi. S. of Lompoc	•	•	•	•	•	•		•				•		•		•	•	•	•	•	•	•

THE DUNES CENTER: The visitor center, in a 1910 house, includes a library and displays with information on plant and animal species of the dunes and the geology, ecology, and geography of the area. The center promotes conservation of the dunes ecosystem and offers docent-led walks. Open Fri.-Sun. noon-4 PM. For information, call: (805) 343-2455.

RANCHO GUADALUPE DUNES PRESERVE (GUADALUPE ENTRANCE): Managed by the Center for Natural Lands Management. The Guadalupe Entrance to the preserve provides access to the Guadalupe Dunes and Mussel Rock Dunes. Access is also provided to the Guadalupe-Nipomo National Wildlife Refuge just north of the Santa Maria River. Call: (805) 343-2354.

POINT SAL STATE BEACH: Secluded, undeveloped beach 8.6 miles west of Hwy. 1 at the foot of the Casmalia Hills; access is via the long, rough dirt and tarmac Point Sal Road. The road is unmaintained, often impassable, and subject to closure during the winter or during missile launchings at Vandenberg Air Force Base. Steep paths lead to the beach from the dirt parking area. Point Sal, which is a harbor seal haul-out and seabird roosting area, is within walking distance. For information, call: (805) 733-3713.

VANDENBERG AIR FORCE BASE FISHING ACCESS: Limited public access along the blufftop from Purisima Point south for approximately five miles; trails lead down the steep bluffs to pocket beaches. The Air Force does not permit swimming, surfing, diving, or abalone gathering and may impose some seasonal closures to protect snowy plover and least tern and during rocket launches. Visitors must first obtain a pass from the base Wildlife Warden. Access is limited to 50 persons per day, weekends and holidays only. To obtain a pass, visitors must call the warden's office by Friday at 12:00 PM to reserve a pass for the subsequent weekend. Call: (805) 606-6804.

OCEAN BEACH COUNTY PARK: The 28-acre park, next to a broad sandy beach and Santa Ynez River lagoon is habitat for several endangered species such as the California brown pelican, California least tern, western snowy plover and salt marsh bird's beak. A 1/4-mile-long path leads beneath the railroad trestle and to Vandenberg Air Force Base's beach. Call: (805) 934-6148.

SURF RAILROAD DEPOT: An operating railroad station provides parking and access to Vandenberg Air Force Base beaches. The Depot provides beach access across the railroad tracks with warning lights and bells; visitors should use caution when crossing.

VANDENBERG AIR FORCE BASE BEACH ACCESS: Public beach access is from 1.1 miles north of the Santa Ynez River to 3.5 miles south, and includes the beach west of Ocean Park and Surf Depot. The beach provides habitat for the western snowy plover. During nesting season, March 1 to September 30, most of the beach is closed to public use, but the Air Force allows public access to a ½-mile stretch of beach. Access to this open area is available from Surf Depot and Ocean Beach County Park (along a ½-mile long trail just west of the railroad tracks). To protect the snowy plover, please respect all posted restrictions. The beach may also be closed prior to and during rocket launches from the base. Call: (805) 606-6804.

JALAMA BEACH COUNTY PARK: 28-acres; broad sandy and rocky beach with a blufftop campground; 112 campsites, store, snack stand, and wheelchair-accessible restrooms. Overnight and day-use fees. Hazardous rip currents. Beach wheelchair available. Call: (805) 736-3504. Vandenberg Air Force Base also provides one mile of unlimited beach access northwest of Jalama Beach.

0   20

Miles

177

Amtrak Coast Starlight

The Amtrak Coast Starlight, a passenger train that runs daily between Seattle, Washington and Los Angeles, California, provides panoramic views of much of the California coast between Monterey and Ventura counties. Some portions of the coast along the train's route are not readily accessible by any other means. Points of interest along the Coast Starlight's route include Elkhorn Slough, the sand dunes along the San Luis Obispo and Santa Barbara County coasts, Vandenberg Air Force Base, and the Santa Barbara Channel coast. For the best views of the coast, passengers should sit in the upper deck of one of the Starlight's observation cars.

Approximately one hour out of the San Jose station going south (about 10 minutes out of Salinas going north), the Coast Starlight begins its run close to the Monterey County coast at the Elkhorn Slough estuary, one of the richest wildlife habitats in California. The origin of the slough is not completely certain. One theory is that during the Pleistocene period (10,000 to 3 million years ago), the slough was the point where runoff from the Sacramento and San Joaquin valleys emptied into the Pacific Ocean. Elkhorn Slough was fed regularly with flows from the Salinas River until the early 1900s, when a new river mouth was formed by extensive flooding caused by heavy storms; the old river channel that ran to the slough was subsequently diked.

Today the slough provides valuable habitat for harbor seals, sea otters, and more than 70 species of fish and 90 species of birds; as many as 20,000 birds may be seen nesting and

feeding in the slough at one time. Elkhorn Slough meets Monterey Bay at Moss Landing, which is the location of a Pacific Gas and Electric Company electric power generating station. The power plant's twin smokestacks can be seen rising above the horizon beyond the slough. The agricultural fields seen along the slough are used primarily for growing strawberries.

After leaving Salinas, the Coast Starlight travels south through the Salinas River Valley on the train's way to the city of San Luis Obispo. As the Starlight descends the Cuesta Pass into San Luis Obispo, a string of eroded, conical-shaped peaks come into view. These peaks, or volcanic plugs, are the remnants of volcanic activity that occurred 11-25 million years ago. The peaks lie in a line from the city of San Luis Obispo to Morro Bay. The westernmost peak (not visible from the train), which is 576 feet high, is Morro Rock, situated at the mouth of Morro Bay. Bus transportation to Hearst Castle, which is located on the coast at San Simeon north of Morro Bay, is available to passengers disembarking in the city of San Luis Obispo.

About 15 minutes out of San Luis Obispo going south (two hours and 15 minutes out of the Santa Barbara station going north), the Starlight reaches the coastline at Pismo Beach, long known as a summer seaside resort and famous for Pismo clams. Pismo Beach is the northern end of a 40-mile stretch of coast that is characterized by extensive sand dunes. The Coast Starlight travels along the inland side of the dune fields; at times the shoreline is visible beyond the expanse of bare and vegetated dunes.

Thirty minutes out of San Luis Obispo going south (two hours out of the Santa Barbara station going north), the Coast Starlight crosses the Santa Maria River and enters Santa Barbara County at Guadalupe, a small town that serves as a shipping point for agricultural products from the fertile Santa Maria Valley. Twenty minutes beyond Guadalupe (one hour out of the Santa Barbara station going north), the Coast Starlight enters Vandenberg Air Force Base.

Vandenberg has been used as a launch site for satellites and missile tests, and is the West Coast launch site for NASA's Space Shuttle; train passengers can see launch pad towers scattered throughout the dunes of the 35-mile-long base.

Point Conception, about 1-1/2 hours out of San Luis Obispo going south (50 minutes out of Santa Barbara going north), marks the point where the coastline changes from a north-south orientation and continues to Santa Barbara along a west-to-east line parallel to the Channel Islands. This shoreline orientation, combined with the prevailing wave direction, makes this portion of the coast one of the world's best surfing spots.

Petroleum deposits lie beneath the ocean floor within the Santa Barbara Channel; a number of offshore oil platforms, and the Channel Islands beyond them, are visible from the Coast Starlight. Continuing south to the city of Santa Barbara, the Starlight passes a number of oil storage stations, pumping facilities, and other shoreside facilities that support offshore oil operations.

The Coast Starlight crosses the Santa Barbara-Ventura County line 20 minutes out of Santa Barbara (30 minutes out of Oxnard going north) at Rincon Point, a popular surfing area. The Starlight parallels the Old U.S. Highway 101 from the Rincon area into Ventura; during periods when waves are high, spray from waves breaking on the rocky shoreline may carry over the highway to the railroad tracks.

After crossing the Ventura River, about 30 minutes out of Santa Barbara, the Starlight crosses to the inland side of Highway 101. The masts of boats in the Ventura Marina are the last bit of coastal scenery one can see from the Coast Starlight before it leaves the farms of the Oxnard Plain for the Los Angeles metropolitan area.

For reservations, scheduling, fares, or other information on the Coast Starlight, call Amtrak at 1-800-872-7245 or see www.amtrak.com.

Santa Barbara County

POINT CONCEPTION TO NAPLES

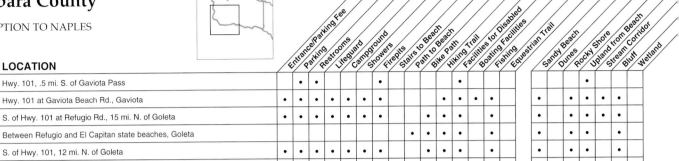

NAME	LOCATION	Entrance/Parking Fee	Parking	Restrooms	Lifeguard	Campground	Showers	Firepits	Stairs to Beach	Path to Beach	Bike Path	Hiking Trail	Facilities for Disabled	Boating Facilities	Fishing	Equestrian Trail	Sandy Beach	Dunes	Rocky Shore	Upland from Beach	Stream Corridor	Bluff	Wetland
Gaviota Rest Area	Hwy. 101, .5 mi. S. of Gaviota Pass		•	•				•					•							•			
Gaviota State Park	Hwy. 101 at Gaviota Beach Rd., Gaviota	•	•	•	•	•	•	•			•	•	•	•			•		•	•	•	•	
Refugio State Beach	S. of Hwy. 101 at Refugio Rd., 15 mi. N. of Goleta	•	•	•	•	•	•		•	•	•		•				•		•	•		•	
Bike Path and Ramp to Beach	Between Refugio and El Capitan state beaches, Goleta								•	•	•		•									•	
El Capitan State Beach	S. of Hwy. 101, 12 mi. N. of Goleta	•	•	•	•	•	•			•			•				•		•	•		•	
El Capitan Ranch Park	11560 Calle Real, N. of El Capitan State Beach, Goleta	•	•	•		•	•	•					•							•	•		

GAVIOTA REST AREA: The rest area is one-half mile from the entrance to Gaviota State Park and is situated in a narrow canyon along Hwy. 101. Both the northbound and southbound stops have 30 parking spaces, restrooms, public telephones, and picnic tables.

GAVIOTA STATE PARK: The 2,776-acre park includes 5.5 miles of shoreline, a day-use picnic area, a campground, and a fishing pier. Camping facilities include a store, 39 trailer sites, 20 tent sites, and 20 sites for self-contained R.V.'s; no hookups or sanitary station. Overnight and day-use fees. Wheelchair-accessible restrooms are available. The park extends on both sides of Hwy. 101 with primary access at the campground near the ocean. The Gaviota Pier has a three-ton boat launch; bait and tackle are available. Lifeguards on duty during the summer. Access to the upland wilderness area is via Hwy. 1 north toward Lompoc; turn right at the stop sign. Hiking trails lead from the dirt parking lot to a small hot spring (approximately body temperature) and the Los Padres National Forest; no overnight parking permitted in the parking lot. Call: (805) 968-1033.

REFUGIO STATE BEACH: The 90-acre park includes a sandy beach with rocky shore and tidepool habitats. The campground has 85 campsites, additional sites for self-contained R.V.'s, and a snack stand. Lifeguards on duty during the summer. Day-use parking, wheelchair-accessible restrooms, fishing licenses, and bait are available. Overnight and day-use fees. Call: (805) 968-1033.

BIKE PATH AND RAMP TO BEACH: A bike path, which is also open to hikers, runs seaward of the railroad tracks and Hwy. 101 between Refugio and El Capitan state beaches. A wheelchair-accessible ramp, located at the south end of the oil facilities tunnel leading to Las Flores Canyon, leads from the path to the two state beaches.

EL CAPITAN STATE BEACH: The 133-acre park includes a narrow sandy beach, and upland day-use and camping facilities. A hiking trail leads through a grassy picnic area along the rocky shoreline. The campground includes 142 campsites and a snack stand; no hook-ups or sanitary station. Day-use parking and enroute campsites also available. Overnight and day-use fees. Some restrooms are wheelchair accessible. El Capitan Point is a noted surfing area. Lifeguards on duty during the summer. For information, call: (805) 968-1033.

EL CAPITAN RANCH PARK: Privately managed campground, located north of Hwy. 101 in a riparian canyon. Facilities include 96 cabins, 26 furnished tents, and 6 meeting spaces, store, laundry, snack bar, and game arcade. Overnight fee. For information and reservations, call or write: 11560 Calle Real, Goleta 93117; (805) 968-2214.

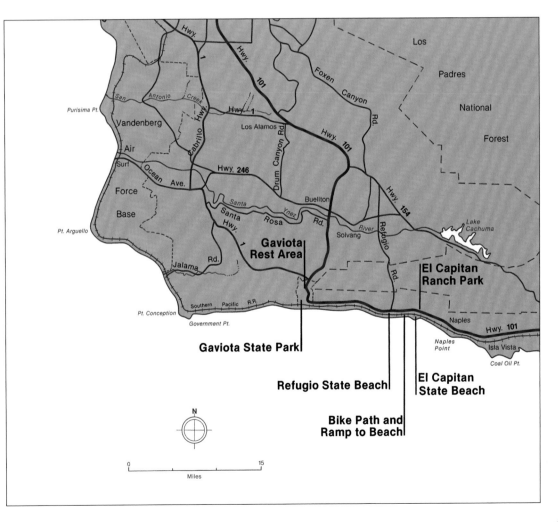

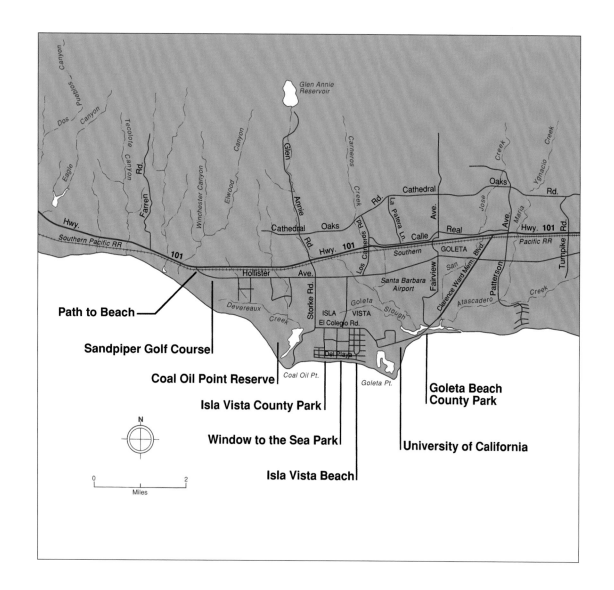

Path to Beach

Sandpiper Golf Course

Coal Oil Point Reserve

Isla Vista County Park

Window to the Sea Park

Isla Vista Beach

Goleta Beach County Park

University of California

Isla Vista County Park

Santa Barbara County

ISLA VISTA / GOLETA

FACILITIES — ENVIRONMENT

NAME	LOCATION	Entrance/Parking Fee	Parking	Restrooms	Lifeguard	Campground	Showers	Firepits	Stairs to Beach	Path to Beach	Bike Path	Hiking Trail	Facilities for Disabled	Boating Facilities	Fishing	Equestrian Trail	Sandy Beach	Dunes	Rocky Shore	Upland from Beach	Stream Corridor	Bluff	Wetland
Sandpiper Golf Course	8000 Hollister Ave., Goleta	•	•	•																•		•	
Path to Beach	8301 Hollister Ave., Goleta		•							•							•						
Coal Oil Point Natural Reserve	End of entrance road off Storke Rd. and Colegio Rd., Isla Vista		•									•			•		•	•	•	•	•	•	•
Isla Vista County Park	Del Playa Dr. at Camino Corto, Isla Vista		•						•	•					•		•					•	
Window to the Sea Park	Del Playa Dr., just S. of Camino del Sur, Isla Vista		•																	•		•	
Isla Vista Beach	S. of Del Playa Dr., Isla Vista		•						•						•		•		•			•	•
University of California at Santa Barbara	Off Clarence Ward Memorial Blvd., Goleta, and on El Colegio Rd., Isla Vista	•	•	•					•	•	•				•		•	•	•	•		•	•
Goleta Beach County Park	5990 Sandspit Rd., Goleta		•	•	•			•					•	•	•		•		•	•			•

SANDPIPER GOLF COURSE: Full 18-hole golf course located on a blufftop overlooking the Santa Barbara Channel. Privately owned, open to the public; the facility includes a pro shop and snack stand. Open at 7:00 AM. Entrance fee. For information, call: (805) 968-1541.

PATH TO BEACH: At Bacara Resort and Spa; parking, trail to beach.

COAL OIL POINT NATURAL RESERVE: The reserve is part of the University of California's statewide Natural Reserve System. Pedestrian access to the reserve is restricted to unposted areas; do not disturb vegetation or other resources. Posted areas are sensitive habitats and use is limited to official scientific study; do not trespass. Devereaux Slough, located within the reserve, provides several coastal lagoon habitats for numerous birds. A two-mile-long self-guided walking tour around the reserve perimeter begins at the reserve gate. The beach below the dunes is open to the public. This beach supports snowy plovers, a threatened species, and there may be some restrictions on public access in this area to protect the habitat. Call: (805) 893-4127.

ISLA VISTA COUNTY PARK: The 1.4-acre grassy blufftop park overlooks the sandy Isla Vista Beach. Facilities include picnic tables, blufftop benches, and a sand volleyball court. A stairway to the beach is west of the grassy area. On-street parking only.

WINDOW TO THE SEA PARK: The blufftop overlook features a very small landscaped park with a swinging bench. No access to the beach. On-street parking only. Call: (805) 968-2017.

ISLA VISTA BEACH: There are stairways to the sandy beach and tidepool areas at the ends of Camino Majorca, Camino del Sur, and Camino Pescadero, and a paved ramp to the beach at the end of El Embarcadero. Blufftop overlooks are at the ends of Camino Majorca, Camino Pescadero, and El Embarcadero. On-street parking only.

UNIVERSITY OF CALIFORNIA AT SANTA BARBARA: Includes the main campus at Goleta Point and the west campus at Coal Oil Point. Pedestrian and bicycle access are permitted through both campuses. Vehicle access through the west campus is limited primarily to those with university business. Visitors may drive onto the main campus when parking is available; parking fee. Paths and stairs that lead to the beach adjacent to the main campus are located near the Santa Cruz dormitory, and near the south and east portions of the Campus Lagoon. Call: (805) 893-2487 or 893-8000.

GOLETA BEACH COUNTY PARK: The 29-acre park features a wide, sandy beach, a grassy picnic area, a fishing pier, and a children's play area. Facilities include volleyball courts, a restaurant and snack stand, parking, and wheelchair-accessible restrooms. There are three group picnic areas; for reservations, call: (805) 568-2465. A beach wheelchair is available. The 1,450-foot Goleta Pier has a four-ton capacity boat hoist (fee). Goleta Slough, comprising over 350 acres of wetland area, is popular for canoeing and bird watching. Call: (805) 967-1300.

Coal Oil Point Reserve

Devereaux Lagooon, Coal Oil Point Reserve

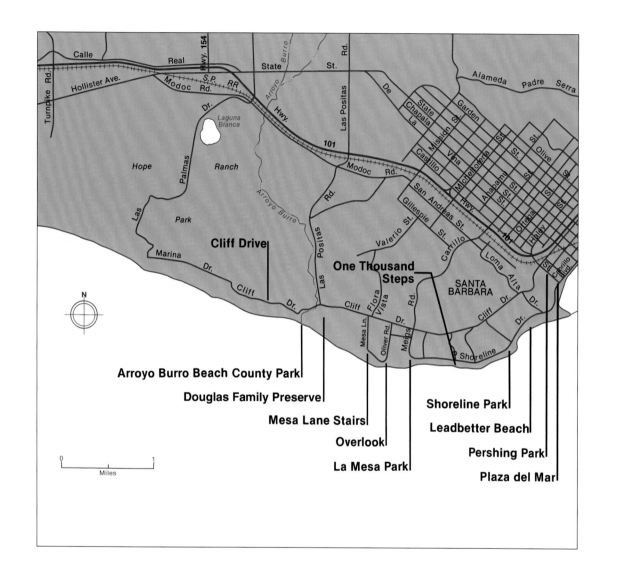

Cliff Drive

One Thousand
Steps

SANTA
BARBARA

Arroyo Burro Beach County Park

Douglas Family Preserve

Mesa Lane Stairs

Overlook

La Mesa Park

Shoreline Park

Leadbetter Beach

Pershing Park

Plaza del Mar

One Thousand Steps

Santa Barbara County

CITY OF SANTA BARBARA

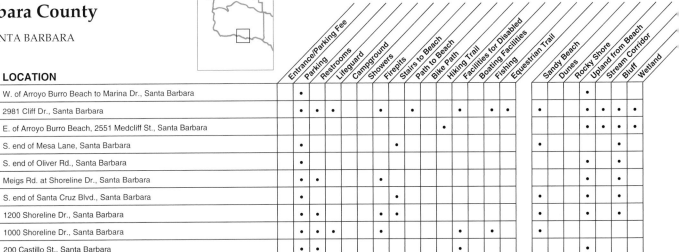

NAME	LOCATION	Entrance/Parking Fee	Parking	Restrooms	Lifeguard	Campground	Showers	Firepits	Stairs to Beach	Path to Beach	Bike Path	Hiking Trail	Facilities for Disabled	Boating Facilities	Fishing	Equestrian Trail	Sandy Beach	Dunes	Rocky Shore	Upland from Beach	Stream Corridor	Bluff	Wetland
Cliff Drive	W. of Arroyo Burro Beach to Marina Dr., Santa Barbara	•																				•	
Arroyo Burro Beach County Park	2981 Cliff Dr., Santa Barbara	•	•	•			•			•			•		•	•	•		•	•	•	•	
Douglas Family Preserve	E. of Arroyo Burro Beach, 2551 Medcliff St., Santa Barbara											•							•	•	•	•	•
Mesa Lane Stairs	S. end of Mesa Lane, Santa Barbara	•							•								•					•	
Overlook	S. end of Oliver Rd., Santa Barbara	•																		•		•	
La Mesa Park	Meigs Rd. at Shoreline Dr., Santa Barbara	•	•					•												•		•	
One Thousand Steps	S. end of Santa Cruz Blvd., Santa Barbara	•							•								•		•			•	
Shoreline Park	1200 Shoreline Dr., Santa Barbara	•	•						•	•							•		•			•	
Leadbetter Beach	1000 Shoreline Dr., Santa Barbara	•	•	•			•							•	•		•						
Pershing Park	200 Castillo St., Santa Barbara	•	•										•							•			
Plaza del Mar	Castillo St. at Cabrillo Blvd., Santa Barbara	•	•										•							•			

Leadbetter Beach

CLIFF DRIVE: Scenic drive with views of the Santa Barbara Channel and Channel Islands; there is a turnout near Yankee Farm Rd.

ARROYO BURRO BEACH COUNTY PARK: The six-acre park contains a sandy beach, a natural stream habitat area, and coastal bluffs. Facilities include volleyball courts, a restaurant and snack stand, a grassy picnic area, and wheelchair-accessible restrooms. The parking lot is often crowded on summer days. A short equestrian trail leads from Cliff Dr. to the sandy beach. Lifeguards on duty during the summer. Call: (805) 687-3714.

DOUGLAS FAMILY PRESERVE: The 70-acre open space commands views of the Santa Barbara Channel and the Channel Islands, as well as the Santa Ynez Mountains. The property contains valuable coastal resources including important wetland, oak woodland, coastal sage scrub, and monarch butterfly habitats. Several types of wetlands have been located on the preserve which is accessible by trail from Hwy. 101 at the south end of Las Positas Road.

MESA LANE STAIRS: A stairway descends a steep cliff to a beach that is popular with surfers and sunbathers; limited on-street parking.

OVERLOOK: The undeveloped blufftop area provides views of the Channel Islands. Beach access from the bluff is hazardous; use the Mesa Lane Stairs to the west.

LA MESA PARK: The nine-acre neighborhood park overlooks the Santa Barbara Channel from La Mesa Bluff above the beach. Facilities include picnic tables, a barbecue area, a playground, and restrooms; on-street parking only.

ONE THOUSAND STEPS: The blufftop overlook provides a stairway to the beach.

SHORELINE PARK: The 15-acre grassy park features a picnic area, a children's play area, and a blufftop overlook of the Santa Barbara Harbor and Channel. The park is popular for walking, kite-flying, frisbee, and picnics. Stairs lead to the shore.

LEADBETTER BEACH: The wide, sandy beach is located along a shallow cove. Facilities include a landscaped picnic area, snack stand, and restrooms. The parking lot is shared with the harbor.

PERSHING PARK: Five-acre park with a lighted baseball diamond and two softball fields used for team sports. The Santa Barbara City College tennis courts are in the park and are open for public use on weekends.

PLAZA DEL MAR: Five-acre park with a shaded, grassy picnic area and a view of the harbor. Summertime concerts are sometimes held at the band pavilion.

Harbor Seals

The harbor seal *(Phoca vitulina)* differs significantly from the other marine mammals found in the waters off the California coast. The subspecies of harbor seal that inhabits the waters offshore the Santa Barbara and Ventura coasts has a black or brown coat with silver, white, or yellow spots; because of the spots, the harbor seal is sometimes called the leopard seal. As the name suggests, harbor seals are most often found in bays, harbors, and river mouths and are frequently seen hauling out on the sand in these areas. However, harbor seals have also been observed offshore on San Nicolas Island, one of the Channel Islands.

Unlike fur seals and sea lions, harbor seals are unable to turn their hind flippers forward for movement on land. Whereas fur seals and sea lions use their front flippers for swimming, harbor seals rely primarily on their hind flippers. The harbor seals' lack of external ears further distinguishes them from fur seals and sea lions. Male harbor seals average five to six feet in length, with a weight of 275 pounds when fully grown; females are somewhat smaller.

Harbor seals have an interesting diving technique. The seal floats vertically with only its head above the water's surface; when it wants to dive, it drops straight down until totally submerged and then turns horizontal to swim.

Harbor seals feed on fish, squid, octopus, and some shellfish. These seals do not breed in colonies. Males do not have harems, which is characteristic of most pinniped species, but they do mate with more than one female. Pups are born from the end of May through July, usually on land.

Harbor seals are protected by state and federal laws.

Santa Barbara County

SANTA BARBARA HARBOR AREA

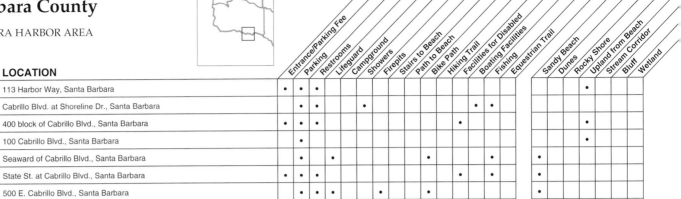

NAME	LOCATION	Entrance/Parking Fee	Parking	Restrooms	Lifeguard	Campground	Showers	Firepits	Stairs to Beach	Path to Beach	Bike Path	Hiking Trail	Facilities for Disabled	Boating Facilities	Fishing	Equestrian Trail	Sandy Beach	Dunes	Rocky Shore	Upland from Beach	Stream Corridor	Bluff	Wetland
Santa Barbara Maritime Museum	113 Harbor Way, Santa Barbara	•	•	•															•				
Santa Barbara Harbor	Cabrillo Blvd. at Shoreline Dr., Santa Barbara		•	•			•							•	•								
Los Baños del Mar	400 block of Cabrillo Blvd., Santa Barbara	•	•	•									•						•				
Ambassador Park	100 Cabrillo Blvd., Santa Barbara		•																•				
West Beach	Seaward of Cabrillo Blvd., Santa Barbara		•		•						•		•				•						
Stearns Wharf	State St. at Cabrillo Blvd., Santa Barbara	•	•	•									•		•		•						
Chase Palm Park	500 E. Cabrillo Blvd., Santa Barbara		•	•	•		•						•				•						
Santa Barbara International Tourist Hostel	134 Chapala St., Santa Barbara	•	•	•			•										•						

SANTA BARBARA MARITIME MUSEUM: Located in the Waterfront Center, houses Chumash displays, waterfront, whaling, and ship model exhibits, children's area, and a Channel Islands exhibit. Open Thurs.-Tues. 11AM-5PM. For information, call: (805) 962-8404.

SANTA BARBARA HARBOR: Harbor facilities include a marina with more than 1,000 boat slips, marine specialty shops, boat repair services, restaurants, sport fishing excursions, boat rentals, private floating dry dock, boat hoists (1,000 and 3,000 lb. cap., fee), launching ramp (fee), bait and tackle shop, and yacht club. Fuel dock open 8 AM-5 PM. Guest slips are available through the Harbormaster; call: (805) 963-1737. Scenic views from the walkway along the breakwater; good surfing conditions outside the harbor.

LOS BAÑOS DEL MAR: This municipal swimming pool, located at West Beach, offers lessons, and competition and recreational swimming. Parking is available at the harbor. For information on hours and fees, call: (805) 966-6119.

AMBASSADOR PARK: The half-acre park, located on the north side of Cabrillo Blvd., is a State Historic Landmark and is the site of an old Indian village. The grassy open area bordered by palm trees provides a view of the harbor.

WEST BEACH: The 11.5-acre sandy beach is located between Stearns Wharf and the harbor and has a boardwalk for pedestrians, bicyclists, and roller skaters.

STEARNS WHARF: Originally constructed in 1872, Stearns Wharf historically served the city of Santa Barbara as a seaport for cargo, passenger, and fishing ships. A major portion of the wharf is open for fishing. There are also restaurants, shops, and a seafood market. The Sea Center, under renovation, will have exhibits on the marine resources of Channel Islands National Park. The Nature Conservancy offices offer exhibits of its California preserves. The wharf is open 24 hours. Fee parking on the wharf; free parking is available in the city lots at Cabrillo Blvd. and Santa Barbara Street. Call: (805) 564-5518.

CHASE PALM PARK: The park has a mile-long stretch of wide, sandy beach with a strip of turf, lined with palm trees and picnic areas, located along East Cabrillo Boulevard. Community arts and craft show every Sunday. Facilities include a pedestrian walkway, Skater's Point, bike path, carousel, and restrooms. Palm Park Cultural Center is used for community meetings. For information on the Cultural Center or craft show, call: (805) 962-8936.

SANTA BARBARA INTERNATIONAL TOURIST HOSTEL: Located across from Amtrak station, 60 beds, laundry facilities, kitchen, showers. Open 8-11:30AM, 3:30-11:30PM. For information, call: (805) 963-3586.

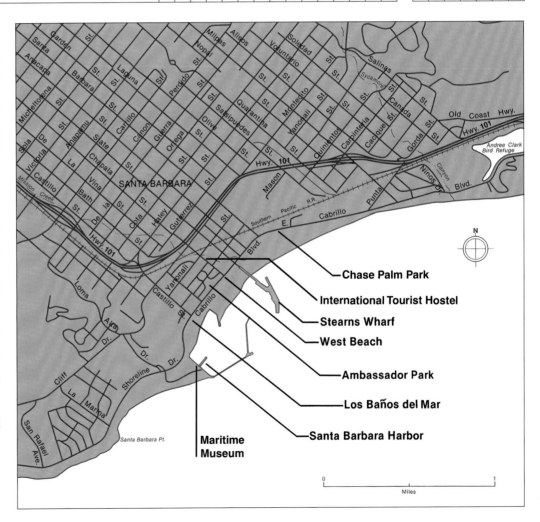

Chase Palm Park
International Tourist Hostel
Stearns Wharf
West Beach
Ambassador Park
Los Baños del Mar
Santa Barbara Harbor
Maritime Museum

0   1   Miles

Seals and Sea Lions

Pinnipedia is the Latin name for the group of marine mammals commonly known as seals and sea lions; pinnipedia means "feather feet," which describes the animals' flipper-like limbs. Like cetaceans (whales and dolphins), pinnipeds rely on fat, or blubber, to keep warm. Fur seals and sea lions belong to the *Otariidae* family of the pinnipeds. The characteristic features of this family include small external ears, hind flippers that turn forward for moving on land, and large fore flippers that are used chiefly for swimming.

A number of species of fur seals and sea lions have been observed off the California coast and some species have established breeding grounds, or rookeries, on the Channel Islands. Seals and sea lions may also be seen resting on many other offshore rocks and islands along the coast.

The northern, or Alaska, fur seal *(Callorhinus ursinus)*, which has established rookeries in the Channel Islands (on San Miguel Island and just off the island on Castle Rock), almost became extinct as a result of extensive fur trading during the 18th and 19th centuries. The fur seals' breeding habits also aided in their near extinction. Northern fur seals usually return to breed at the same rookery where they were born, regardless of any harassment or danger posed by intruders; therefore, once hunters located a fur seal rookery, they were assured of finding seals there year after year. Fur seals and all other marine mammals are now protected by state and federal laws.

Male northern fur seals grow to up to eight feet in length and weigh up to 700 pounds. Mature females are much smaller; their average length is four to five feet, and they weigh approximately 125 pounds. Adult males have a dark brown coat, while females and pups are grayish. Along the California coast, the northern fur seal diet consists primarily of squid, anchovies, and hake (a non-commercial fish). Fur seals usually stay well offshore, except during the breeding season; they are known to spend months at a time in the water without coming ashore at all.

People throughout the world have probably seen the California sea lion *(Zalophus californianus)* more than any other pinniped, because this species is used almost exclusively in the trained seal acts of circuses and marine animal parks. The most likely place to observe them in a natural environment is off the Santa Barbara County coast. The California sea lion territory ranges from British Columbia to Baja California; most are located between San Francisco and Baja, with the largest rookeries on San Miguel and San Nicolas islands.

Male California sea lions have a pronounced ridge running down the middle of their skulls, which distinguishes them from females and other species of sea lions. A characteristic that helps distinguish both sexes of this species from other species is their almost constant barking. The average California sea lion is comparable to the northern fur seal in size, although the largest of the male sea lions can weigh up to 1,000 pounds. Males have a dark brown color; females are lighter brown. The California sea lion diet consists primarily of octopus, squid, and many species of non-commercial fish.

The Steller, or northern, sea lion *(Eumetopias jubatus)* is one of the largest pinnipeds found in the waters off the California coast. Males measure up to 13 feet long and weigh as much as one ton; adult females average nine feet in length, with a weight of 600 pounds. The Steller sea lion's larger size and lighter color distinguish it from the California sea lion. In addition, Steller sea lions do not bark as regularly as California sea lions.

Stellers breed during June and July. Rookeries are typically located on the rocky parts of islands; the largest Steller breeding colonies are on the Farallon and Año Nuevo islands. Large numbers of Stellers also haul out, or rest, on Seal Rocks offshore from the Cliff House in San Francisco. However, a small percentage of the Steller sea lion population does breed as far south as the Channel Islands.

Santa Barbara County

CITY OF SANTA BARBARA / MONTECITO

NAME	LOCATION	Entrance/Parking Fee	Parking	Restrooms	Lifeguard	Campground	Showers	Firepits	Stairs to Beach	Path to Beach	Bike Path	Hiking Trail	Facilities for Disabled	Boating Facilities	Fishing	Equestrian Trail	Sandy Beach	Dunes	Rocky Shore	Upland from Beach	Stream Corridor	Bluff	Wetland
Cabrillo Ball Field	Milpas St. at Punta Gorda St., Santa Barbara		•																	•			
East Beach	1100 E. Cabrillo Blvd., Santa Barbara		•	•	•		•			•			•		•		•						
Dwight Murphy Field	Ninos Dr. and Por La Mar, Santa Barbara		•	•			•													•			
A. Child's Estate Zoological Gardens	1300 E. Cabrillo Blvd., Santa Barbara	•	•	•			•			•										•			
Andree Clark Bird Refuge	E. Cabrillo Blvd. at Hwy. 101, Santa Barbara		•								•									•			•
Stairways to Beach	Along Channel Dr., Montecito		•						•								•		•	•		•	
Hammonds Beach	End of Eucalyptus Lane, Montecito		•					•		•							•			•			
Miramar Beach	End of Eucalyptus Lane, Montecito		•						•								•						

CABRILLO BALL FIELD: The five-acre field has a single softball diamond, bleachers, and lights. For group reservations of the field, call: (805) 564-5422.

EAST BEACH: The 44-acre park with a half-mile wide, sandy beach includes a grassy picnic area with barbecues, a wading pool, children's play equipment, and beach volleyball courts; annual state and local volleyball tournaments are held here. The Cabrillo Pavilion, built in 1925, includes a youth recreation center and public bathhouse; call: (805) 897-1982. The Cabrillo Arts Center offers art shows, lectures, and movies; call: (805) 962-8956.

DWIGHT MURPHY FIELD: The 10.5-acre park, located across from A. Child's Zoo, is used primarily for softball and soccer. Facilities include lighted fields, children's play area, picnic tables, and barbecues. For group reservations of the field, call: (805) 564-5422.

A. CHILD'S ESTATE ZOOLOGICAL GARDENS: A beautifully landscaped 81-acre zoo and park. The gardens occupy a prominent knoll that overlooks East Beach and the harbor area. Facilities include a nature theater, children's train ride, play area, snack stand, and picnic area. Entrance fee. Call: (805) 962-5339.

ANDREE CLARK BIRD REFUGE: 42-acre refuge near the zoo and East Beach with an enclosed saltwater marsh. The refuge is a habitat for a variety of birds, including herons, egrets, cormorants, ducks, and geese. Popular bike path and grassy area are along the south and east shores. Limited parking available on Los Patos Way.

STAIRWAYS TO BEACH: Three public stairways lead to the sandy beach south of Channel Drive. They are located at the end of Butterfly Lane, and along the boardwalk seaward of the Biltmore Hotel. Channel Dr. provides on-street parking and is a scenic route with views of the Santa Barbara Harbor and Channel.

HAMMONDS BEACH: Several trails lead to the sandy beach. The trail system will eventually be linked by a footbridge over Montecito Creek. The Peter Bakewell Trail leads from a small parking area at the foot of Eucalyptus Lane west to Montecito Creek and follows the east bank of the creek to the beach. The Surfrider Trail leads from the south side of the Southern Pacific railroad tracks near Danielson Rd. to the northwest corner of Hammonds Meadow, located adjacent to the beach. Planned facilities include benches and a shower.

MIRAMAR BEACH: The narrow beach, accessible via a ramp and trail, is shared with the adjacent hotel. Limited on-street parking available.

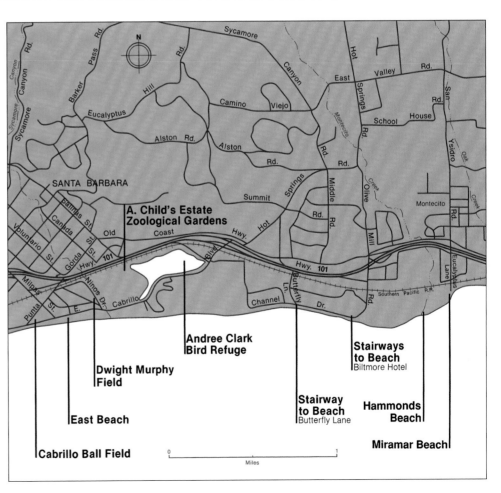

Coastal Crops

artichoke

Moderate weather conditions, fertile coastal valleys and terraces, and a long growing season make California's coastal strip well suited to agriculture. Specialty crops such as artichokes and Brussels sprouts, as well as a number of other food and ornamental crops, may be seen growing in the cultivated fields along the coast.

The artichoke plant is a member of the thistle family; the portion of the plant that is harvested for eating is its unopened flower head. If allowed to mature, the head of the artichoke will develop into an attractive purple thistle flower. The lower portion of each "petal," which is in fact a bract, is edible, as is the fleshy heart, which is found near the flower's base, beneath the inedible "choke." Artichokes are usually prepared for eating by boiling or steaming them until the leaves are tender. Artichoke hearts can be pickled, deep fried, or used to make soup.

The coastal areas of San Mateo and Santa Cruz counties produce the majority of this country's Brussels sprouts. Brussels sprouts are members of the mustard family. The edible sprouts are small cabbage-like heads that grow on a vertical stem beneath the leafy portion of the plant. Raw Brussels sprouts are quite tough, but when cooked make a tasty and very nutritious side dish.

Citrus fruits, avocados, and kiwi fruit are lucrative crops produced in some coastal areas of Santa Barbara and Ventura counties. Since their establishment during California's mission period (1769-1833), lemons and oranges have been a major cash crop in Ventura County; in addition, citrus packing is a major industry for the area. The avocado (also known as the alligator pear) is a greenish, pear-shaped fruit that grows on trees; these trees are members of the laurel family. Avocados have become an increasingly popular food item in the west, and are used most often as an ingredient in salads, sandwiches, and dips. In the 2000-2001 season 213,000 tons of avocados were produced in the state of California.

The kiwi fruit, introduced into California from New Zealand, is an egg-sized, fuzzy, brown-skinned fruit that grows on a vine. The edible portion of the fruit is the green, fleshy interior, which is similar in texture and taste to the strawberry. In 1981 production amounted to 6,000 tons; by 1989, production had increased to 40,000 tons. Kiwi fruit grown in California represents 95 percent of the total U.S. crop. The kiwi fruit is enjoyed by many Californians in pies, fruit salads, beverages, and eaten out of hand.

Other food crops grown in coastal areas include strawberries, raspberries, blackberries, tomatoes, and leafy vegetables such as lettuce, spinach, and cabbage. Several coastal areas also produce ornamental crops such as nursery stock, seeds, and flowers. Lilies are grown on the Smith River coastal plain in Del Norte County, and other flowers are grown in San Mateo County near Half Moon Bay, in Santa Barbara and Ventura counties between Carpinteria and Oxnard, and along the northern San Diego coast.

Brussels sprouts

kiwi fruit

avocados

strawberries

Santa Barbara County

SUMMERLAND / CARPINTERIA

NAME	LOCATION	Entrance/Parking Fee	Parking	Restrooms	Lifeguard	Campground	Showers	Firepits	Stairs to Beach	Path to Beach	Bike Path	Hiking Trail	Facilities for Disabled	Boating Facilities	Fishing	Equestrian Trail	Sandy Beach	Dunes	Rocky Shore	Upland from Beach	Stream Corridor	Bluff	Wetland
Lookout County Park	Lookout Park Rd., Summerland		•	•	•		•		•								•		•			•	
Carpinteria Salt Marsh Nature Park	Ash Ave. and 3rd St., Carpinteria												•										•
Carpinteria City Beach	End of Linden Ave. and Ash Ave., Carpinteria		•						•						•		•						
Carpinteria State Beach	End of Palm Ave., Carpinteria	•	•	•	•	•	•	•	•	•			•		•		•	•		•	•	•	•
Tarpits Park	E. end of campground access road, Carpinteria State Beach, Carpinteria	•	•							•	•	•					•					•	
Santa Monica Creek Trail	Via Real to Foothill Rd., Carpinteria											•								•	•		
Carpinteria Bluffs Nature Park	Carpinteria Ave. at Bailard Ave., Carpinteria		•	•						•	•	•										•	
Rincon Beach County Park	Bates Rd. and Hwy. 101, Carpinteria		•	•					•	•					•		•		•			•	

LOOKOUT COUNTY PARK: Small blufftop park with a grassy picnic area, volleyball court, and children's play area; a paved ramp leads to a sandy beach. Lifeguards service on weekends during the summer. Call: (805) 969-1720.

CARPINTERIA SALT MARSH NATURE PARK: The restored salt marsh area is located at the eastern end of the Carpinteria Marsh. There are interpretive panels, an amphitheater, an overlook, and nature trail at the 15-acre park. The site is served by the Seaside Shuttle. Call: (805) 684-5405.

CARPINTERIA CITY BEACH: One-quarter-mile of narrow, sandy beach. There is limited parking at the ends of Ash, Holly, Elm, and Linden avenues. The beach is served by the Seaside Shuttle.

CARPINTERIA STATE BEACH: Area encompasses 50 acres, including a narrow beach bordered by dunes on the east side and by a bluff on the west. Facilities include a Chumash Indian interpretive display, 101 tent and trailer campsites, 160 motor home sites, a trailer sanitation station, a grassy picnic area, and day-use parking. Some campsites and restrooms are wheelchair accessible. Overnight and day-use fees. Lifeguards on duty during the summer. Known as the "safest beach on the coast" because of the shallow offshore shelf that prevents rip currents. A popular surfing area, called the "tarpits," is located off the south end of the beach. For information, call: (805) 684-2811.

TARPITS PARK: Nine-acre beachfront bluff area with trails and overlooks; Chumash Indian historical site with natural asphaltum seeps. Enter through Carpinteria State Beach.

SANTA MONICA CREEK TRAIL: Hiking trail along the east bank of the creek north of Hwy. 101 between Via Real and Foothill Road. The graded trail follows a flood control easement and parallels the channelized creek.

CARPINTERIA BLUFFS NATURE PARK: The 53-acre bluff top park has trails and a coastal overlook; primarily natural open space with a 6-acre ballfield adjacent to Carpinteria Avenue.

RINCON BEACH COUNTY PARK: A wooden stairway, located near the picnic area and paved parking lot north of Bates Rd., leads down the steep bluff to a sandy beach. Provides access to Rincon Point, one of the most popular surfing areas along the California coast. Additional parking and beach access is available south of Bates Rd. at Rincon Point in Ventura County.

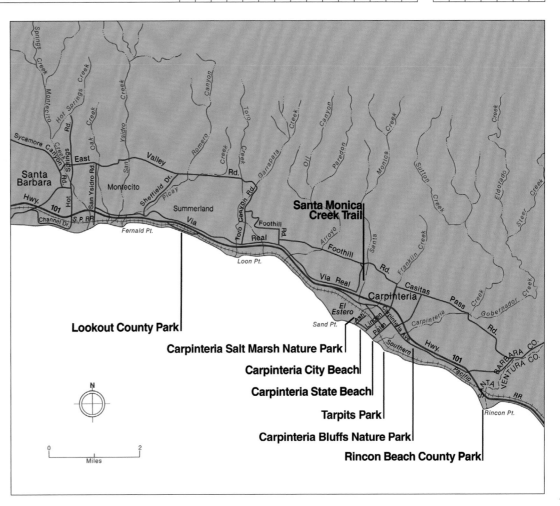

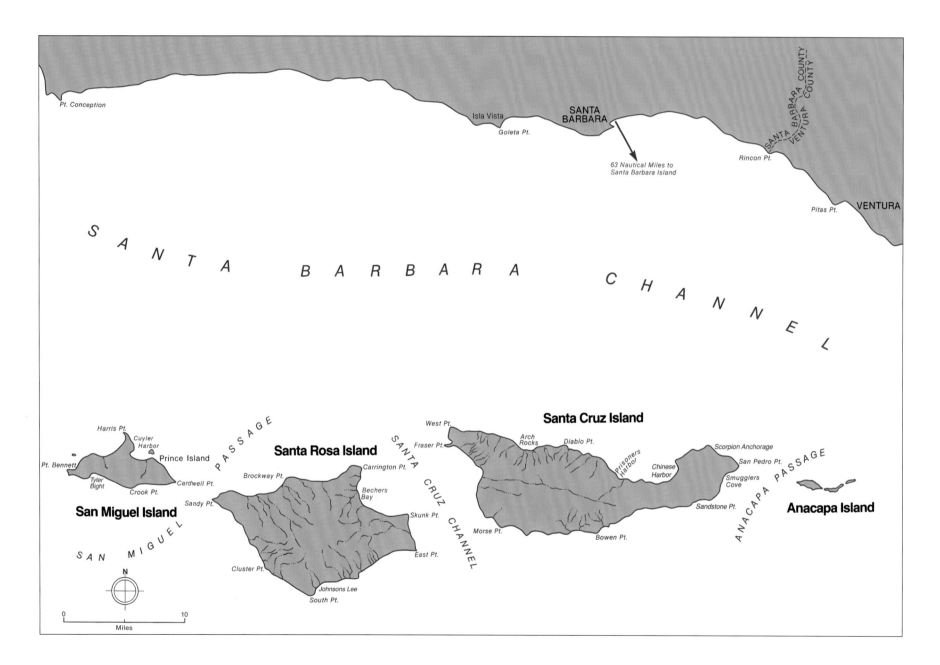

Pt. Conception

Isla Vista

Goleta Pt.

SANTA
BARBARA

SANTA
BARBARA COUNTY
VENTURA COUNTY

*63 Nautical Miles to
Santa Barbara Island*

Rincon Pt.

Pitas Pt.

VENTURA

S A N T A   B A R B A R A   C H A N N E L

Harris Pt.

Cuyler
Harbor

Prince Island

Pt. Bennett

Tyler
Bight

Crook Pt.

Cardwell Pt.

San Miguel Island

S A N   M I G U E L

P A S S A G E

Brockway Pt.

Sandy Pt.

Cluster Pt.

Santa Rosa Island

Carrington Pt.

Bechers
Bay

Skunk Pt.

East Pt.

Johnsons Lee

South Pt.

S A N T A   C R U Z   C H A N N E L

West Pt.

Fraser Pt.

Arch
Rocks

Diablo Pt.

Santa Cruz Island

Prisoners
Harbor

Chinese
Harbor

Morse Pt.

Bowen Pt.

Sandstone Pt.

Scorpion Anchorage

San Pedro Pt.

Smugglers
Cove

A N A C A P A   P A S S A G E

Anacapa Island

N

0   10

Miles

NAME	LOCATION	Entrance/Parking Fee	Parking	Restrooms	Lifeguard	Campground	Showers	Firepits	Stairs to Beach	Path to Beach	Bike Path	Hiking Trail	Facilities for Disabled	Boating Facilities	Fishing	Equestrian Trail	Sandy Beach	Dunes	Rocky Shore	Upland from Beach	Stream Corridor	Bluff	Wetland
San Miguel Island	Offshore, 38 nautical miles S.W. of Santa Barbara		•			•						•		•					•			•	
Santa Rosa Island	Offshore, 28 nautical miles S.W. of Santa Barbara		•			•	•					•	•	•			•		•			•	•
Santa Cruz Island	Offshore, 22 nautical miles S.W. of Santa Barbara		•			•		•				•		•			•		•			•	

The Channel Islands are made up of eight islands off the coast of Southern California. The four northern Channel Islands–San Miguel, Santa Rosa, Santa Cruz, and Anacapa–form a chain that is the southern boundary of the Santa Barbara Channel. The southern Channel Islands are Santa Barbara, Santa Catalina (see Los Angeles County), San Nicolas, and San Clemente.

Channel Islands National Park, established in 1980, is composed of five of the eight Channel Islands–San Miguel, Santa Rosa, Santa Cruz, Anacapa, and Santa Barbara. The Channel Islands National Marine Sanctuary extends six nautical miles around each of the five islands; boating, diving, and fishing are regulated within these waters. Santa Rosa, Anacapa, and Santa Barbara islands are administered and managed by the National Park Service. San Miguel Island is also managed by the National Park Service but is owned by the U.S. Navy. The western 76 percent of Santa Cruz Island is owned by The Nature Conservancy, the eastern 24 percent by the National Park Service, and the island is administered jointly.

For additional information on the Channel Islands , contact Channel Islands National Park Headquarters and Visitor Center, 1901 Spinnaker Dr., Ventura 93001; (805) 658-5730. Island Packers is the park's concessionaire for boat travel departing from Ventura and Channel Island harbors to all the islands. For information, rates, or reservations, write or call Island Packers, Inc., 1867 Spinnaker Dr., Ventura 93001; (805) 642-1393. Reservations should be made well in advance throughout the year. Truth Aquatics in Santa Barbara offers organized tours and multi-day, live-aboard cruises to the northern Channel Islands; for information, call: (805) 963-3564.

SAN MIGUEL ISLAND: The westernmost Channel Island, eight miles long and four miles wide, totals 14 square miles. The island's natural features include a caliche forest, formed by a combination of calcium carbonate and rainwater that calcifies on the roots of trees. Depending on the time of year, California sea lions, Steller sea lions, northern elephant seals, harbor seals, northern fur seals, and Guadalupe fur seals can be seen at Point Bennett, a 15-mile round-trip hike from Cuyler Harbor. Landing is permitted only at Cuyler Harbor. Beach day use at Cuyler Harbor and along the trail to the campground and ranger station does not require a permit; for free backcountry hiking permits, call: (805) 658-5711. Ten-acre Prince Island just outside of Cuyler Harbor is off limits to visitors due to seabird nesting. Camping on San Miguel is by permit only, obtainable by calling 1-800-365-2267. Primitive campsites are one mile from the landing and must be reserved in advance; camping fee. Pit toilets, windbreaks and picnic tables are available; campers need to bring their own stoves and water.

SANTA ROSA ISLAND: The second-largest island in the northern Channel Islands is 15 miles long and 10 miles wide; it was added to Channel Islands National Park in 1986. The island features high mountains with deeply cut canyons, gentle rolling hills, and flat marine terraces. Extensive grasslands cover about 85 percent of the island and support large populations of bird species. A number of rare or endangered plant and animal species and several endemic species live here. The island is surrounded by expanses of kelp beds which provide habitat for numerous marine organisms. There are many undisturbed Chumash Indian archaeological sites on the island. A ranger station is located near the landing and anchorage at Bechers Bay near Carrington Point. Primitive campsites 1.5 miles from the landing have windbreaks, tables, pit toilets, a cold shower, and water; campers need to bring their own stoves. Advance reservations and camping fee are required; call: 1-800-365-2267. Boaters can obtain free permits to camp on the beach; call: (805) 658-5711. Tours and restrooms are available for visitors in wheelchairs. For air travel to Santa Rosa Island, call: (805) 987-1301.

SANTA CRUZ ISLAND: The largest and most diverse of the islands within the national park, Santa Cruz Island measures 96 square miles. At 2,470 feet, the highest of all Channel Islands mountains is found here. The island features large sea caves, beaches, tidepools, hiking trails, and rare or endangered species including the endemic island fox. The western 76 percent of Santa Cruz Island is owned by the Nature Conservancy and is managed as the Santa Cruz Island Preserve. For a permit to land on the island west of the property line between Prisoners Harbor and Valley Anchorage, contact the Santa Cruz Island Preserve, 1901 Spinnaker Dr., Ventura, CA 93001, or call: (949) 263-0933 ext. 306.

No permit is needed to land on the island east of the property line between Prisoners Harbor and Valley Anchorage. Camping facilities include the Del Norte backcountry site on the north side of the island (accessible by hiking 3.5 miles), and a primitive campground for up to 240 visitors at Scorpion Valley with water available. At both sites, a permit and camping fee are required, tables and pit toilets are available, and campers need to bring their own stoves. Call 1-800-365-2267 for advance reservations.

Santa Cruz Island

Cathedral Cove, East Anacapa Island

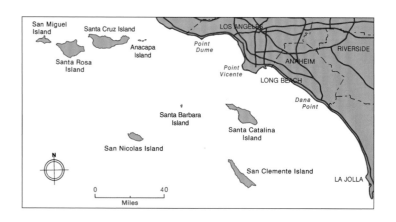

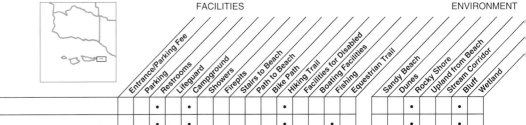

NAME	LOCATION	Entrance/Parking Fee	Parking	Restrooms	Lifeguard	Campground	Showers	Firepits	Stairs to Beach	Path to Beach	Bike Path	Hiking Trail	Facilities for Disabled	Boating Facilities	Fishing	Equestrian Trail	Sandy Beach	Dunes	Rocky Shore	Upland from Beach	Stream Corridor	Bluff	Wetland
Anacapa Island	Offshore, 14 nautical miles S.W. of Ventura		•	•								•							•			•	
Santa Barbara Island	Offshore, 46 nautical miles S. of Ventura		•	•								•			•				•			•	

ANACAPA ISLAND: The closest island to the mainland, 700-acre Anacapa Island is actually three separate islets which are collectively five miles long: West Anacapa, Middle Anacapa, and East Anacapa. The Anacapas are landscaped by rugged cliffs and edged with sea caves. Kayaking and diving are popular here; several submerged shipwrecks still exist. There are guided walks on East Anacapa during the summer, as well as underwater viewing using a video monitor. Sheep were herded on the islets in the early 1900s but removed in 1938 when Anacapa became part of the Channel Islands National Monument, a predecessor to the Channel Islands National Park.

East Anacapa has primitive campsites near the ranger station with tables and outhouses; campers need to bring their own water and stoves. Reservations and fee are required and a permit must be obtained in advance by calling 1-800-365-2267. There is a long stairway and .5-mile walk from the landing cove to the campsites; light packs are recommended. Trail guides for the 1.5-mile self-guided nature trail are available at the visitor center, which has a small museum. The island is noted for Arch Rock, brown pelicans, harbor seals and sea lions, and the springtime blooming of giant coreopsis (tree sunflower). Migrating California gray whales often pass close to Anacapa between January and March. Visitors should note that the high-intensity foghorn at the automated lighthouse on the east island can permanently damage hearing within 100 yards. Beaches on East Anacapa are not accessible, but visitors can swim in the landing cove on calm days.

The slopes of West Anacapa Island are the primary West Coast nesting site for the endangered California brown pelican. To protect the pelican rookery, West Anacapa has been designated a Research Natural Area that is closed to the public; the pelican breeding season is approximately between March 1 and July 31. Except at Frenchy's Cove, no landings are permitted on West or Middle Anacapa without written permission from the park superintendent. The beach at Frenchy's Cove has tidepools and a snorkeling area.

SANTA BARBARA ISLAND: This island is approximately one square mile and lies far south of the four northern Channel Islands. Its steep cliffs rise to a marine terrace that is topped by two peaks, the highest point being Signal Peak at 635 feet. There are 5.5 miles of trails, including the self-guided Canyon View Nature Trail near the ranger station and campground. Camping is allowed only in the campground; reservations and fee are required and a permit must be obtained in advance by calling 1-800-365-2267. Campers must bring their own food, stove, and water; picnic tables and outhouses are available. There is a visitor center with a museum. The island is accessible via a steep climb up the trail from the beach at the northeast landing cove below the ranger's quarters.

In 1846, goats were pastured on Santa Barbara Island. During the 1920s, farming, grazing, intentional burning by island residents, and the introduction of rabbits severely damaged the native vegetation. Today, with the rabbits now removed, the native vegetation is recovering, and there are stands of giant coreopsis, a species of sunflower. Western gulls nest in abundance here. California sea lions frequently haul out along the western rocky shore and elephant seals are sometimes seen on the northwest shore of Webster's Point. Disturbing the marine mammals is prohibited. The island night lizard, a threatened species, inhabits the island.

Island Packers is the park's concessionaire for boat travel departing from Ventura and Channel Island harbors to all the islands. For information, rates, or reservations, write or call Island Packers, Inc., 1867 Spinnaker Dr., Ventura 93001; (805) 642-1393. Reservations should be made well in advance throughout the year.

The southern Channel Islands, San Nicolas and San Clemente, are military reservations and are not open to the public. However, boating and diving are allowed offshore.

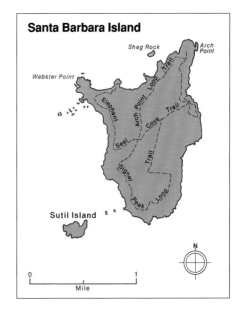

Santa Barbara Island

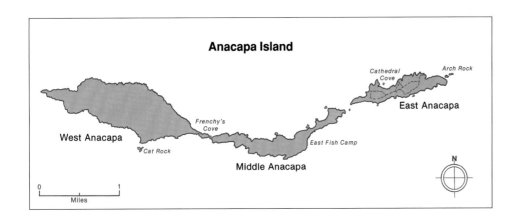

Anacapa Island

San Buenaventura State Beach

Ventura County

Neighboring the large Los Angeles metropolis to the south, Ventura County's 43-mile-long coastline offers numerous sandy beaches. In the north and south county areas, the coastal ranges of the Santa Ynez and Santa Monica Mountains border the shoreline, providing a scenic, rural mountain backdrop to the narrow beaches found here. Dividing these ranges in the central county is the large, flat Oxnard Plain, which contains the cities of Ventura, Oxnard, and Port Hueneme. The Oxnard Plain, the former delta area of the Santa Clara River, is noted for its agricultural fertility and the adjoining long stretches of wide, sandy beach that are characteristic of the county's central coast.

Offshore of Ventura, between 10 and 45 miles away, Channel Islands National Park includes the islands of San Miguel, Santa Rosa, Santa Cruz, Santa Barbara, and Anacapa. These islands are open to the public by permit, and provide excellent areas for hiking, diving, and fishing. Anacapa is the only island actually within Ventura County, while the other four are within Santa Barbara County. The Channel Islands National Park Headquarters is located at Ventura Harbor.

The northern Channel Islands off Ventura and Santa Barbara were formed approximately 14 million years ago by volcanic activity. Isolated from the mainland, many endemic species of flora and fauna have evolved and survived on the islands. Archaeological evidence suggests that one of the larger northern Channel Islands, Santa Rosa, may be one of the earliest sites of human occupation in the Americas.

Portuguese navigator Juan Rodríguez Cabrillo, during his exploration of the California coast for Spain, recorded in 1542 the first sighting of the Channel Islands, then inhabited by the Chumash Indians. Cabrillo died a year later from injuries sustained in a fall, and is believed to be buried on San Miguel Island.

The Chumash of the islands and mainland were decimated by diseases introduced by the European explorers and settlers. In the early 1800s, most of the remaining Chumash were inducted into the Spanish missions in the cities of Santa Barbara and Ventura, and later became local ranch hands. The deserted islands were subsequently inhabited by sea otter and seal hunters, who remained until the mid-1800s, when the islands were used for cattle and sheep ranching. Anacapa and Santa Barbara islands were proclaimed a national monument by Franklin Roosevelt in 1938, and San Miguel Island was acquired by the military and at one time used as a bombing range and missile tracking station. Today, the islands, including Santa Cruz, are all within the Channel Islands National Park boundaries.

The military also owns the Point Mugu Lagoon along the south coast, one of Ventura County's most productive wildlife habitats. Although the lagoon is entirely within the Navy's Pacific Missile Test Center at Point Mugu, a significant portion of the wetlands can be observed from the Point Mugu Wildlife Sanctuary overlook on Highway 1 near Point Mugu Rock. A total of 191 species of birds have been sighted here; approximately 10,000 birds annually winter at the lagoon.

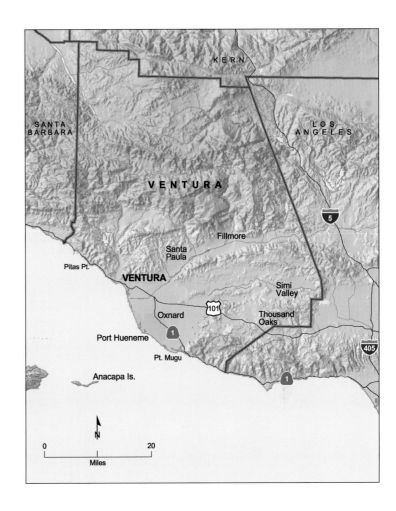

For additional information about Ventura County's coast, contact the Greater Ventura Chamber of Commerce, 785 S. Seaward Avenue, Ventura 93003, (805) 648-2875, or the Oxnard Chamber of Commerce, 400 S. "A" Street, Oxnard 93030, (805) 385-8860 or see www.oxnardchamber.org. For transit information, contact: South Coast Area Transit (SCAT), 301 E. 3rd Street, Oxnard 93032, (805) 643-3158 or 487-4222 or see www.scat.org.

Offshore Oil

California is one of the nation's leading producers of oil. While most of the oil comes from land within the state, approximately 20 percent of it comes from offshore. California ranks fourth in overall oil production among the states, behind Louisiana, Texas and Alaska. Oil has been produced from offshore wells in California since 1896, when over 400 wells were drilled from wooden piers extending into the ocean off the Santa Barbara coast. These wells were approximately 600 feet deep, and produced between one and two barrels of oil each day. Today, the wells in the Santa Barbara Channel are drilled in water up to 900 feet deep, 8,000-9,000 feet beneath the ocean floor, and produce an average of 145,000 barrels a day.

In the 1950s, as the first leases for exploring and drilling on state-owned tidelands were issued, exploration and development of oil proceeded rapidly. In 1954 the first artificial offshore island for oil drilling was built, and in 1957 the first offshore platform, Hazel, was built off Summerland in Santa Barbara County. The federal government began leasing its offshore lands on the Outer Continental Shelf (those lands seaward of three nautical miles offshore) in 1966.

Santa Barbara Channel, Platform Hope

Offshore oil production is concentrated in Southern California. In 1998 about 67.4 million barrels of oil were produced in state and federal waters offshore California. California oil production from all sources, both land and offshore, has continued to decline since the peak year of 1985. The oil reserves are trapped in areas of subsurface irregularities, such as salt plugs or domes, buried reefs, faults, and folds, which create a trap in which oil accumulates. Various sources of data are used to determine where oil may lie; magnetic electrical, seismic, and drilling information all help determine the relative amounts and locations of oil. The number of wells drilled and where the wells are located depends on the estimated boundaries of the field.

Exploratory wells are usually drilled from floating vessels such as drillships that are moored above the site, or anchored semisubmersibles that float above the sea bottom. Exploration of oil offshore continues until enough oil is found to be economically worthwhile to remove, or until a series of dry wells discourages further exploration. Exploratory activity may last several years after a lease sale. If any oil and/or gas discoveries warrant development and the Development and Production Plan (DPP) has been approved by all applicable agencies, the oil company will proceed in constructing offshore platforms and onshore processing facilities.

Offshore oil production facilities include platforms and artificial islands. Each platform contains living quarters for the crew, work decks, a heliport, and a derrick, used to drill the oil wells. The derrick is the tower that stands above the platform and is the most visible portion of the platform from shore; on a clear day, the platforms can be seen 10-11 miles offshore. Artificial islands, such as those off Long Beach, have the same function as platforms; however, the islands have been camouflaged, soundproofed, and landscaped with trees and shrubs so that they resemble offshore high-rise developments.

Once the well is drilled, the flow of oil is controlled by a system of valves and gauges called a "christmas tree." Water, steam, or gas is sometimes injected into the well to maintain the pressure that forces the oil out. The oil is then transferred to shore for processing and refining. In California, oil and/or gas is transported to onshore processing facilities or market destinations either by pipeline or by tanker. Transportation of oil and gas by pipeline is the environmentally preferred method because it minimizes the risk of catastrophic oil spills and release of air pollutants.

Ventura County

NORTHERN VENTURA COAST

NAME	LOCATION	Entrance/Parking Fee	Parking	Restrooms	Lifeguard	Campground	Showers	Firepits	Stairs to Beach	Path to Beach	Bike Path	Hiking Trail	Facilities for Disabled	Boating Facilities	Fishing	Equestrian Trail	Sandy Beach	Dunes	Rocky Shore	Upland from Beach	Stream Corridor	Bluff	Wetland
Rincon Point	Off Hwy. 101, S. of Bates Rd.		•	•					•						•		•		•				
La Conchita Beach	Along Hwy. 101 between Rincon Point and Mussel Shoals														•		•						
Mussel Shoals Beach	W. of old Pacific Coast Hwy., Mussel Shoals		•														•		•				
Seacliff Beach	N. end of old Pacific Coast Hwy., S. of Mussel Shoals		•																				
Hobson County Park	Off old Pacific Coast Hwy., just S. of Seacliff	•	•	•		•	•	•					•		•					•			
Rincon Parkway North	Along old Pacific Coast Hwy., between Hobson and Faria County Parks	•	•	•		•				•					•		•		•				

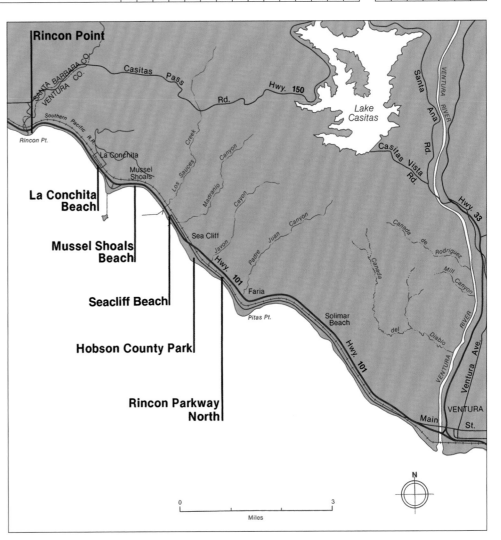

RINCON POINT: Parking lot is south of Rincon Point Parkway and adjacent to the Hwy. 101 southbound entrance ramp. Short dirt pathway to the predominantly cobble beach begins at the south end of the lot; restrooms are at the north end. Heavily used surfing area. Private beach property is adjacent to the north; do not trespass.

LA CONCHITA BEACH: Shoreline access is along the Hwy. 101 shoulder down a riprap revetment extending from the sign indicating the end of the freeway south to Mussel Shoals. Small sandy beach area is just north of Mussel Shoals. No parking or pedestrians allowed along the freeway shoulder to the north. The Mussel Shoals community is private; do not trespass.

MUSSEL SHOALS BEACH: Access to a sandy public beach fronting the Mussel Shoals community is at the north end of Breakers Way and at the west end of Ocean Avenue. Parking on the old Pacific Coast Highway; no parking in residential area. The extent of public and private rights is undetermined and subject to further investigation for the beach above the mean high tide line from the north side of the foot of Ocean Ave. to approximately 160 feet north. A rocky beach with tidepools is just south of the point.

SEACLIFF BEACH: Wide sandy beach and rocky seawall adjacent to the former Mobil Oil piers north of Seacliff. Primarily used for surfing. A small portion of the old pier is now available as a public viewing platform. Main parking area is along the old Pacific Coast Highway east of Hwy. 101; do not block service access to the beach. Pedestrian tunnel to the beach beneath Hwy. 101.

HOBSON COUNTY PARK: Small day-use and overnight camping area next to the ocean. Picnic sites, barbecue pits, concession stand, showers, wheelchair-accessible restrooms, and 31 trailer and tent campsites; overnight fee. The park is periodically closed during heavy storms. For information and reservations, call: (805) 654-3951.

RINCON PARKWAY NORTH: Roadside parking area along the seawall of the old Pacific Coast Highway. Overnight camping permitted in designated areas for self-contained recreational vehicles; overnight fee, 127 sites, no reservations accepted. Picnic tables and pit toilets are available. For information, call: (805) 654-3951.

Seaside Wilderness Park

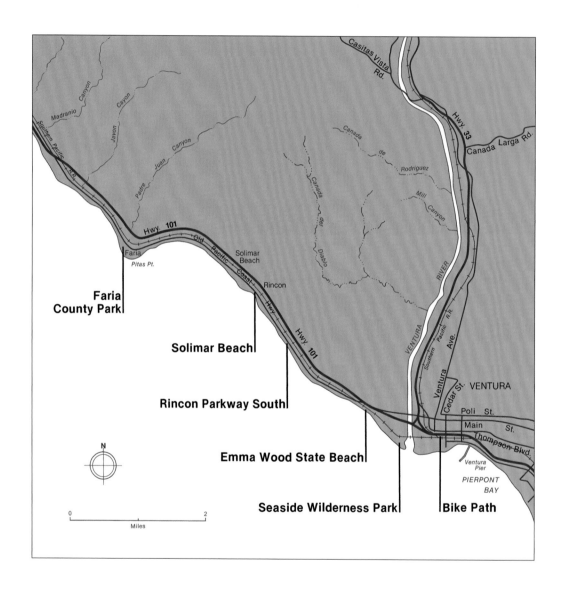

Faria
County Park

Solimar Beach

Rincon Parkway South

Emma Wood State Beach

Seaside Wilderness Park

Bike Path

N

0   2
          Miles

Ventura County

PITAS POINT TO THE CITY OF VENTURA

NAME	LOCATION	Entrance/Parking Fee	Parking	Restrooms	Lifeguard	Campground	Showers	Firepits	Stairs to Beach	Path to Beach	Bike Path	Hiking Trail	Facilities for Disabled	Boating Facilities	Fishing	Equestrian Trail	Sandy Beach	Dunes	Rocky Shore	Upland from Beach	Stream Corridor	Bluff	Wetland
Faria County Park	Off old Pacific Coast Hwy. at Pitas Point	•	•	•		•	•	•	•				•		•		•		•				
Solimar Beach	W. of old Pacific Coast Hwy., Solimar Beach		•							•							•						
Rincon Parkway South	Along old Pacific Coast Hwy., N. and S. of Solimar Beach	•	•			•			•						•		•						
Emma Wood State Beach	S. end of old Pacific Coast Hwy., N. of West Main St., Ventura	•	•	•	•	•	•		•		•				•		•						
Seaside Wilderness Park	W. of Hwy. 101, N. of the Ventura River, Ventura		•														•	•	•				
Bike Path	Along Main St. and the Ventura River, Ventura										•											•	

FARIA COUNTY PARK: Stairs lead to sandy beach and rocky shoreline. There are 40 tent and trailer campsites with picnic tables located adjacent to the ocean; overnight fee. Reservations are recommended and can be made for group sites from September through May. Facilities include firepits, showers, concession stand, and wheelchair-accessible restrooms. Park may be closed during heavy storms. Call: (805) 654-3951.

SOLIMAR BEACH: A public sandy beach fronts the private community of Solimar Beach. The beach is accessible from paths located approximately 450 feet northwest and 75 feet southeast of the community or from Faria County Park or Rincon Parkway. Parking along the Rincon Parkway.

RINCON PARKWAY SOUTH: Roadside parking areas along the seawalls of the old Pacific Coast Highway. Day-use areas are just north of Solimar Beach across from Dulah Rd., and just north of Emma Wood State Beach. Overnight camping for R.V.'s is permitted south of Solimar Beach in 103 designated spaces; overnight fee. No reservations. A stairway south of Solimar Beach provides beach access at low tide. For information, call: (805) 654-3951.

EMMA WOOD STATE BEACH: The Ventura River Group Camp (formerly Emma Wood Group Camp) is 1/4-mile from the southern portion of the beach. There are four 30-person sites with restrooms, camp stoves, picnic tables, cold outdoor showers, and a hiker/bicyclist camp with a two-night limit. Overnight and day-use fees; for reservations, call: 1-800-444-7275. Lifeguards available by request for large groups only. For beach information, call: (805) 899-1400. The northern part of the sandy beach includes a camping area with 61 individual sites and 2 group sites.

SEASIDE WILDERNESS PARK: Only access is from Emma Wood State Beach to the north, along the beach and sand dunes west of the railroad tracks; 3/4-mile walk. Small, undeveloped park of Monterey pines and palm trees adjacent to the mouth of the Ventura River. Good bird-watching area.

BIKE PATH: Path begins east of railroad tracks at southbound Hwy. 101 entrance ramp near Emma Wood State Beach and extends south to Oxnard State Beach. The path runs along Main St. in Ventura, along the east side of the Ventura River, and seaward of the Ventura Fairgrounds passing through Promenade Park and San Buenaventura State Beach.

Faria County Park

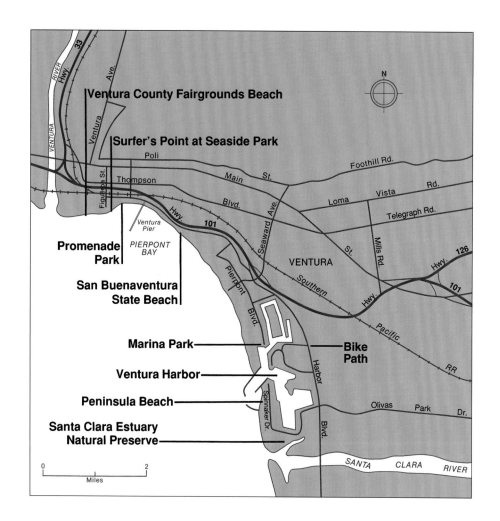

Ventura County Fairgrounds Beach

Surfer's Point at Seaside Park

Promenade Park

San Buenaventura State Beach

Marina Park

Ventura Harbor

Peninsula Beach

Santa Clara Estuary Natural Preserve

Bike Path

Poli

Foothill Rd.

Thompson

Main St.

Blvd.

Loma Vista Rd.

Telegraph Rd.

Ventura Pier

PIERPONT BAY

VENTURA

126

Pacific

Olivas Park Dr.

SANTA CLARA RIVER

0 2
Miles

Ventura County Fairgrounds Beach

Ventura County

CITY OF VENTURA

NAME	LOCATION	Entrance/Parking Fee	Parking	Restrooms	Lifeguard	Campground	Showers	Firepits	Stairs to Beach	Path to Beach	Bike Path	Hiking Trail	Facilities for Disabled	Boating Facilities	Fishing	Equestrian Trail	Sandy Beach	Dunes	Rocky Shore	Upland from Beach	Stream Corridor	Bluff	Wetland
Ventura County Fairgrounds Beach	Between the Ventura River and Surfer's Point at Seaside Park, Ventura									•									•				
Surfer's Point at Seaside Park	Foot of Figueroa St., Ventura		•	•			•			•			•				•						
Promenade Park	From foot of Figueroa St. E. to San Buenaventura State Beach, Ventura		•							•							•						
San Buenaventura State Beach	From the Ventura Pier S. to Marina Park, Ventura	•	•	•	•		•	•		•			•				•						
Marina Park	S. end of Pierpont Blvd., Ventura		•							•	•			•			•	•					
Ventura Harbor	W. of Harbor Blvd. at Anchors Way Dr., Ventura		•	•									•	•	•								
Bike Path	Along Arundell Barranca, Ventura										•											•	
Peninsula Beach	Along the W. end of Spinnaker Dr., Ventura		•	•									•				•						
Santa Clara Estuary Natural Preserve	S. of Spinnaker Dr. at the Santa Clara River, Oxnard		•																			•	•

VENTURA COUNTY FAIRGROUNDS BEACH: The rocky cobble beach is accessible from Surfer's Point downcoast. The Ventura County Fair runs annually for approximately one week at the beginning of October. Other events are held almost every weekend. For information, call: (805) 654-3951.

SURFER'S POINT AT SEASIDE PARK: Offshore of the sandy beach is a popular surfing area. Facilities include parking, outdoor showers, and wheelchair-accessible restrooms.

PROMENADE PARK: Long, narrow park with a concrete walkway and bike path along the edge of the beach. Facilities include benches, tables, playground equipment, volleyball standards, and wide ramps to the beach. Parking at Surfer's Point or at the lot west of the Ventura Pier.

SAN BUENAVENTURA STATE BEACH: Park Headquarters, parking lot, main entrance, and facilities are off San Pedro St. at Pierpont Boulevard. Additional access is at the Ventura Pier area, along Harbor Blvd. between park headquarters and the pier, and at 24 residential street ends between Marina Park and San Pedro Street. Public parking for cars and bicycles is available at the lot on the southeast corner of Seaward Ave. and Zephyr Court.

The wide, sandy beach protected by breakwaters is noted for good swimming; lifeguards on duty daily during the summer and on weekends in spring and fall. Facilities include picnic areas, a snack bar, beach equipment rentals, volleyball standards, outdoor showers, dressing rooms, and wheelchair-accessible restrooms. Park open 7 AM-sunset. 1,700-foot-long Ventura Pier at the northwest end of the park has a restaurant, bait shop, and snack bar, and is wheelchair accessible. Bike path begins at the park entrance and runs north along the edge of the beach. Parking fee at the main entrance. For information, call: (805) 654-4610.

MARINA PARK: Located in the sand dunes of the north peninsula of Ventura Harbor. Facilities include a short bike path, playground equipment, basketball court, sheltered picnic sites, ocean beach access, and a small boat dock. Sailing classes are available year-round from the Ventura City Parks and Recreation Department; call: (805) 652-4550.

VENTURA HARBOR: Extensive inland harbor with two marinas, boat charters, fuel dock, launching ramp (no fee), boat storage and repair, small sailboat rentals (14- and 21-foot), bait shop, fishing equipment, and wheelchair-accessible restrooms. Car and boat trailer parking off Anchors Way Dr. near the boat ramp.

Two public golf courses are east of Harbor Blvd. on Olivas Park Dr., (805) 642-4303 or 642-2231. For whale-watching trips and Channel Islands tours; call Island Packers: (805) 642-1393. For harbor information, call: (805) 642-8538.

The Channel Islands National Park Headquarters and Visitor Center, located at the harbor at 1901 Spinnaker Dr., is open daily from 8:30 AM-5 PM. The Visitor Center provides park information and contains exhibits of the islands' flora and fauna; call (805) 658-5730.

BIKE PATH: Public bikeway along Arundell Barranca from Ventura Harbor, inland to Arundell Ave. then to Main Street.

PENINSULA BEACH: Municipal sandy beach on the south peninsula of Ventura Harbor. The small Marina Cove play area has restrooms and limited paved parking at the north end of Spinnaker Dr.; undeveloped parking area at the south end. The swimming area at Marina Cove is protected by breakwaters.

SANTA CLARA ESTUARY NATURAL PRESERVE: The sensitive marshland wildlife habitat at the mouth of the Santa Clara River is accessible from Peninsula Beach to the north or McGrath State Beach to the south. The wastewater treatment ponds, located north of the river along Spinnaker Dr., provide habitat for migratory waterfowl and are a popular bird-watching area.

Ventura Harbor

Channel Islands Harbor

Ventura County

OXNARD / PORT HUENEME

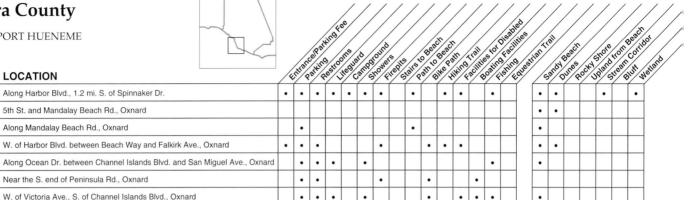

NAME	LOCATION	Entrance/Parking Fee	Parking	Restrooms	Lifeguard	Campground	Showers	Firepits	Stairs to Beach	Path to Beach	Bike Path	Hiking Trail	Facilities for Disabled	Boating Facilities	Fishing	Equestrian Trail	Sandy Beach	Dunes	Rocky Shore	Upland from Beach	Stream Corridor	Bluff	Wetland
McGrath State Beach	Along Harbor Blvd., 1.2 mi. S. of Spinnaker Dr.	•	•	•	•	•	•	•		•		•	•		•		•	•			•		•
Mandalay County Park	5th St. and Mandalay Beach Rd., Oxnard																•	•					
Oxnard Shores	Along Mandalay Beach Rd., Oxnard	•							•								•	•					
Oxnard State Beach	W. of Harbor Blvd. between Beach Way and Falkirk Ave., Oxnard	•	•	•			•			•	•	•					•	•					
Hollywood Beach	Along Ocean Dr. between Channel Islands Blvd. and San Miguel Ave., Oxnard	•	•	•	•		•							•			•						
Peninsula Park	Near the S. end of Peninsula Rd., Oxnard	•	•				•			•			•				•						
Channel Islands Harbor	W. of Victoria Ave., S. of Channel Islands Blvd., Oxnard	•	•	•			•			•	•		•	•			•						
Silver Strand Beach	Along Ocean Dr., San Nicholas Ave. to Sawtelle Ave., Port Hueneme	•	•	•	•										•		•						
Port Hueneme Beach Park	Along W. end of Surfside and Ocean View drives, Port Hueneme	•	•				•			•		•	•		•		•			•	•		
Bubbling Springs Park	E. of Park Ave. and Ventura Rd., Port Hueneme	•					•				•												
Ormond Beach	Foot of Perkins Rd., S. of Port Hueneme	•							•								•	•					•

McGRATH STATE BEACH: The 295-acre state beach stretches along two miles of shoreline south of the Santa Clara River and includes dunes, a sandy beach, a marsh area at the river mouth, and a campground with 174 semi-private sites with tables and stoves. A nature trail begins north of the entrance kiosk. Lifeguards on duty daily during summer and on weekends in spring and fall. Parking and wheelchair-accessible restrooms are available. McGrath Lake wildlife area is at the southern end of the park; public access to the small freshwater lake is along the beach and dunes west of the Chevron Oil Company facilities. Overnight and day-use fees. Call: (805) 654-4610.

MANDALAY COUNTY PARK: 104 acres of undeveloped beach and dunes. Access is from 5th St. to the south. The power plant facility to the north is private; do not trespass.

OXNARD SHORES: A partially developed subdivision located west of Mandalay Beach Rd. and Capri Way between W. Fifth St. and Amalfi Way. Public beach access exists seaward of the residential parcels; there are nine unmarked public accessways, located between residential parcels within the subdivision, that lead from the road to the beach. In addition, there are two public beach areas in the subdivision: at the north end between W. Fifth St. and Channel Way, and near the south end at Neptune Square. A small park with playground equipment and picnic tables is at Neptune Square.

OXNARD STATE BEACH: 62-acre park with a dune trail system and sandy beach. Facilities include a group picnic area (100 persons), athletic field, covered pavilion with barbecue grills, pedestrian/bicycle path, parking, and wheelchair-accessible restrooms. In addition, there are 14 individual/family barbecue units with wind screens, picnic tables, and grills. Call: (805) 385-7950.

HOLLYWOOD BEACH: Access is via street ends along Ocean Drive. Restrooms with outdoor showers and a small parking lot are at the corner of Ocean Dr. and La Brea Street. Lifeguards on duty summer only. Volleyball standards and nets are available.

PENINSULA PARK: The park has a sandy playground area, boat dock, grassy picnic areas with day-use cooking grills, two tennis courts, parking, and restrooms.

CHANNEL ISLANDS HARBOR: Harbor facilities include boat ramps, hoists, berths, fuel dock, and boat trailer parking. Channel Islands Harbor Park is situated along most of the inner perimeter of the harbor; it includes grassy picnic sites, bicycle and pedestrian paths along the waterfront, restrooms with outdoor showers, and a beach area with lifeguards (summer only) at the corner of Anacapa and Victoria avenues. For information: Harbormaster Office, 3900 Pelican Way; (805) 385-8697.

The Channel Islands Harbor Visitor's Center located at 2741 South Victoria Ave. at Fisherman's Wharf is open daily 10 AM-4 PM. For information, call: (805) 985-4852. The Maritime Museum located at 2731 South Victoria Ave. has ship models dating from the 17th century. Open daily 11 AM-5 PM; for information, call: (805) 984-6260.

SILVER STRAND BEACH: The beach is west of Ocean Dr.; main access, restrooms, and parking are at the foot of San Nicholas Avenue. Lifeguards (summer only) and volleyball standards. Enroute camping available. At the south end of the beach, the rock-filled shipwreck of the S.S. *La Jenelle* is now a fishing jetty; access and parking at the end of Sawtelle Avenue. The jetty may be hazardous during heavy surf.

PORT HUENEME BEACH PARK: Broad sandy beach with playground facilities, protected picnic sites with barbecue grills, grassy area, and a snack bar. A pedestrian/bike path runs along the edge of the beach. Parking and wheelchair-accessible restrooms are available. The 1,240-foot-long Port Hueneme Pier, open 24 hours and lighted, has cutting tables with sinks and a bait shop with fishing supplies. For information, call: (805) 986-6555.

BUBBLING SPRINGS PARK: Grassy inland park with picnic tables, firepits, and a playground area. The park includes a 1.5-mile-long bike path that runs along a landscaped drainage channel to Port Hueneme Beach Park. Parking lots are off Park Ave. at the Community Center, and at Ventura Rd. and Bard Road. The bike path passes Moranda Park, which contains tennis courts, a baseball field, and a playground.

ORMOND BEACH: Low dunes with wide sandy beach, most of which is publicly owned. A path leads from the parking lot at the end of Perkins Rd. to the beach. Additional access is off Hueneme Rd. at the end of Arnold Road.

The California Coastal Trail

This trail system is a work in progress, with the goal of providing trail access to and along California's 1,100 mile long coast. When complete, the Coastal Trail will be continuous along or near the coast, linking the Oregon border to the Mexican border. The vision of trail planners is for a system that connects parks, beaches, bicycle routes, hostels and other statewide trail networks. The Coastal Trail will also connect to inland trails. Currently, the planned trail is about 65% complete. Both the state of California and the federal government have identi-fied completion of the Coastal Trail as a high state and national priority.

In addition to shoreline trails, California's coastal mountain ranges provide a variety of opportunities for hiking and backpacking in wilderness areas, many of which are accessible from Highway One. California's mild climate permits hiking and backpacking year-round; however, during the summer, early morning and late afternoon fog are prevalent near the coast, and during the rainy season (November through March), hikers should be prepared for cold, wet weather.

Some of the more popular areas for hiking and backpacking near the coast are Del Norte Coast Redwoods State Park in Del Norte County, King Range Conservation Area in Humboldt County, Point Reyes National Seashore in Marin County, Ventana Wilderness in Monterey County, and the Santa Monica Mountains National Recreation Area in Ventura and Los Angeles counties.

Ventura County

SOUTHERN VENTURA COAST

NAME	LOCATION	Entrance/Parking Fee	Parking	Restrooms	Lifeguard	Campground	Showers	Firepits	Stairs to Beach	Path to Beach	Bike Path	Hiking Trail	Facilities for Disabled	Boating Facilities	Fishing	Equestrian Trail	Sandy Beach	Dunes	Rocky Shore	Upland from Beach	Stream Corridor	Bluff	Wetland
Mugu Lagoon	Pacific Missile Test Center, N. of Point Mugu Rock		•																				•
Point Mugu Beach	At Point Mugu Rock, S. of the Navy Firing Range		•														•	•				•	
La Jolla Valley (Unit of Point Mugu State Park)	E. of Hwy. 1 and Thornhill Broome Beach	•	•	•		•						•								•	•		
Thornhill Broome Beach (Unit of Point Mugu State Park)	W. of Hwy. 1, 1.5 mi. N. of Sycamore Cove	•	•	•	•	•	•	•					•				•						
Sycamore Canyon Campground (Unit of Point Mugu State Park)	E. of Hwy. 1 and Sycamore Cove	•	•	•		•	•	•												•	•		
Sycamore Cove Beach (Unit of Point Mugu State Park)	9000 Pacific Coast Hwy.	•	•	•	•		•	•					•				•						
County Line Beach	W. of Hwy. 1 at Yerba Buena Rd., Malibu		•	•					•	•					•		•		•			•	
Staircase Beach	40000 Pacific Coast Hwy., Malibu		•						•								•					•	

MUGU LAGOON: The 1,800-acre saltwater lagoon, located within the Navy's Pacific Missile Test Center, is the largest coastal estuary-lagoon between Morro Bay and San Diego County. Wildlife viewing is possible from the small pavilion on Hwy. 1, 1/2-mile north of Point Mugu Rock. Public access within the lagoon is restricted to weekday group interpretive tours, available by reservation only, made at least one month in advance; groups must consist of 15 to 25 persons, adults only. Call the P.M.T.C. Public Affairs Office: (805) 989-1704.

POINT MUGU BEACH: The sandy beach is located below Hwy. 1 between Point Mugu Rock and the Navy Firing Range to the north. There is blufftop off-road parking on the west side of Hwy. 1, both north and south of Point Mugu Rock.

The following are administered by the State Department of Parks and Recreation as part of Point Mugu State Park: La Jolla Valley, Thornhill Broome Beach, Sycamore Canyon Campground, and Sycamore Cove Beach. For information, call: (818) 880-0350.

LA JOLLA VALLEY: A small walk-in campground and group camp are located east of Hwy. 1 in La Jolla Canyon. A trailhead provides access into the 13,000-acre Point Mugu State Park within the Santa Monica Mountains. The La Jolla Valley Natural Preserve is adjacent to this trailhead, and the Boney Mountain Wilderness area is inland; both are popular for hiking. Two-mile walk to 12 campsites; tap water and picnic tables are available. Trailer camping in the parking lot area. No open fires; the area may be closed during periods of high fire danger. The gates to the park close at sunset. Overnight fee. For camping reservations: 1-800-444-7275.

THORNHILL BROOME BEACH: Rocky and sandy beach, formerly called La Jolla Beach, offers 80 primitive unprotected trailer and tent campsites directly on the beach, adjacent to Hwy. 1; overnight fee. Picnic tables, firepits with grills, cold outdoor showers, wheelchair-accessible pit toilets, and sanitary tank disposal; summertime lifeguards. Dogs on leashes only.

SYCAMORE CANYON CAMPGROUND: The campground has a trailhead for Point Mugu State Park, 50 developed drive-in campsites, and a special hiker/bicyclist group campsite; overnight fee. Reservations: 1-800-444-7275. Restrooms, picnic tables, pay showers, and barbecue grills are available.

SYCAMORE COVE BEACH: Headquarters for Point Mugu State Park, this developed park area with a sandy beach contains picnic tables, grassy areas, cooking grills, and restrooms. Wheelchair-accessible facilities include a wooden walkway that leads out to the beach, outdoor showers, and chemical toilets. Day-use fee; summertime lifeguards. There are several signed coastal access points with narrow sandy beaches and blufftop pull-outs to the south along Hwy. 1, including one at Deer Creek Rd. with stairway beach access.

COUNTY LINE BEACH: Undeveloped beach area, also known as Yerba Buena Beach; popular surfing area. Chemical toilets and a blufftop dirt parking area are available.

STAIRCASE BEACH: The beach is the northernmost undeveloped portion of Leo Carrillo State Beach. It is accessible from a path at the public dirt parking lot of the state park rangers' residence at 40000 Pacific Coast Highway, and at .2 miles and .5 miles south; additional access is from the North Beach Campground of Leo Carrillo State Beach to the south in Los Angeles County. No open fires allowed. Call: (818) 880-0350.

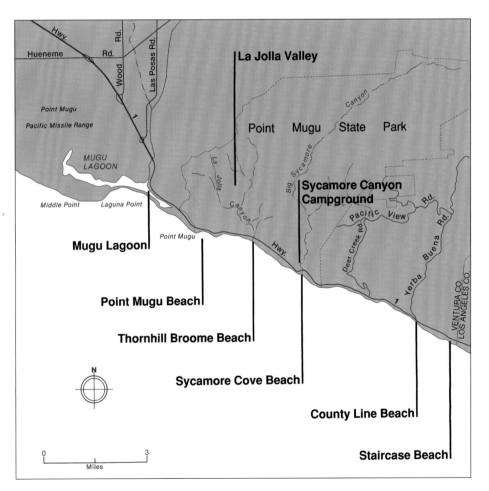

Marina del Rey

Los Angeles County

I n the book *Two Years Before the Mast*, written in 1840, author Richard Henry Dana wrote of Los Angeles: "In the hands of an enterprising people, what a country this might be." Since that time, "enterprising people" have transformed Los Angeles County from an arid coastal basin inhabited by Chumash and Gabrielino Indians into one of the country's largest metropolitan regions, with industries ranging from motion pictures and television to oil refining and aerospace systems. Throughout this transformation, the county's 74 miles of coastline have been a significant attraction and source of recreation; today the coast is visited more than 62 million times a year by tourists and the county's residents.

Los Angeles's Mediterranean climate has always had a major influence on the county's development. Coastal air and water temperatures are warm; summer air temperatures average in the 80s, while water temperature is a comfortable 67 degrees. Night and early morning fog or low clouds are prevalent in many areas, but the overcast usually burns off by mid-morning.

In the summer of 1769, a Spanish missionary named Juan Crespí, one of the first non-natives to visit Los Angeles, described the area as a beautiful and perfect place for a settlement. Descriptions by subsequent visitors and entrepreneurs continued to praise the climate and other virtues of the Los Angeles area. Railroad lines eventually connected east with west and spawned a series of development booms. The beach community of Venice, which was built in 1892 on the sand dunes and marshes at the mouth of Ballona Creek, was one of these boom developments. It was patterned after the Italian city of the same name, and included a series of interconnecting canals complete with gondolas. Today, although many canals are filled in, Venice is a bustling artist's community and popular tourist spot.

The topography of the county's coastline varies dramatically. The Santa Monica Mountains, at the northwest end of the county, drop sharply into the ocean at Malibu, resulting in a series of rocky coves, headlands, points, and sandy pocket beaches. The shore from Malibu to the west is the most rural portion of the coastline. The shoreline of Santa Monica Bay consists of a string of wide, sandy beaches backed by urban development resting on a coastal plain called the Los Angeles Basin. Oil was discovered beneath the basin in 1891, and from the turn of the century to the present, oil derricks, well pumps, and refineries have been a noticeable landscape feature from Venice south.

The Palos Verdes Peninsula, 15 miles of rocky shoreline with tidepools and excellent diving and fishing spots, separates Santa Monica Bay from San Pedro Bay. Catalina Island, located 22 miles offshore, is visible from the south end of the peninsula at Point Fermin. The island is largely a rural area with rugged mountain terrain, and was a hideout for buccaneers during the 17th and 18th centuries; since the resort town of Avalon was developed in 1887, Catalina has become a popular visitor destination.

A little more than half of the county's coastline is in public ownership. The diverse shoreline topography and facilities available to the public allow people to enjoy virtually every type of beach recreation, including sunbathing, swimming, surfing, fishing, and diving. A number of communities along the coast have fishing piers, boat launches, and marinas; Marina del Rey, located just south of Venice along Ballona Creek, is the largest artificial small pleasure-craft harbor in the world.

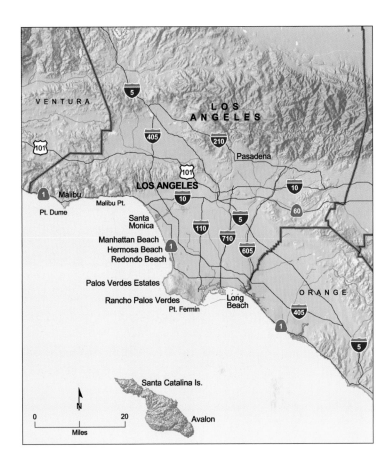

For more information on the Los Angeles County coast, call or write the Chamber of Commerce office in the area you plan to visit, or the Greater Los Angeles Visitors and Convention Bureau, Visitor Information Center, 685 Figueroa Street, Los Angeles 90017, (213) 698-8822. For transit information, call the Southern California Metropolitan Transportation Authority (MTA) (213) 626-4455 or see www.mta.net.

Other transit services within Los Angeles County include Santa Monica Municipal Bus Lines, call: (310) 451-5444 or see www.bigbluebus.com; Torrance Transit System, (310) 618-6266; Long Beach Public Transit, (562) 570-2301 or www.lbtransit.com.

El Pescador State Beach

Leo Carrillo State Beach

El Matador State Beach

Los Angeles County

WESTERN MALIBU

NAME	LOCATION	Entrance/Parking Fee	Parking	Restrooms	Lifeguard	Campground	Showers	Firepits	Stairs to Beach	Path to Beach	Bike Path	Hiking Trail	Facilities for Disabled	Boating Facilities	Fishing	Equestrian Trail	Sandy Beach	Dunes	Rocky Shore	Upland from Beach	Stream Corridor	Bluff	Wetland
Leo Carrillo State Beach	36000 block of Pacific Coast Hwy., Malibu	•	•	•	•	•	•	•	•	•		•	•		•		•		•	•	•	•	
Nicholas Canyon County Beach	Pacific Coast Hwy., about 1 mi. S. of Leo Carrillo, Malibu	•	•	•	•			•	•	•							•					•	
Charmlee County Park	Encinal Canyon Rd., N. of Pacific Coast Hwy., Malibu		•	•								•	•			•				•			
El Pescador State Beach	32900 Pacific Coast Hwy., Malibu	•	•	•	•					•			•				•					•	
La Piedra State Beach	32700 Pacific Coast Hwy., Malibu	•	•	•	•					•			•				•					•	
El Matador State Beach	32350 Pacific Coast Hwy., Malibu	•	•	•	•				•	•			•				•					•	
Lechuza Beach	Broad Beach Rd. at East and West Sea Level Drives, Malibu		•						•	•							•						

LEO CARRILLO STATE BEACH: 3,000-acre park at the west end of Malibu, named after L.A.-born television actor Leo Carrillo. Good surfing, swimming, diving, and camping; nature and hiking trails, tidepools, and rock formations. Day-use and camping fees. The 6,600-foot-long beach is divided into two areas by Sequit Point, which contains sea caves and a natural tunnel. Lifeguards year-round. Migrating gray whales may be seen from the beach November-May.

The park campground has both inland and beach sites. The canyon campground, with 127 sites for tents and vehicles, has wheelchair-accessible restrooms and hot pay showers, and a group campground. The beach campground has 32 sites for self-contained R. V.'s, and provides wheelchair-accessible restrooms and outdoor showers; campground is accessible only to vehicles less than 8' in height. Firepits and grills available for both camp areas. A camp store is open seasonally. Visitor Center open weekends, 10 AM-3 PM. Call: (818) 880-0350.

NICHOLAS CANYON COUNTY BEACH: Across Pacific Coast Hwy. from the Malibu Riding and Tennis Club. The fee parking lot is on the bluff, and has pit toilets and a small paved observation area; a stairway and wheelchair-accessible path lead down to the 23-acre sandy beach, also accessible from Leo Carrillo to the west. Cliffs are highly eroded; surfing and diving at the beach. Lifeguard. Call: (310) 305-9503.

CHARMLEE COUNTY PARK: A 520-acre park in a natural setting with free parking, picnic tables, wheelchair-accessible restrooms, and a view of the ocean. A nature center has live animals and interpretive displays; 15 miles of multi-use trails. Spring wildflower displays; full-moon hikes every month. Open sunrise to sunset. For information, call: (310) 457-7247.

The following are units of Robert H. Meyer Memorial State Beaches, which are administered by the California Department of Parks and Recreation: El Pescador State Beach, La Piedra State Beach, and El Matador State Beach. Steep cliffs; stay on trails. There is private property adjacent to each beach; do not trespass. Lifeguards on duty during summer only. For information, call: (818) 880-0350 or 457-8143.

EL PESCADOR STATE BEACH: Ten acres; fee parking lot, wheelchair-accessible pit toilet, and picnic tables on the bluff; a pedestrian trail, not wheelchair accessible, leads down the bluff to the narrow, sandy beach. Beach open 8:00 AM to sunset.

LA PIEDRA STATE BEACH: Nine acres, with a fee parking lot, picnic tables, a wheelchair-accessible pit toilet, and picnic tables on the bluff; a pedestrian trail, not wheelchair accessible, leads down the bluff to the narrow, sandy beach.

EL MATADOR STATE BEACH: Eighteen acres; facilities include a large, fee parking lot, wheelchair-accessible pit toilet, and picnic area. Beach access is via trails, not wheelchair-accessible, and stairway down the bluff to the narrow, sandy beach. Scenic sea stacks.

LECHUZA BEACH: Parking along Broad Beach Road. Pedestrian gates located at the intersection of Broad Beach Rd. and West Sea Level Dr. and East Sea Level Drive. Walk down to wide sandy beach; gates open to the public dawn to dusk.

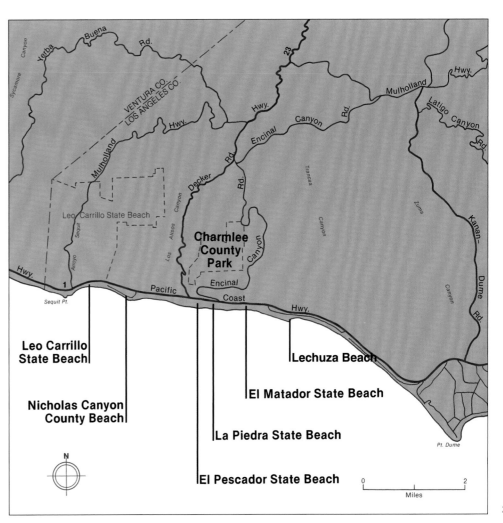

Rip Currents

A nearshore current called a rip current (commonly known as rip tide) is often formed along the coast and presents a potential hazard to the uninformed swimmer. When higher than average waves pass over an offshore sand bar and break on the beach in rapid succession, water is piled up in the surf zone faster than it can recede back to the open ocean. In order to recede, the water breaches the bar in various locations. Thus, rip channels are created as strong currents of water rush through the channels directly seaward. After the volume of water passes beyond the sand bar, the current dissipates.

Rip currents move away from the shoreline faster than most people can swim. Therefore, ocean swimmers should look for rip current warnings and avoid swimming in these areas. Rip currents can often be identified by the sight of converging waves which seem to arrive from several different directions at once. Fortunately, rip currents are confined to narrow channels, so if a swimmer does get caught in the current, he or she can get out of it by swimming parallel to or at a slight diagonal toward the shoreline for a short distance. It is important to note that although a rip current may be strong, it will generally not pull a swimmer far offshore before losing its energy.

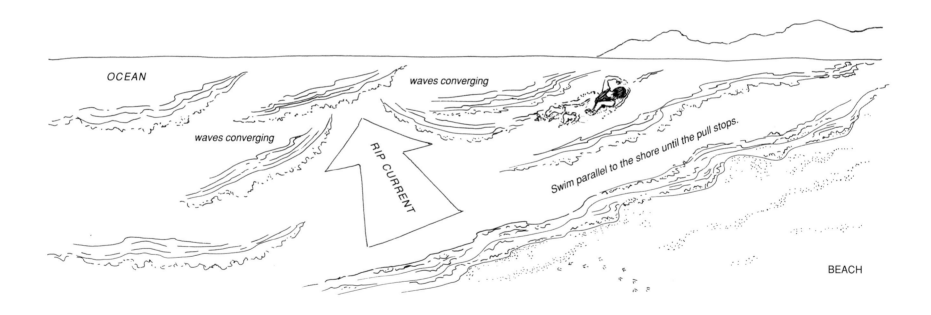

OCEAN

waves converging

waves converging

RIP CURRENT

Swim parallel to the shore until the pull stops.

BEACH

Los Angeles County

POINT DUME / MALIBU

NAME	LOCATION	Entrance/Parking Fee	Parking	Restrooms	Lifeguard	Campground	Showers	Firepits	Stairs to Beach	Path to Beach	Bike Path	Hiking Trail	Facilities for Disabled	Boating Facilities	Fishing	Equestrian Trail	Sandy Beach	Dunes	Rocky Shore	Upland from Beach	Stream Corridor	Bluff	Wetland
Stairway to Beach	31344 Broad Beach Rd., Malibu								•								•						
Walkway and Steps to Beach	31200 Broad Beach Rd., Malibu								•								•						
Zuma Beach County Park	30000 block of Pacific Coast Hwy., Malibu	•	•	•	•		•				•		•				•						
Point Dume Whale Watch	Corner of Westward Beach Rd. and Birdview Ave., Malibu																		•			•	
Point Dume State Beach	S. end of Westward Beach Rd., Malibu	•	•	•	•		•				•		•				•	•				•	
Point Dume State Preserve	Cliffside Drive, Malibu		•						•			•					•		•				

Los Angeles County-operated public accessways and beaches may be recognized by the County-maintained garbage cans located at accessway entrances or along the beaches themselves. Rules and regulations are also posted.

STAIRWAY TO BEACH: A public stairway leads to the beach at 31344 Broad Beach Rd.; open sunrise to sunset. Private property on either side; do not trespass.

WALKWAY AND STEPS TO BEACH: A cement walkway and steps lead to the beach at 31200 Broad Beach Road. Private property on either side; do not trespass. Limited street parking.

ZUMA BEACH COUNTY PARK: Wide, sandy beach on the west side of Point Dume outside Santa Monica Bay; L.A.'s largest county-owned beach. Facilities include swings, outdoor showers, dressing rooms, restrooms, food concessions, sand volleyball courts, and a large parking lot; fee. Rough surf; surfing and diving. Two beach wheelchairs are kept at the beach, call for availability. Zuma Lifeguard Headquarters: (310) 457-2525.

POINT DUME WHALE WATCH: A stairway at Westward Beach Rd. and Birdview Rd. leads up to the whale watch, which has benches and provides a view of the ocean. Stairway begins next to the restaurant.

POINT DUME STATE BEACH: Good swimming, surfing, diving, and tidepools; sandstone cliffs. Beach closes at dark. Fee parking lot, outdoor shower, and restrooms are located at the southeast end of the beach. Access is also available from Zuma Beach to the northwest. Beach is adjacent to the Point Dume Headlands, where there are informal trails.

POINT DUME STATE PRESERVE: Headlands provide outstanding views, particularly during whale migration. Designated parking spaces along Cliffside Drive. Shuttle available from City of Malibu. Trail system preserves the bluff/sand dune habitat; stairway leads down to Little Dume Beach.

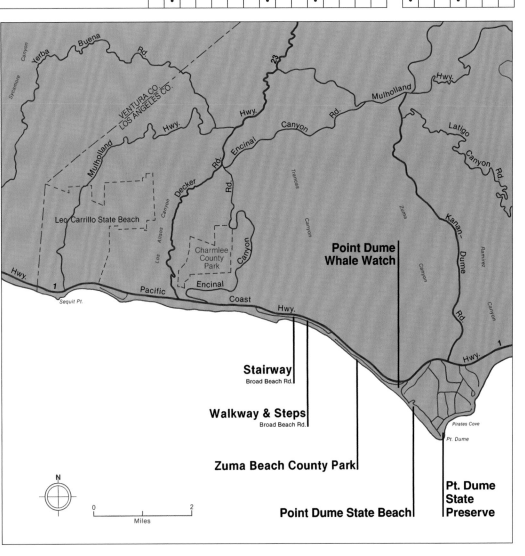

Sport Fishing

There are many methods of fishing the California coast, such as pier fishing, poke pole fishing in tidepools, fishing from shore, which includes surf casting and rock fishing, and trolling and deep sea fishing (bottom fishing) from party boats and skiffs offshore. The type of gear, bait, and method of fishing vary with the species of fish.

Steelhead trout and king salmon, for example, are usually caught by trolling, while rockfish are caught by bottom fishing, surfperch by surf fishing, and monkeyface and rock eels by poke-poling.

Steelhead trout and king salmon are two of the most important Northern California species; striped bass and lingcod are also important. Bonito, yellowtail, and barracuda, which are found off-shore around kelp beds, are significant Southern California fish, as are Pacific mackerel, jack mackerel, albacore, bluefin tuna, and blue shark. Halibut, rockfish, and surfperch, other popular catches, are found along most of the coast.

A fishing license is required if you are 16 years of age or older in order to take any fish, mollusk, invertebrate, amphibian, or crustacean in California, except if you fish from a public pier along the coast. There are fees for resident and non-resident licenses. A fishing license stamp is also required for trout, salmon, and steelhead fishing, and an inland water license stamp is required in applicable circumstances; these stamps must be permanently affixed to the licenses.

Licenses are good for one year; there are also special one-day licenses for non-residents and vacationers. Fishing licenses are available at most bait and tackle shops, sporting goods stores, and diving shops, or from the Department of Fish and Game.

The *California Sport Fishing Regulations*, published by the Department of Fish and Game, includes such information as the type of bait that may be used for some of the common species, the type of gear permitted, seasons and hours when each species may be taken, and the legal limit per day for each species. Marine Protected Areas where sport fishing is not permitted have been designated at various locations along the California coast, in an effort to help fish populations stabilize. In addition, fishing for certain species may be subject to closure at certain times of the year or in certain locations; anglers are advised to keep up on current rules. You can get a copy of the current *Marine Sport Fishing Regulations* or *Fresh Water Sport Fishing Regulations* where you buy your license.

For further information, contact any of the Marine Region offices of the Department of Fish and Game:

20 Lower Ragsdale Dr., Suite 100
Monterey 93940 (831) 649-2870

350 Harbor Blvd.
Belmont 94002 (650) 631-7730

4665 Lampson Ave., Suite C
Los Alamitos 90720 (562) 342-7100

Ocean salmon hotline: (707) 431-4341
www.dfg.ca.gov

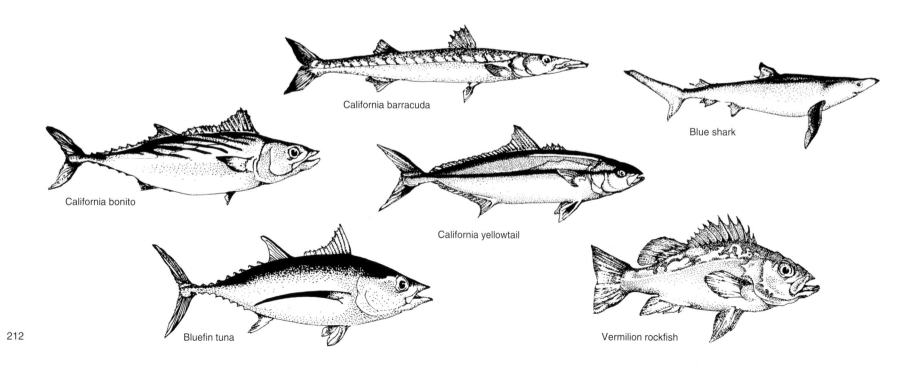

California barracuda

Blue shark

California bonito

California yellowtail

Bluefin tuna

Vermilion rockfish

Los Angeles County

MALIBU

NAME	LOCATION	Entrance/Parking Fee	Parking	Restrooms	Lifeguard	Campground	Showers	Firepits	Stairs to Beach	Path to Beach	Bike Path	Hiking Trail	Facilities for Disabled	Boating Facilities	Fishing	Equestrian Trail	Sandy Beach	Dunes	Rocky Shore	Upland from Beach	Stream Corridor	Bluff	Wetland
Paradise Cove	28128 Pacific Coast Hwy., Malibu	•	•	•									•		•		•					•	
Escondido Beach	27420, 27400, and 27150 Pacific Coast Hwy., Malibu	•							•						•		•			•			
Stairs to Beach	Latigo Shore Dr. and Pacific Coast Hwy., Malibu								•								•		•				
Dan Blocker County Beach	26000 block of Pacific Coast Hwy., Malibu		•	•		•									•		•						
Malibu Beach R.V. Park	25801 Pacific Coast Hwy., Malibu	•	•	•		•	•	•					•							•			
Stairways to Beach	Along Malibu Rd., Malibu								•								•						
Malibu Bluffs State Park	Malibu Canyon Rd. and Pacific Coast Hwy., Malibu		•	•					•	•			•							•		•	

PARADISE COVE: Private fee beach and fishing pier; fee parking. Open sunrise to sunset. No surfing or surf fishing. Wheelchair-accessible restrooms, full-service restaurant, and summer snack bar. Call: (310) 457-2511.

ESCONDIDO BEACH: There are three stairways on Pacific Coast Hwy. between Paradise Cove and Malibu Cove Colony Dr. that lead to Escondido Beach; look for the County- maintained garbage can and the Caltrans Coastal Access sign. Diving area. Street parking on highway shoulder; two off-highway spaces. For information, call the Mountains Recreation Conservation Authority: (310) 456-7049.

STAIRS TO BEACH: Latigo Shore Drive stairway, near the intersection with Pacific Coast Highway, provides access to Latigo Beach. Good surfing and tidepooling.

DAN BLOCKER COUNTY BEACH: Also known as Solstice Beach. Narrow, sandy beach with .7 mile of ocean frontage; lifeguard at peak season; diving and surfing. Outdoor shower at lifeguard station. Limited roadside parking only. For information, call: (310) 457-2525.

MALIBU BEACH R.V. PARK: 120 trailer sites with hookups, and 50 tent sites. Provides coin-operated laundry facilities, propane, picnic tables, barbecue grills, showers, video game room, spa, and wheelchair-accessible restrooms. Overnight fee. Call: (310) 456-6052.

STAIRWAYS TO BEACH: Public stairways at 25118, 24714, 24602, 24434, and 24318 Malibu Rd. lead to the beach; stairways are marked by County-maintained garbage cans. There is limited off-street parking near the 25118 stairway. Private property adjoins each accessway; do not trespass. Stairways close at dusk.

MALIBU BLUFFS STATE PARK: An upland park between Pacific Coast Hwy. and Malibu Rd.; a 1.5-mile-long dirt path leads from the park to the beach stairways on Malibu Road. Facilities include wheelchair-accessible restrooms, free parking, grassy picnic area, playing field, and a paved pathway. Enter from Malibu Canyon Rd., directly opposite Pepperdine University. For information, call: (310) 317-1364.

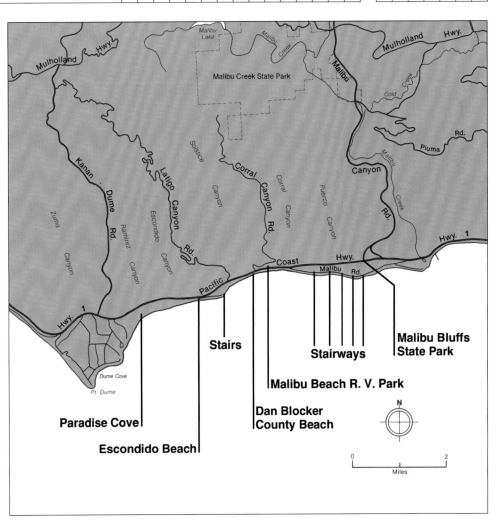

Paradise Cove

Escondido Beach

Stairs

Dan Blocker County Beach

Malibu Beach R. V. Park

Stairways

Malibu Bluffs State Park

Wildfires

Numerous wildfires occur annually throughout California. Most of these fires are extinguished quickly, before causing much damage. However, almost every year there are uncontrollable wildfires that burn extensive areas, resulting in millions of dollars in damages to homes, property, and natural resources. Many of the largest wildfires occur in Southern California, where there are long arid summers, seasonal dry winds, and extensive areas of dry brushland and rugged terrain.

The most critical periods of wildfire danger occur when vegetation is dry, temperatures are hot, and relative humidities are low. In California, which has a Mediterranean climate, these conditions occur every summer and fall, primarily because of seasonal weather patterns that produce very little rainfall from May to November. Drought conditions, a regular part of California's weather system, exacerbate the critical periods.

Wildfire conditions become extremely hazardous in the months following the hot, dry summer when strong winds blow from the east for several days at a time. These winds, called foehn winds (locally known in Southern California as Santa Ana winds), occur when a strong high-pressure system forms in the Great Basin with the development of a corresponding low-pressure system off the Southern California coast. The low-pressure system draws desert air over the Southern California mountains, creating hot dry winds with speeds of up to 60 to 90 miles per hour. Under these conditions, wildfires are extremely difficult to contain; most of the devastating fires in Southern California burn during periods of Santa Ana winds.

The areas near the coast in Southern California that are most susceptible to large wildfires are the vast coastal sage scrub and chaparral brushlands. Over the past 100,000 years, fires started by lightning or possibly by Native Americans have periodically burned through chaparral areas. Many chaparral species are remarkably well adapted to deal with these recurring fires; shrubs can sprout from burned-over stumps, and some chaparral seeds, which are capable of remaining dormant for long periods of time between fires, are resistant to heat and may in fact even require heat to germinate.

In the early 1900s, the U.S. Forest Service and the California Department of Forestry adopted policies of prohibiting fires on certain designated public and private lands. However, the practice of complete fire exclusion has allowed some brushlands to reach over-maturity, creating conditions where historically unburned areas contain high amounts of very old dead and dry brush. Fires occurring in these areas, especially during periods of Santa Ana winds, are particularly catastrophic.

To prevent the spread of wildfires, the U.S. Forest Service, the California Department of Forestry, and private landowners have established numerous fuel and fire breaks, particularly where wildland areas adjoin communities and other developments. These fuel and fire breaks are wide swaths, sometimes several miles long, where vegetation is partially or completely removed. Other methods of fire prevention include the use of controlled or "prescribed" burns in the early spring or late fall. These carefully planned burns during wet months reduce the amount of flammable vegetation that would normally exist during peak fire hazard seasons.

Although arson accounts for the origin of many California wildfires, most fires on public lands are ignited by lightning or by accidental causes such as downed or shorted power lines, sparks from railroad trains, campfires, the use of machines or firearms, and discarded cigarettes and matches. During periods of high fire danger, some California park lands may be closed, or campfires prohibited. Because wildfires are easiest to extinguish while they are still small, any fires should be reported immediately to local fire prevention authorities.

Los Angeles County

MALIBU PIER TO TOPANGA BEACH

NAME	LOCATION	Entrance/Parking Fee	Parking	Restrooms	Lifeguard	Campground	Showers	Firepits	Stairs to Beach	Path to Beach	Bike Path	Hiking Trail	Facilities for Disabled	Boating Facilities	Fishing	Equestrian Trail	Sandy Beach	Dunes	Rocky Shore	Upland from Beach	Stream Corridor	Bluff	Wetland
		FACILITIES															**ENVIRONMENT**						
Malibu Lagoon State Beach	23200 block of Pacific Coast Hwy., Malibu	•	•	•	•		•			•	•						•			•		•	
Malibu Pier	23000 block of Pacific Coast Hwy., Malibu	•	•										•	•	•								
Zonker Harris Accessway	22670 Pacific Coast Hwy., Malibu									•							•						
Stairway to the Beach	20350 Pacific Coast Hwy., Malibu								•								•						
Stairway to the Beach	20000 Pacific Coast Hwy., Malibu								•								•		•				
Las Tunas State Beach	19400 block of Pacific Coast Hwy., Malibu		•		•										•		•		•			•	
Topanga State Beach	18500 block of Pacific Coast Hwy., Malibu	•	•	•	•		•			•					•		•		•		•	•	

MALIBU LAGOON STATE BEACH: The 167-acre state beach includes Malibu Lagoon, a small brackish lagoon at the mouth of Malibu Creek; the Malibu Lagoon Museum; the Malibu Pier; and Surfrider Beach, a very popular surfing and swimming beach. Beach facilities include restrooms, volleyball, outdoor showers, and lifeguard. Fee parking lots; some free roadside parking. Beach is open sunrise to sunset. Call: (818) 880-0350. Facilities at the lagoon include wheelchair-accessible restrooms and picnic tables, and interpretive trails. The Malibu Lagoon Museum, open Wed.-Sat. 11 AM-3 PM, is located about 300 yards west of the pier at 23200 Pacific Coast Highway. Free tours are available; call: (310) 456-8432.

MALIBU PIER: 700-foot-long wooden pier built in 1903 and reconstructed in 1946 is a unit of Malibu Lagoon State Beach.

ZONKER HARRIS ACCESSWAY: Ten-foot-wide cement walkway leads to the beach; sign on street indicates beach access. Named after Zonker Harris, created by Gary Trudeau as a part of the Doonesbury comic strip. Private property on either side; do not trespass.

STAIRWAYS TO THE BEACH: Public stairways lead to the beach, each marked by County-maintained garbage cans. Private property on either side; do not trespass. Stairways are locked after dark.

LAS TUNAS STATE BEACH: Narrow sandy beach, sometimes rocky, below the bluff. Swimmers and divers should beware of the rusted metal groins in the water that hold the sand. Unpaved parking area. For information, call: (310) 305-9546.

TOPANGA STATE BEACH: Narrow sandy beach, sometimes rocky, below the bluff, with 1.1 miles of ocean frontage; wheelchair-accessible restrooms; beach wheelchairs are available. There is a popular surfing area near Topanga Creek. Fee parking lot. Separate lot at beach level for vehicles carrying wheelchairs. For information, call: (310) 305-9503.

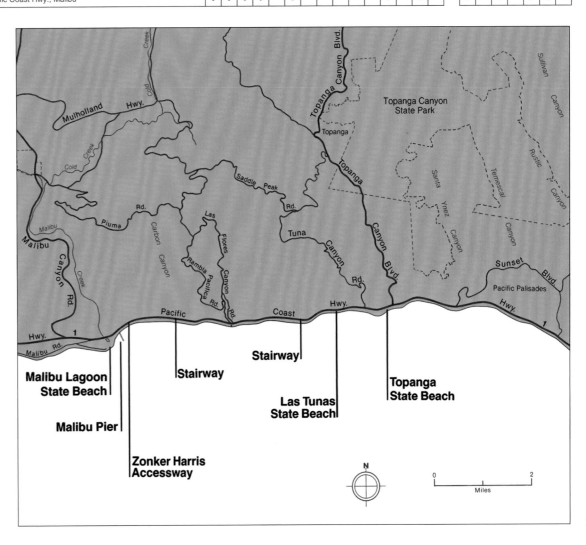

Santa Monica Mountains, Malibu Creek State Park

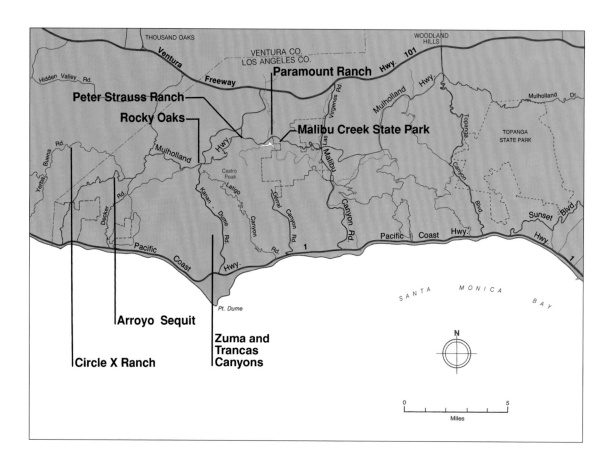

Los Angeles County

SANTA MONICA MOUNTAINS WEST

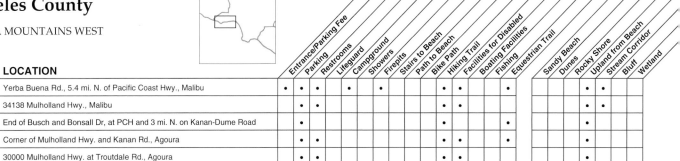

NAME	LOCATION	FACILITIES — Entrance/Parking Fee	Parking	Restrooms	Lifeguard	Campground	Showers	Firepits	Stairs to Beach	Path to Beach	Bike Path	Hiking Trail	Facilities for Disabled	Boating Facilities	Fishing	Equestrian Trail	ENVIRONMENT — Sandy Beach	Dunes	Rocky Shore	Upland from Beach	Stream Corridor	Bluff	Wetland
Circle X Ranch	Yerba Buena Rd., 5.4 mi. N. of Pacific Coast Hwy., Malibu	•	•	•		•		•				•	•			•				•	•		
Arroyo Sequit	34138 Mulholland Hwy., Malibu		•	•								•	•							•	•		
Zuma and Trancas Canyons	End of Busch and Bonsall Dr, at PCH and 3 mi. N. on Kanan-Dume Road		•									•			•					•			
Rocky Oaks	Corner of Mulholland Hwy. and Kanan Rd., Agoura		•	•								•	•							•			
Peter Strauss Ranch	30000 Mulholland Hwy. at Troutdale Rd., Agoura		•	•								•	•							•			
Paramount Ranch	Cornell Rd. at Mulholland Hwy., 2.5 mi. S. of Ventura Fwy.		•	•								•	•		•					•			
Malibu Creek State Park	28754 Mulholland Dr., Calabasas	•	•	•		•	•	•				•	•		•					•	•		

The Santa Monica Mountains run from the Oxnard Plain east to the Los Angeles River. The dramatic contrast between the natural and scenic environment of the mountains and the intensely developed urban areas of Los Angeles and the San Fernando Valley makes the Santa Monicas an extremely valuable resource, particularly as a recreational area.

The mountains offer panoramic views of the ocean from the summits, some of which are as high as 3,000 feet, and provide large open spaces; much of the Santa Monica Mountains is still wild and undeveloped. Mulholland Drive and Highway follow the crest of the mountains for nearly 50 miles, from Hollywood to the ocean at Leo Carrillo State Beach, providing a major scenic access route. For general information on parks in the Santa Monica Mountains, call: (818) 597-1036.

The Santa Monica Mountains National Recreation Area (SMMNRA), consisting of 150,000 acres within the mountains and along the coast, is an area of the National Park System established in November 1978 to preserve the mountains' natural, cultural, and scenic resources and to provide recreational and educational opportunities for the public.

Several areas within the SMMNRA are currently open to the public for hiking and picnicking; there are also free ranger-led hikes in some of these areas. The National Park Service hopes to acquire more public land within the SMMNRA and to develop additional recreational facilities. For information, write or call the National Park Service at 401 West Hillcrest Drive, Thousand Oaks, CA 91630, (805) 370-2300 or see www.nps.gov/samo/. To obtain the Park Service's free quarterly calendar of hikes and other events in the mountains, write, Attention: Calendar Coordinator.

One important feature of the SMMNRA is its extensive public trail system, the largest unit of which is the Backbone Trail Corridor, which runs 55 miles along the crest of the mountains from Point Mugu State Park to Will Rogers State Historic Park.

The Santa Monica Mountains Conservancy was established by the California State Legislature in 1980 to acquire land and operate programs for park, recreation, and conservation purposes in the Santa Monica Mountains area. Their headquarters is at 5750 Ramirez Canyon Road, Malibu 90265; for information, call: (310) 589-3200.

The California Department of Parks and Recreation owns and operates several state parks in the mountains. Their Santa Monica Mountain Area Headquarters is at 1925 Las Virgenes Road, Calabasas, 91302. For information, call: (818) 880-0350.

In addition to the park units listed below, several other areas in the Santa Monica Mountains afford hiking and riding opportunities. For information on Red Rock Canyon, Corral Canyon, Malibu Springs, and Diamond X Ranch, call: (805) 370-2300.

CIRCLE X RANCH: 1,655-acre park with ridges, woodlands, streams, and the highest peak in the Santa Monica Mountains, Sandstone Peak (elevation 3,111 feet). Facilities include two backpacking campgrounds with chemical toilets, picnic areas with barbecue pits, a nature center, restrooms, and over 30 miles of hiking/equestrian trails; wheelchair-accessible restroom at ranger station. Call: (805) 370-2300, ext.1702.

ARROYO SEQUIT: 155-acre park, open daily 8 AM-sunset, with scenic streamside trails, picnic tables, nature center, and wheelchair-accessible restrooms. The Educational Building is available for classroom and other educational uses; Santa Monica City College offers classes in nature study. Main entrance is off Mulholland Hwy., six miles north of Pacific Coast Highway. Call: (805) 370-2300.

ZUMA AND TRANCAS CANYONS: Trailheads located at ends of Busch Drive and Bonsall Drive and at Kanan-Dume Road, 3 miles N. of Pacific Coast Highway. Several loop trails traverse riparian habitat, chaparral, and coastal ridges with ocean views. There is a trail to Upper Zuma Falls.

ROCKY OAKS: 198 acres, with hiking, equestrian, and nature trails; facilities include picnic tables, wheelchair-accessible restrooms, and parking. Rangers lead hikes on selected weekends. Open dawn till dusk. Call: (805) 370-2300.

PETER STRAUSS RANCH: Provides hiking, picnicking, nature center, wheelchair-accessible restrooms, and parking. Concerts and plays in outdoor amphitheater; main building serves as classroom, conference room, and art gallery. Park open daily 8 AM-sunset. Rangers and docents lead walks on selected dates. Call: (805) 370-2300.

PARAMOUNT RANCH: Attractions include Old Western Town, used as a movie set, and major public events held in the large meadow. 336 acres, with hiking/equestrian trails, picnic areas, and wheelchair-accessible restrooms; nature trail up Coyote Canyon, and a five-kilometer running path. Rangers and docents lead hikes on selected dates; free parking. Call: (805) 370-2300.

MALIBU CREEK STATE PARK: 5,000 acres; includes Malibu Creek, Century Lake, woodlands, canyons, waterfall areas, and Kaslow, Udell, and Liberty Canyon natural preserves. Migratory waterfowl are seen in winter at freshwater marsh. Park provides hiking and equestrian trails, a self-guided Braille trail, barbecue pits, wheelchair-accessible restrooms, and visitor center with maps, books, and information on natural and cultural history. Campground has 62 sites with solar pay showers; no hook-ups; one car per site only. Group campsite for up to 60 people is available, with wheelchair-accessible restrooms and solar showers. Day-use and camping fees. For camping reservations, call: 1-800-444-7275. Main entrance is off Malibu Canyon Rd., .25 mile south of Mulholland Highway. Call: (818) 880-0367 or 880-0350.

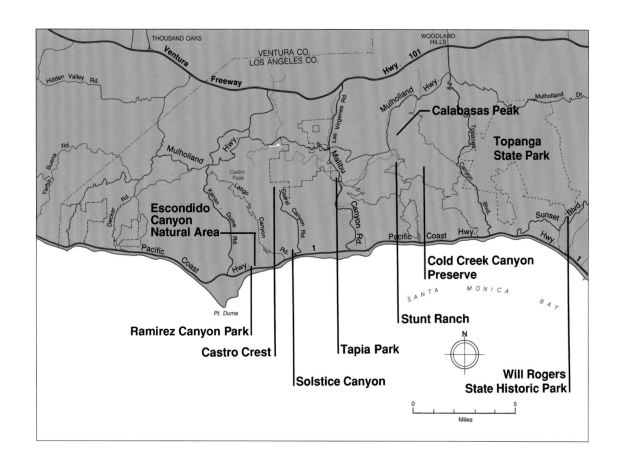

THOUSAND OAKS
VENTURA CO.
LOS ANGELES CO.
WOODLAND HILLS

Ventura
Freeway

Hwy. 101

Hidden Valley Rd.

Mulholland Hwy.

Mulholland Dr.

Calabasas Peak

Yerba Buena Rd.

Mulholland

Castro Peak

Decker Rd.

Kanan Dume Rd.

Latigo

Las Virgenes Rd.

Corral Canyon Rd.

Malibu Canyon Rd.

Topanga Canyon Blvd.

Sunset Blvd.

Topanga State Park

Escondido Canyon Natural Area

Canyon Rd.

Pacific Coast Hwy.

1

Pacific Coast Hwy.

Hwy.

1

Pt. Dume

SANTA   MONICA   BAY

Cold Creek Canyon Preserve

Ramirez Canyon Park

Stunt Ranch

Castro Crest

N

Tapia Park

Solstice Canyon

Will Rogers State Historic Park

0   Miles   5

Santa Monica Mountains

Los Angeles County

SANTA MONICA MOUNTAINS EAST

NAME	LOCATION	Entrance/Parking Fee	Parking	Restrooms	Lifeguard	Campground	Showers	Firepits	Stairs to Beach	Path to Beach	Bike Path	Hiking Trail	Facilities for Disabled	Boating Facilities	Fishing	Equestrian Trail	Sandy Beach	Dunes	Rocky Shore	Upland from Beach	Stream Corridor	Bluff	Wetland
Ramirez Canyon Park	5750 Ramirez Canyon Rd., Malibu		•	•								•								•			
Escondido Canyon Natural Area	Off Pacific Coast Hwy. 1 mi. up Winding Way, Malibu		•	•								•								•			
Castro Crest	Off Pacific Coast Hwy., at the end of Corral Canyon Rd., Malibu	•	•									•				•				•			
Solstice Canyon	3700 Solstice Canyon Rd., off Corral Canyon Rd., Malibu		•	•								•	•			•				•	•		
Tapia Park	Las Virgenes Rd. and Malibu Canyon Rd., Calabasas	•	•	•				•				•	•			•				•	•		
Calabasas Peak	Stunt Rd., off Mulholland Hwy., Calabasas		•	•								•								•			
Stunt Ranch	Stunt Rd., 1.3 mi. S. of Mulholland Hwy., Calabasas		•	•								•								•	•		
Cold Creek Canyon Preserve	Stunt Rd., 1.1 mi. S. of Mulholland Hwy., Calabasas											•				•				•	•		
Topanga State Park	20825 Entrada Rd., off Topanga Canyon Blvd., Topanga	•	•	•		•						•	•			•				•	•		
Will Rogers State Historic Park	14253 Sunset Blvd., Pacific Palisades	•	•	•								•	•			•				•			

RAMIREZ CANYON PARK: Headquarters for the Santa Monica Mountains Conservancy. For information regarding access to hiking trails and other park activities, call: (310) 589-3200.

ESCONDIDO CANYON NATURAL AREA: From parking lot, 4.2 mile-long trail crosses Escondido Creek several times leading to the spectacular 150 foot high Escondido Falls with deep pool at base.

CASTRO CREST: 800-acre park with sweeping views; provides parking, picnicking, and hiking and equestrian trails, including four-mile loop trail and portions of the Backbone Trail Corridor, which is planned to connect all the parks in the Santa Monica Mountains National Recreation Area. Call: (805) 370-2300.

SOLSTICE CANYON: 556 acres, with views, sycamore groves, and year-round stream; open daily 8 AM-6 PM summer, 8 AM-5 PM winter. Provides hiking and equestrian trails, picnic tables, and wheelchair-accessible restrooms. Visitors center provides exhibits, maps, and books. Main entrance off Corral Canyon Rd. just north of Pacific Coast Highway. Call: (805) 370-2300.

TAPIA PARK: A sub-unit of Malibu Creek State Park, located adjacent to Malibu Creek, five miles south of Hwy. 101 on Las Virgenes Road. The 94.5-acre park provides picnic areas with grills, a sports field, hiking and equestrian trails, and wheelchair-accessible restrooms. No bicycles allowed; no dogs on trails. Parking fee. Call: (818) 880-0367.

CALABASAS PEAK: 40 acres of parkland accessible by a trail that starts on fire road off Stunt Rd.; 1.8-mile hike uphill. Regular full-moon hikes and other group events. Parking lot on Stunt Road.

STUNT RANCH: 243 acres are open to the public. The Stunt High Trail crosses through Stunt Ranch, connects to the Backbone Trail and continues up to Saddle Peak. Dirt parking lot at northern end of Stunt High Trail. In addition, 67 acres of the ranch are managed as the UCLA Stunt Ranch Santa Monica Mountains Reserve; access to this area by permit only. For information, call: (310) 206-3887.

COLD CREEK CANYON PRESERVE: 600-acre preserve containing numerous native plant and animal species. Provides nature trail and hiking/equestrian trails; one trail connects to the Backbone Trail. Docents conduct cultural and natural history programs. Access by permit only. Call: (818) 346-9675.

TOPANGA STATE PARK: 13,000 pristine acres with meadows, woodlands, and a small intermittent stream; 35 miles of trails and fire roads, picnic areas, walk-in and equestrian campsites with drinking water, and wheelchair-accessible restrooms near parking areas. No fires allowed; no dogs; bicycles on fire roads only. Panoramic views from Eagle Rock. Day-use fee. Call: (310) 455-2465.

WILL ROGERS STATE HISTORIC PARK: 186-acre park; former ranch of the late Will Rogers, famous cowboy humorist. The main house displays possessions and mementos. The visitor center shows a 12-minute film of Will Rogers' life. Hiking and equestrian trails; bicycles on fire roads only; no fires allowed; no dogs. Restrooms are wheelchair accessible. Park open daily 8 AM—dusk. Parking fee. Call: (310) 454-8212.

Santa Monica Mountains

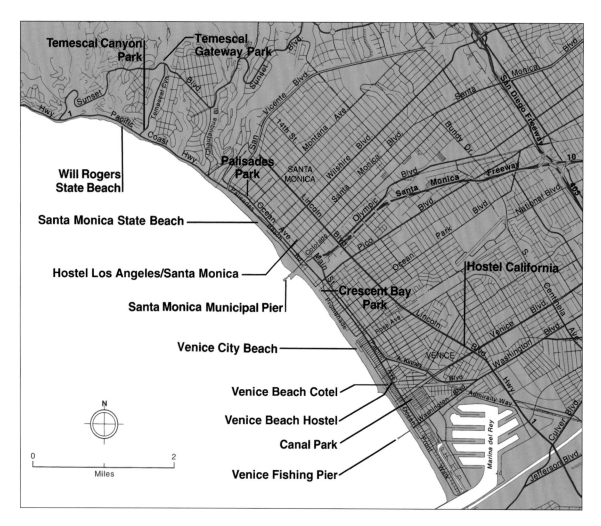

Temescal Canyon Park

Temescal Gateway Park

Will Rogers State Beach

Palisades Park

SANTA MONICA

Santa Monica State Beach

Hostel Los Angeles/Santa Monica

Hostel California

Santa Monica Municipal Pier

Crescent Bay Park

Venice City Beach

VENICE

Venice Beach Cotel

Venice Beach Hostel

Canal Park

Marina del Rey

Venice Fishing Pier

N

0   2
Miles

Los Angeles County

PACIFIC PALISADES / SANTA MONICA / VENICE

NAME	LOCATION	Entrance/Parking Fee	Parking	Restrooms	Lifeguard	Campground	Showers	Firepits	Stairs to Beach	Path to Beach	Bike Path	Hiking Trail	Facilities for Disabled	Boating Facilities	Fishing	Equestrian Trail	Sandy Beach	Dunes	Rocky Shore	Upland from Beach	Stream Corridor	Bluff	Wetland
Will Rogers State Beach	16000 block of Pacific Coast Hwy., Pacific Palisades	•	•	•	•		•			•			•		•		•						
Temescal Canyon Park	Temescal Canyon Rd. and Pacific Coast Hwy., Pacific Palisades		•	•																•			
Temescal Gateway Park	End of Temescal Canyon Rd., N. of Sunset Blvd., Pacific Palisades	•	•	•								•	•							•			
Palisades Park	Ocean Ave., from Colorado Ave. to Adelaide Dr., Santa Monica		•	•								•	•							•		•	
Santa Monica State Beach	W. of Pacific Coast Hwy., Santa Monica	•	•	•	•		•			•			•		•		•						
Hostel Los Angeles/Santa Monica	1436 2nd St., Santa Monica	•		•			•						•								•		
Santa Monica Municipal Pier	Bicknell and Ocean avenues, Santa Monica	•	•	•					•				•		•								
Crescent Bay Park	1800 Promenade, Santa Monica	•	•						•														
Venice City Beach	1531 Ocean Front Walk, Venice	•	•	•	•		•			•			•				•						
Venice Fishing Pier	End of Washington Blvd., Venice	•	•	•	•		•						•		•								
Venice Beach Cotel	25 Windward Ave., Venice	•	•	•			•														•		
Venice Beach Hostel	1515 Pacific Ave., Venice	•	•																		•		
Canal Park	Linnie Canal and Dell Ave., Venice		•																		•		
Hostel California	2221 Lincoln Blvd., Venice	•		•			•														•		

WILL ROGERS STATE BEACH: Sandy beach, several miles long, popular for swimming, diving, and surfing. Provides volleyball courts, food facilities, wheelchair-accessible restrooms, beach wheelchairs, outdoor showers, and lifeguard service. Named for the cowboy humorist Will Rogers. Fee parking. The South Bay Bicycle Trail starts here and extends over twenty miles south to Torrance County Beach. For information, call: (310) 305-9545.

TEMESCAL CANYON PARK: The park, also referred to as Lower Temescal, is located on both sides of Temescal Canyon Rd., and extends one mile up the canyon to Pacific Palisades High School. The park features dirt paths, playground, and restrooms; view of the ocean; picnicking. Free roadside parking.

TEMESCAL GATEWAY PARK: Also referred to as Upper Temescal. Provides trail access into the Santa Monica Mountains National Recreation Area. No bicycles. Provides fee parking and wheelchair-accessible restrooms. Call: (310) 454-1395 x103. The Palisades Malibu YMCA pool is located within the park; call: (310) 454-3514.

PALISADES PARK: The park is located on the bluff above Pacific Coast Hwy. overlooking Santa Monica Beach. Facilities include benches, shuffleboard courts, a walkway, and restrooms. The Senior Recreation Center on Ocean Blvd. is open Mon.-Fri. 9 AM-4 PM and Sat. and Sun. 11 AM-4 PM, and features a Camera Obscura; call: (310) 458-8644. Several bicycle/pedestrian overpasses lead from the park down to the beach. A Visitor Assistance Stand near Arizona St. has free maps, brochures, and tour information.

SANTA MONICA STATE BEACH: Extremely wide, sandy beach below the bluff, divided by the Santa Monica Pier into north and south beaches. Facilities include picnic tables, playground, wheelchair-accessible restrooms, outdoor showers, equipment rentals, and volleyball nets; lifeguard. Popular surfing area. Fee parking. Access to the beach is from the parking lots or via bicycle/pedestrian overpasses from Palisades Park; there is also access from the foot of Idaho Avenue. A promenade extends the length of the south beach and part of the north beach. The South Bay Bicycle Trail runs along the beach. Call: (310) 305-9545. The nearby California Heritage Museum, located at 2612 Main St., has historical exhibits, decorative arts, and a photo archive. Open Wed.-Sun. 11 AM-4 PM, call: (310) 392-3537.

HOSTEL LOS ANGELES / SANTA MONICA: A 280-bed hostel located two blocks from the Santa Monica Pier. Facilities include lounge room, self-service kitchen, library, washer/dryer, locked storage areas, and lockers. Open 24 hours. The hostel is fully wheelchair accessible. Overnight fee. Call: (310) 393-9913.

SANTA MONICA MUNICIPAL PIER: Entrance at Colorado and Ocean avenues. Pier features amusement arcade, giant Ferris wheel, roller coaster, restaurants and food concessions, fishing equipment, shops, and the famous carousel. For information, call: (310) 458-8900. Fee parking in the beach lots. Wheelchair-accessible restrooms. Two stairways, one on the north side and one on the south, and one graded ramp lead from the pier to the beach. The Ocean Discovery Center, located at the pier, is an aquarium and marine science education center; open on weekends; admission fee; for information, call: (310) 393-6149. Carousel Park, just east of the pier, is wheelchair accessible and provides a children's play area and a viewing area.

CRESCENT BAY PARK: Small, grassy park with benches, picnic tables, and gazebo south of the pier with stairs to parking lot below; metered parking on the street or in the beach lot.

VENICE CITY BEACH: Wide, sandy beach stretching to Marina del Rey Harbor entrance channel. Grunion catching in season. Provides outdoor showers, wheelchair-accessible restrooms and a beach wheelchair; a playground and picnic area are located near the Venice Fishing Pier. Fee parking. For information, call: (310) 305-9545. The Ocean Front Walk, adjacent to the beach, has shops, restaurants, street merchants, artists, and performers, and is popular with rollerskaters, bicyclists, and skateboarders. The Muscle Beach area has fitness equipment. The nearby recreation center has basketball, racquetball, picnic areas, and restrooms.

VENICE FISHING PIER: 1,300-foot long pier has fish cleaning stations, restrooms, showers, and fee parking at the base. Entrance at foot of Washington Blvd. Fully accessible; open 6 AM-midnight.

VENICE BEACH COTEL: Located on the Venice Beach boardwalk; sleeps 100 in both shared and private rooms; both private and shared baths available. Most rooms have ocean views. Facilities include common room, lockers, and small parking lot. Open 24 hours; overnight fee. Call: (310) 399-7649.

VENICE BEACH HOSTEL: Shared rooms sleep 40 people in addition to 12 private rooms; facilities include kitchen, laundry room, common room, and storage lockers. Overnight fee; open 24 hours. Call: (310) 452-3052.

CANAL PARK: Small, grassy mini-park with a playground along Linnie Canal.

HOSTEL CALIFORNIA: Open year-round; sleeps 66 in two dorm rooms, also has four private rooms. Facilities include kitchen, laundry, and showers. Overnight fee. 15 minute walk to beach. For information, call: (310) 305-0250.

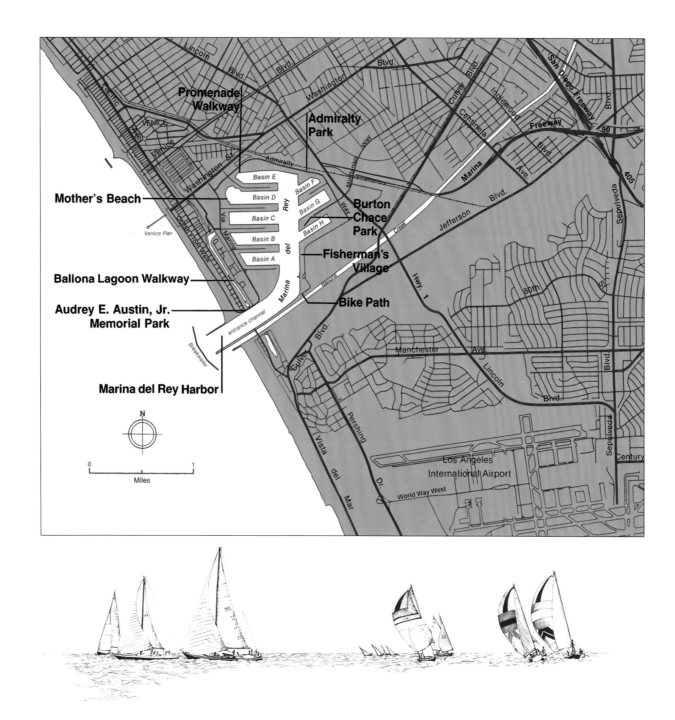

Promenade Walkway

Admiralty Park

Lincoln Blvd.
Washington Blvd.
Culver Blvd.
Centinela
Inglewood
San Diego Freeway
Freeway
90
Blvd.

VENICE
Venice
Washington St.
Admiralty
Marina
Blvd.
405
Sepulveda

Basin E
Basin F
Basin D
Basin G
Basin C
Basin H
Basin B
Basin A

Mother's Beach

Rey
del
Marina

Way
Creek

Burton Chace Park

Jefferson
Blvd.
80th
St.

Venice Pier
Ocean Front Walk
Via Marina

Fisherman's Village

Ballona Lagoon Walkway

Bike Path

Hwy. 1
Ballona

Audrey E. Austin, Jr. Memorial Park

entrance channel

Breakwater

Culver
Blvd.

Manchester Ave.
Lincoln
Blvd.
Blvd.
Century

Marina del Rey Harbor

N

Pershing
Dr.

Vista del Mar

Los Angeles International Airport

World Way West

Sepulveda

0   1
Miles

222

Los Angeles County

MARINA DEL REY

NAME	LOCATION	Entrance/Parking Fee	Parking	Restrooms	Lifeguard	Campground	Showers	Firepits	Stairs to Beach	Path to Beach	Bike Path	Hiking Trail	Facilities for Disabled	Boating Facilities	Fishing	Equestrian Trail	Sandy Beach	Dunes	Rocky Shore	Upland from Beach	Stream Corridor	Bluff	Wetland
Ballona Lagoon Walkway	E. side of Ballona Lagoon									•										•			•
Marina del Rey Harbor	S. of Venice, Marina del Rey	•	•	•			•						•	•	•					•			
Audrey E. Austin, Jr. Memorial Park	S. end Pacific Ave., Marina del Rey	•	•										•							•			
Mother's Beach	End of Basin D, Panay Way, Marina del Rey	•	•	•	•		•	•		•			•	•			•						
Promenade Walkway	In front of Marina City Club, Marina del Rey																			•			
Admiralty Park	Admiralty Way, Marina del Rey	•	•							•										•			
Burton Chace Park	Basin H, W. end of Mindanao Way, Marina del Rey	•	•	•			•	•					•	•	•					•			
Fisherman's Village	Basin H, Fiji Way, Marina del Rey	•	•	•										•	•					•			
Marina del Rey Bike Path	Around perimeter of Harbor, Marina del Rey										•									•			

BALLONA LAGOON WALKWAY: A dirt public walkway along the east side of Ballona Lagoon; can be reached from several street ends in the Silver Strand subdivision, from a pedestrian bridge at Lighthouse St., and from Audrey E. Austin, Jr. Memorial Park. Good bird watching opportunities.

MARINA DEL REY HARBOR: Built in 1957; the world's largest small craft harbor, with more than 5,000 private pleasure craft. Facilities include public boat slips, marine supplies, hoists, fuel docks, pumpout stations, haul-out yards, and charter boats. There are also shops, restaurants, hotels, and six yacht clubs.

A public boat launch ramp at Basin H, Fiji Way, is ten lanes with floating docks; open 24 hours. Fee for parking and use of ramp. Adjacent to the ramp is a public dry-storage boating facility; fee. There are also private facilities for boat rentals and mast-up dry boat storage. Special events include the Christmas Boat Parade, the annual boat show at Burton Chace Park, summer concerts, and the California Cup Race.

More than 1,900 feet of docking space for guest berths with water, electricity, restrooms, and hot showers available at Basin H; report to the Community Building in Burton Chace Park for berth assignment. Advance reservations only for cruising groups. One week limit on stay. Mailing address: L.A. County Dept. of Beaches and Harbors, 13837 Fiji Way, Marina del Rey 90292. The Marina del Rey Visitor Information Center is at Admiralty Way and Mindanao Way: 4701 Admiralty Way, Marina del Rey 90292. For information, call: (310) 305-9545.

There is public access along most of the bulkheads in the harbor. Other accessways are listed below.

AUDREY E. AUSTIN, JR. MEMORIAL PARK: Provides paved walkway with benches and view piers along the north side of the entrance channel of the harbor. Provides access to the south end of Venice Beach via a paved path along the north jetty of the entrance channel, and access to the Ballona Lagoon walkway. Metered parking.

MOTHER'S BEACH: Officially called Marina del Rey Public Swimming Beach, the area is called Mother's Beach because of the gentle swimming conditions, however, always check water quality reports before use. There is a launch for non-motorized craft weighing less than 200 lbs. Beach wheelchairs available. A paved ramp provides wheelchair access into the water; a grab-rail on the side of the ramp offers guidance for chairs or support. Facilities include lifeguard, wheelchair-accessible restrooms, outdoor showers, volleyball nets, sheltered picnic tables, and promenade. Good windsurfing. Access is off Palawan Way.

PROMENADE WALKWAY: Concrete public walkway along the bulkhead; starts near fire station on Admiralty Way and continues around boat basin to Palawan Way. Access also through the Ritz Carlton Hotel on Admiralty Way. Walkway open 6 AM-9 PM.

ADMIRALTY PARK: A linear grassy park along Admiralty Way between Lincoln Blvd. and Washington St.; with par course and mile-long jogging loop. Parking fee. The Marina del Rey Bike Path runs along one edge.

BURTON CHACE PARK: Ten-acre park at the west end of Mindanao Way with a panoramic view of the main channel. Provides transient boat docks, fishing dock, fish-cleaning facility, observation deck, snack bar, and wheelchair-accessible restrooms. There are also barbeque grills and large picnic shelters that can be reserved; call: (310) 305-9595. Annual in-the-water boat show is in early June and annual lighted Christmas boat parade is on the second Saturday in December. Parking at the park is free only on weekdays during the daytime.

FISHERMAN'S VILLAGE: Commercial area modeled after New England seaport, with shops, galleries, snack stands, restaurants; sailboat, powerboat, and sportfishing boat charters and rentals; fishing licenses and bait and tackle; and a view of the harbor. For information, call: (310) 823-5411. Validated fee parking. Harbor tours leave from the Boat House in the lighthouse area of Fisherman's Village, 13727 Fiji Way; call: (310) 577-6660.

MARINA DEL REY BIKE PATH: Bike path along the perimeter of the harbor; connects with the South Bay Bicycle Trail, which extends from Will Rogers State Beach to Torrance County Beach.

Bicycling

California's coast is one of the most scenic areas in the state for bicycling. Throughout the year, particularly in summer, bicyclists tour along the coast, or use bicycles simply to get to the beach.

California has extensive bikeway systems along the coast, ranging from the 1,000-mile-long Pacific Coast Bicycle Route to numerous bike paths (paved paths exclusively for the use of bicycles) and bike lanes (special lanes delineated by a painted line on the road) that provide access to and between local beaches and coastal parks. Bicycle racks are frequently provided at beaches and parks.

The Pacific Coast Bicycle Route runs near the coast from the Oregon border to Mexico, offering spectacular vistas and veering inland only when there is no other suitable roadway. The route ranges from freeway shoulders to exclusive bike paths, while the topography varies from steep hills in the north to flat ocean terraces in the south.

The Pacific Coast Bicycle Route passes numerous historic sites, state parks, and beaches, plus several national parks. Many of these parks provide special camping sites for bicyclists, charging a low overnight fee. There are also low-cost accommodations and bicycle facilities at most of the coastal hostels.

For more detailed information and maps
covering the entire route, request a free catalog from:

Adventure Cycling Association
P. O. Box 8308
Missoula, MT 59807   1-800-755-2453

www.adv-cycling.org

For maps and information on local and regional bikeways, contact local bike shops or visitor's centers.

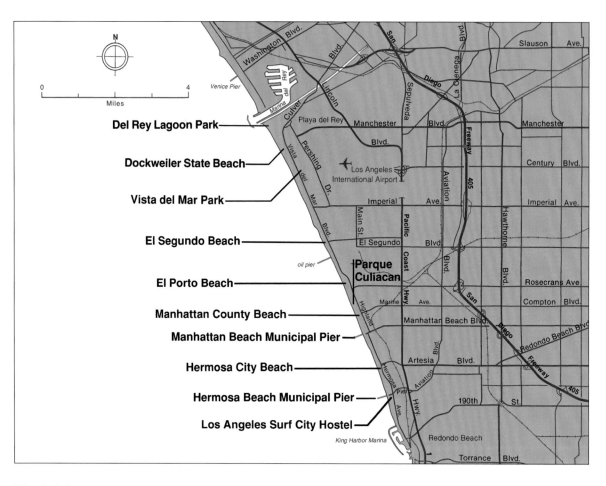

Bicycle Safety:

1. Wear a helmet and protective clothing. A significant number of injuries in bicycle accidents could be reduced or eliminated by wearing a well-fitting cycling helmet. Helmets are widely available at bicycle shops and sporting goods stores. Long sleeves protect from sunburn and minimize abrasions.

2. Ride defensively. Never assume motorists are aware of you; many drivers are so busy watching other cars that they don't even notice bicycles.

3. Obey the same rules and regulations as motorists.

4. Stay as far to the right side of the road as safely possible, except to pass or on one-way streets.

5. Keep an eye out for parked cars with people inside; someone may open a door or the car may pull out into traffic suddenly.

6. Beware of cars turning right across your path; the driver may not know you're there.

7. Let other drivers know what you're doing: use hand signals, and always ride single file.

8. Keep your bicycle in good condition; check tires (pressure and tread), brake shoes, chain, and all cables frequently.

For the complete set of laws pertaining to bicycle operation, refer to section 21200 of the California Vehicle Code.

Los Angeles County

PLAYA DEL REY
MANHATTAN BEACH / HERMOSA BEACH

NAME	LOCATION	Entrance/Parking Fee	Parking	Restrooms	Lifeguard	Campground	Showers	Firepits	Stairs to Beach	Path to Beach	Bike Path	Hiking Trail	Facilities for Disabled	Boating Facilities	Fishing	Equestrian Trail	Sandy Beach	Dunes	Rocky Shore	Upland from Beach	Stream Corridor	Bluff	Wetland
Del Rey Lagoon Park	6660 Esplanade, Playa del Rey		•	•					•				•							•			•
Dockweiler State Beach	Runs from Harbor Channel to Vista del Mar and Grand Avenue, Playa del Rey	•	•	•	•	•	•	•	•		•		•		•		•						
Vista del Mar Park	E. side of Vista del Mar, S. of Sandpiper St., Playa del Rey								•											•			
El Segundo Beach	W. end of Grand Ave., El Segundo	•	•	•	•						•		•				•						
El Porto Beach	W. end of 45th St., Manhattan Beach	•	•	•	•						•		•				•						
Parque Culiacan	26th St. and Manhattan Ave., Manhattan Beach	•	•							•			•									•	
Manhattan County Beach	W. of the Strand, Manhattan Beach	•	•	•	•		•				•		•		•		•						
Manhattan Beach Municipal Pier	Foot of Manhattan Beach Blvd., Manhattan Beach	•	•	•			•						•	•	•								
Hermosa City Beach	W. of the Strand, Hermosa Beach		•	•	•		•				•		•		•		•						
Hermosa Beach Municipal Pier	Foot of Pier Ave., Hermosa Beach	•	•	•									•	•	•								
Los Angeles Surf City Hostel	26 Pier Ave., Hermosa Beach	•		•			•													•			

DEL REY LAGOON PARK: Approximately 13 acres; grassy area around the lagoon. Facilities include restrooms, picnic tables, barbecue grills, playground, baseball diamond, and basketball courts; no swimming. Park is across the street from the northern end of Dockweiler State Beach, and provides access to the south jetty of Marina del Rey Harbor entrance, where there are fishing and observation platforms. Parking on street and along Ballona Creek; fee parking for beach at 62nd Place and Pacific Avenue.

DOCKWEILER STATE BEACH: Extremely wide, sandy beach located just west of Los Angeles International Airport. Main entrance is on Vista del Mar at Imperial Highway. Facilities include a bicycle path, sand volleyball courts, lifeguard service, a snack bar, picnic area, firepits, outdoor showers, beach wheelchairs, and wheelchair-accessible restrooms. Fee parking lots, open 6 AM-10 PM. Paved ramps lead from parking areas to bicycle path. For information, call: (310) 372-2166.

The Dockweiler R.V. campground, located at the south end of the beach, has 118 spaces, wheelchair-accessible restrooms, and a sewage disposal unit; full hookups. Vehicles must check in by 10 PM. Camping fee. Call: (310) 322-4951.

VISTA DEL MAR PARK: Small grassy park across from the beach; provides picnic tables, grills, and playground. On-street parking.

EL SEGUNDO BEACH: Sandy beach at the west end of Grand Ave., with beach volleyball, lifeguard, chemical toilets, and fee parking. The South Bay Bicycle Trail runs adjacent to the beach.

EL PORTO BEACH: Facilities include beach volleyball, bicycle racks, swings, lifeguard service, outdoor shower, beach wheelchair, and wheelchair-accessible restrooms. Metered beach parking. Ramps at 44th and 41st streets lead to the Strand, a cement pedestrian walkway, and to the South Bay Bicycle Trail. Snack bar and equipment rentals are located at the end of 42nd Street.

PARQUE CULIACAN: Three-acre grassy park on the bluff; facilities include a basketball court, benches, wheelchair-accessible restrooms, metered parking, and paved access to the beach from the parking area.

MANHATTAN COUNTY BEACH: Sandy beach popular for swimming and surfing. Provides sand volleyball, a playground, lifeguard service, outdoor shower, beach wheelchairs, and wheelchair-accessible restrooms. Fee parking. The Strand, a cement pedestrian walkway, extends the length of the beach, as does the South Bay Bicycle Trail. For information, call: (310) 372-2166.

MANHATTAN BEACH MUNICIPAL PIER: 900-foot-long pier, open 24 hours. Outdoor shower and wheelchair-accessible restrooms with dressing rooms near pier. Snack bar available. Metered parking nearby; bicycle racks on pier. The Roundhouse Marine Studies Lab, a marine life aquarium, is located at the end of the pier; call: (310) 379-8117.

HERMOSA CITY BEACH: Wide sandy beach, popular for swimming and surfing, with sand volleyball courts and swing sets; wheelchair-accessible restrooms and outdoor showers are located at the base of the Hermosa Beach Pier and along the beach. Lifeguard service; beach wheelchair. Call: (310) 372-2166. The Strand, a pedestrian walkway, and the South Bay Bicycle Trail run the length of the beach. Parking near the pier at the end of 11th Street.

HERMOSA BEACH MUNICIPAL PIER: 900-foot-long pier, open until 11 pm, with restrooms. Pier is wheelchair accessible. For Hermosa Sportfishing, call: (310) 372-2124.

LOS ANGELES SURF CITY HOSTEL: Located on the Strand along Hermosa City Beach near the pier; open 24 hours; sleeps 200. Facilities include kitchen and laundry room. Overnight fee; reservations not required. For information, call: (310) 798-2323 or 1-800-305-2901.

King Harbor

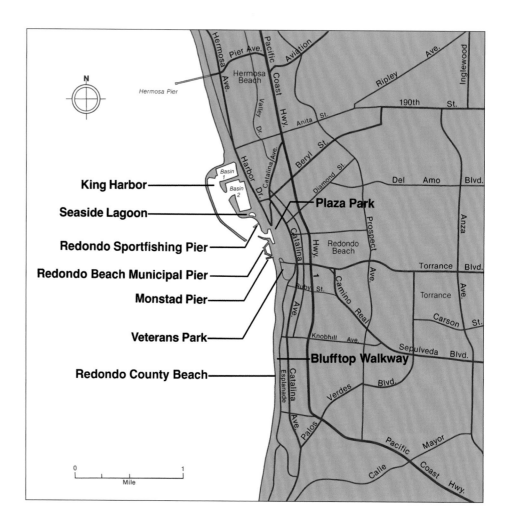

N

Hermosa Pier

King Harbor

Seaside Lagoon

Redondo Sportfishing Pier

Redondo Beach Municipal Pier

Monstad Pier

Veterans Park

Blufftop Walkway

Redondo County Beach

Plaza Park

Hermosa Beach

Basin 1

Basin 2

Redondo Beach

Pier Ave.

Pacific Coast Hwy.

Aviation

Ripley Ave.

Inglewood

190th St.

Anita St.

Beryl St.

Diamond St.

Del Amo Blvd.

Anza

Prospect Ave.

Torrance Blvd.

Catalina Ave.

Camino Real

Torrance Ave.

Carson St.

Ruby St.

Knobhill Ave.

Sepulveda Blvd.

Catalina Ave.

Esplanade

Verdes Blvd.

Palos

Pacific Coast Hwy.

Calle Mayor

Harbor Dr.

Valley Dr.

Hermosa Ave.

0 ——————— 1
Mile

Los Angeles County

KING HARBOR / REDONDO BEACH

NAME	LOCATION	Entrance/Parking Fee	Parking	Restrooms	Lifeguard	Campground	Showers	Firepits	Stairs to Beach	Path to Beach	Bike Path	Hiking Trail	Facilities for Disabled	Boating Facilities	Fishing	Equestrian Trail	Sandy Beach	Dunes	Rocky Shore	Upland from Beach	Stream Corridor	Bluff	Wetland
King Harbor	W. of Harbor Dr., Redondo Beach	•	•	•							•		•	•	•								
Seaside Lagoon	200 Portofino Way, Redondo Beach	•	•	•	•		•	•					•				•						
Redondo Sportfishing Pier	W. of Harbor Dr., N. of Basin #3, Redondo Beach	•	•	•											•								
Plaza Park	Between Harbor Dr. and Catalina Ave., Redondo Beach	•	•						•				•									•	
Redondo Beach Municipal Pier	Foot of Torrance Blvd., Redondo Beach	•	•	•											•								
Monstad Pier	Foot of Torrance Blvd., Redondo Beach	•	•	•									•		•								
Veterans Park	Foot of Torrance Blvd., Redondo Beach	•	•	•									•							•			
Redondo County Beach	W. of Esplanade, Redondo Beach	•	•	•	•		•		•		•		•		•		•						
Blufftop Walkway	W. of Esplanade, Redondo Beach	•	•										•	•								•	

KING HARBOR: Municipal small-craft harbor with three basins, accommodating more than 1,400 boats. For transient berthing call Harbor Patrol: (310) 318-0632. Harbor facilities include slips, marinas, marine supplies, hoists, fuel docks, pumpout station, charter boat rentals, shops, and restaurants. Fee parking. During the summer a small hand-carry boat launching ramp is installed adjacent to Seaside Lagoon. Seasonal whale-watching boat trips available; call: (310) 372-3566.

Basin 3 includes a large commercial-recreational area associated with the Pier Complex, known as the International Boardwalk, which has an amusement arcade. The Pier Complex consists of the Monstad Pier, the Municipal Pier, the International Boardwalk, and a parking structure. Wheelchair-accessible restrooms at entrance of pier and at several locations in the three basins. Public walkways extend around each basin and throughout the harbor and pier area. The South Bay Bicycle Trail runs east of the harbor along Catalina Avenue. For information on King Harbor, call: (310) 376-6926.

SEASIDE LAGOON: Located between Harbor Basins 2 and 3; a 2.5-acre warm saltwater swimming lagoon; picnic tables, volleyball, snack bar, barbeque grills, playground, lawn area, water slides, and diving boards are available. Fee parking; admission fee. Open 10 AM-5:45 PM, Memorial Day-September. Call: (310) 318-0681.

REDONDO SPORTFISHING PIER: 200-foot-long public pier with bait and tackle shop, equipment sales and rentals, sportfishing charters, coffee shop, and snack bar. Fee parking is located near the municipal pier. For information on whale-watching cruises, call: (310) 372-3566. For sportfishing, call: (310) 372-2111.

PLAZA PARK: Also known as Czuleger Park, a small grassy park on a bluff east of Harbor Basin 3 with benches, paved paths, and views; provides access via both stairs and a wheelchair-accessible elevator to pedestrian walkways in King Harbor. A parking structure is located nearby; metered parking on Catalina Avenue.

REDONDO BEACH MUNICIPAL PIER: Also known as Horseshoe Pier, open 24 hours; has commercial facilities such as restaurants and shops. There is a large parking structure located behind the pier at the end of Torrance Blvd.; parking fee.

MONSTAD PIER: This 300-foot-long public pier is one of the best fishing spots in Southern California. Open 24 hours. Pier is wheelchair accessible, with a fishing platform at the seaward end; provides bait and tackle shop, restaurants, and restrooms. Parking fee.

VETERANS PARK: Public park with restrooms, playground, bandshell, picnic area, and paved paths. Also within the park is a senior center and a community center available for rentals. Stairs lead from the park to Redondo Beach. Parking in structure at end of Torrance Boulevard.

REDONDO COUNTY BEACH: Extremely wide, sandy beach with lifeguard, sand volleyball courts, outdoor showers, and wheelchair-accessible restrooms. A walkway and the South Bay Bicycle Trail run the length of the beach. Several overlooks along Esplanade; a number of stairways, walkways, and ramps lead down to the beach from Esplanade. Metered parking along Esplanade. Call: (310) 372-2166.

BLUFFTOP WALKWAY: The walkway runs from the end of Torrance Blvd. south to Knob Hill Ave.; can be reached from Harbor Dr., and from three stairways on Esplanade. At Knob Hill the walkway connects to a sidewalk above the beach that continues south to the Torrance city boundary. Stairs and steep ramps lead from the walkway and sidewalk down to the beach bicycle path. Metered parking.

Pier, Redondo Beach

Point Vicente Fishing Access

Point Vicente Lighthouse

Scenic Overlook, Palos Verdes Estates

Los Angeles County

TORRANCE / PALOS VERDES

NAME	LOCATION	Entrance/Parking Fee	Parking	Restrooms	Lifeguard	Campground	Showers	Firepits	Stairs to Beach	Path to Beach	Bike Path	Hiking Trail	Facilities for Disabled	Boating Facilities	Fishing	Equestrian Trail	Sandy Beach	Dunes	Rocky Shore	Upland from Beach	Stream Corridor	Bluff	Wetland
Torrance County Beach	Along Paseo de la Playa, Torrance	•	•	•	•		•		•	•	•		•		•		•						
Palos Verdes Estates Shoreline Preserve	Entire shoreline of Palos Verdes Estates		•						•						•		•		•	•		•	
Malaga Cove	End of Via Arroyo, off Paseo del Mar, Palos Verdes Estates		•		•					•					•		•					•	
Path to Bluff Cove	At Flat Rock Pt., 600 block of Paseo del Mar, Palos Verdes Estates		•							•					•				•			•	
Overlook at Bluff Cove	1300 block of Paseo del Mar, Palos Verdes Estates		•																			•	
Point Vicente Park	31501 Palos Verdes Dr. West, S. of Hawthorne Blvd., Rancho Palos Verdes		•	•									•									•	
Point Vicente Fishing Access	E. of Point Vicente Park, Rancho Palos Verdes		•	•						•					•				•			•	

TORRANCE COUNTY BEACH: The wide, sandy beach is accessible at the north end by cement ramps leading down from the large fee parking lot on Paseo de la Playa; wheelchair-accessible restrooms, outdoor showers, beach wheelchairs, concession stand, and lifeguard stations. The south end of the beach, a good surfing and diving area, is accessible by walking north from Malaga Cove. Call: (310) 372-2166. The South Bay Bicycle Trail ends at Torrance Beach, having run south for over 20 miles from Will Rogers State Beach. Miramar Park, a grassy park with benches, walkways, and stairs to the beach, is located at Paseo de la Playa and Calle Miramar.

PALOS VERDES ESTATES SHORELINE PRESERVE: Established in 1969; includes 130 acres of city-owned blufftop parkland and submerged lands, running the full length of the city's 4.5-mile-long shoreline. The blufftop parklands are undeveloped and have no facilities; there is ample street parking. Many scenic overlooks and unsigned paths are located along the blufftops. Several footpaths lead down the bluffs to the beach; most are very steep and hazardous. Most of the shoreline is rocky, and there are good tidepools; diving area. Popular surfing spots at Malaga Cove, Bluff Cove, and Lunada Bay. The peninsula shoreline is a marine preserve; no rocks, plants, shells, or animals can be removed.

MALAGA COVE: The primary sandy beach on the Palos Verdes Peninsula; has an outdoor shower, lifeguard, and sand volleyball; a gazebo overlooks the cove. A partially paved road leads down from the parking lot to the beach; several small paths lead from the road to tidepools at the south end. The cove is also known as RAT (Right After Torrance) Beach, and can be reached by foot from Torrance County Beach to the north. A trail in the 500 block of Paseo del Mar at Via Chino leads to the southern end of Malaga Cove, which is a surfing and diving spot known as Haggerty's. For information, call: (310) 378-0383.

PATH TO BLUFF COVE: At Flat Rock Point a broad dirt path leads down to Bluff Cove, a rocky beach with tidepools that is a popular surfing and diving area. Free street parking.

OVERLOOK AT BLUFF COVE: Blufftop parking area that affords views of the Channel Islands. A dirt trail begins here and continues south along the blufftop.

POINT VICENTE PARK: Four-acre blufftop park north of the Point Vicente Lighthouse with parking, a paved blufftop trail, viewing platforms, and picnic tables. The Interpretive Center has wheelchair-accessible restrooms, a gift shop, whale-watching deck, and displays on local history and ecology relating to gray whales; call for hours of operation: (310) 377-5370. The lighthouse was built in 1926 and is open to the public from 10 AM-3 PM on the second Saturday of each month.

POINT VICENTE FISHING ACCESS: A steep dirt path leads down to a rocky beach; popular area for fishing and diving. Blufftop parking lot, where there are restrooms and a paved path that provides views of the shoreline. Lot open dawn till dusk.

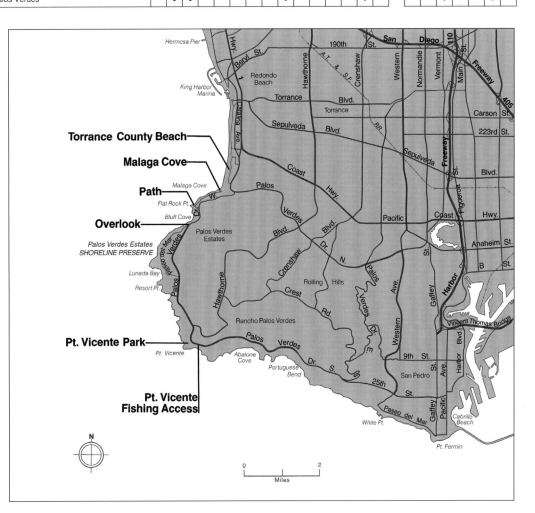

Tidepools

Tidepools are rocky pockets that retain water when the tide recedes; they are found on rocky shores in the intertidal zone, the area between high and low tides. Tidepools are habitat for a myriad of marine life forms that live in the crevices, ledges, and caves of the tidepool, such as barnacles, sea urchins, anemones, sea stars, limpets, mussels, snails, clams, periwinkles, chitons, oysters, scallops, crabs, squid, abalone, shrimp, and other invertebrates, and fish such as gobies and sculpins.

The intertidal marine area is divided into four life zones according to the amount of exposure to air; the ability of each species to live in a particular zone depends upon its relative air tolerance. Other factors include the type of food source, predation by other species, and amounts of required sunlight.

The splash zone, the highest zone, gets the least amount of water and contains such biota as periwinkles and green algae; the upper intertidal zone, which is usually only under water during high tide, supports such organisms as acorn barnacles, limpets, and chitons; the middle intertidal zone, typically covered and uncovered twice a day, contains mussels, gooseneck barnacles, and rockweed; and the low intertidal zone, usually uncovered only by "minus" tides, supports the greatest diversity of life forms including abalone, sea stars, sea urchins, and anemones.

There are four types of food gatherers in tidepools: predators, such as sea stars, sea anemones, and barnacles; filter feeders, such as mussels and clams, which pump water through their filters and digest plankton and bits of plant; algae eaters, such as snails, limpets, and abalone, which feed on kelp and other algae; and scavengers such as crabs, which eat waste food.

Each species in the intertidal zone must be adapted to prevent dehydration and asphyxiation when exposed to air at low tide, and be able to protect itself from wave shock; these adaptations have evolved over millions of years. In the last 50 years people have caused a tremendous amount of change in the environment that these intertidal biota are unable to cope with. Thermal and chemical waste discharges from urban and industrial activities as well as oil spills have severely disrupted many of California's tidepool habitats.

The greatest damage to tidepools is probably from people who trample through them and overturn rocks, exposing the organisms to light, desiccation, and predation. Marine life removed from its natural environment usually cannot survive.

As a result, the Department of Fish and Game has adopted regulations to protect tidal invertebrates and it is now a crime to remove living organisms from tidepools except those species designated in the regulations. A valid California Sport Fishing License is required to take the invertebrates listed, and there is a bag limit of 35 on all organisms for which the take is authorized and for which there is not a bag limit otherwise established.

Tidepool visitors should consult tide tables, which may be found on the weather page of many coastal newspapers, and should keep a close watch on the waves and the tide while observing tidepools; many rocky shore areas along the California coast are hazardous.

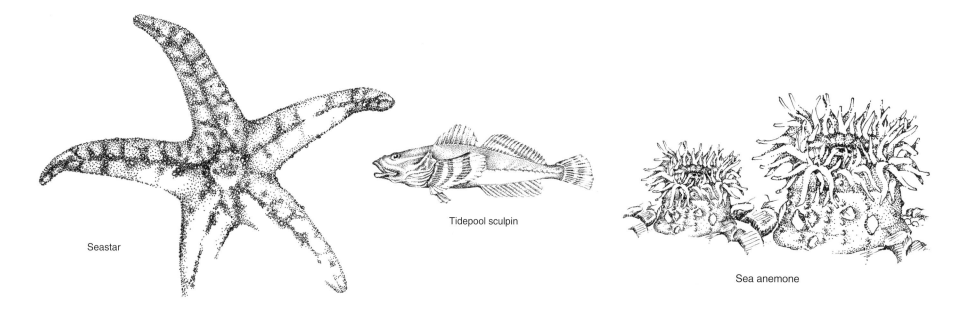

Seastar

Tidepool sculpin

Sea anemone

Los Angeles County

RANCHO PALOS VERDES / POINT FERMIN

NAME	LOCATION	Entrance/Parking Fee	Parking	Restrooms	Lifeguard	Campground	Showers	Firepits	Stairs to Beach	Path to Beach	Bike Path	Hiking Trail	Facilities for Disabled	Boating Facilities	Fishing	Equestrian Trail	Sandy Beach	Dunes	Rocky Shore	Upland from Beach	Stream Corridor	Bluff	Wetland
Long Point	6600 Palos Verdes Dr. South, Rancho Palos Verdes		•							•					•				•			•	
Frank A. Vanderlip, Sr. Park	6500 Seacove Dr., Rancho Palos Verdes																					•	
Abalone Cove Beach	Palos Verdes Dr. South, W. of Narcissa Dr., Rancho Palos Verdes	•	•	•	•		•			•			•		•				•			•	
Ocean Trails	S. of Palos Verdes Dr. South, W. of 25th St., Rancho Palos Verdes		•						•	•	•	•				•						•	
Royal Palms County Beach	Western Ave. and Paseo del Mar, San Pedro	•	•	•	•		•					•	•		•				•			•	
Friendship Community Regional Park	Off Western Ave. and 9th St., San Pedro		•	•				•					•	•						•			
Point Fermin Park and Lighthouse	Paseo del Mar and Gaffey St., San Pedro		•	•				•		•									•	•		•	

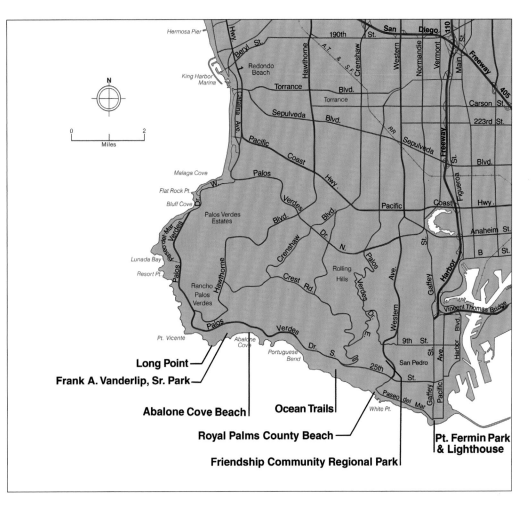

Long Point
Frank A. Vanderlip, Sr. Park
Abalone Cove Beach   Ocean Trails
Royal Palms County Beach
Friendship Community Regional Park
Pt. Fermin Park & Lighthouse

LONG POINT: Allows beach access to site of former Marineland of the Pacific. Free public parking; partially paved pathway leads to rocky shore. Gate is open from one hour before sunrise to one hour after sunset. Good diving.

FRANK A. VANDERLIP, SR. PARK: Located off Seawolf Dr. south of Palos Verdes Drive. Small blufftop park with benches overlooking the ocean; no beach access. Paved pathway leads from street to viewing area at edge of bluff. No facilities. Street parking along Seacove Drive.

ABALONE COVE BEACH: A steep dirt path leads down to the rocky shore; seasonal lifeguard; chemical toilets; cold showers. Area from lifeguard tower east to Inspiration Point is an ecological reserve; all invertebrates are protected and cannot be removed. The beach is a popular surfing and diving spot. Call: (310) 372-2166. Abalone Cove Shoreline Park has views from the blufftop; fee parking lot, picnic area, and restrooms. Wayfarers Chapel, located across Palos Verdes Dr. South from the parking lot, is an unusual and beautiful chapel designed by Lloyd Wright (son of Frank Lloyd Wright) in 1946, made of redwood, stone, and glass with a 50-foot tower.

OCEAN TRAILS: Publicly accessible hiking trails and bikeways with good views lead around golf course, including steep trails to the beach. A nature preserve is along the blufftop. Free parking on side streets and in lots at the ends of Ocean Trails Dr. and La Rotonda Drive. Trails may also be reached from Shoreline Park at the conjunction of W. 25th St. and Palos Verdes Dr. South. Restrooms and some trails are wheelchair-accessible.

ROYAL PALMS COUNTY BEACH: The Royal Palms Hotel was situated here until it was washed out by a storm in the 1920s; majestic palms and garden terraces remain. Popular surfing and diving spot; lifeguards; cold showers. There are steep, rugged cliffs above a rocky shoreline with tidepools. Bluff has a playground, picnic area, wheelchair-accessible nature trail, and some free parking. White Point Nature Preserve is across Paseo del Mar. Fee parking lots on bluff and at the beach are open until sunset; both have restrooms. White Point Beach is directly east. Road and dirt trails lead down from parking lot to both beaches; pedestrian access to White Point Beach also off Weymouth Ave. by steep paved road. For information, call: (310) 372-2166.

FRIENDSHIP COMMUNITY REGIONAL PARK: Facilities include dirt and paved pathways, covered picnic tables, and barbeque grills; views of Los Angeles Harbor. Park is 123 acres and wheelchair accessible. Open 7 AM-dusk. A nature center offers exhibits, classes, and a small gift shop; open 9 AM - 5 PM daily; free admission. Call: (310)-519-6115. The adjacent Bogdanovich Recreation Center includes a playground, baseball diamonds, and restrooms.

POINT FERMIN PARK AND LIGHTHOUSE: 37 landscaped acres on a bluff overlooking the ocean and L.A. Harbor. Facilities include a bandshell, picnic tables, playground, barbecue grills, and restrooms. Migrating whales can be seen offshore. Two steep but maintained trails, one accessible from Barbara St. and the other from Meyler or Roxbury St., lead to the narrow shoreline. Point Fermin Lighthouse, built in 1874, is not open to the public.

Grunion

Of all the thousands of fish that live in the sea, only the grunion lays its eggs on land. Grunion are small, silvery fish, related to topsmelt, that come ashore to spawn from March through August, just after the highest high tide. California grunion are found generally between Morro Bay and Abreojos, Baja California; they live three to four years and reach a length of six inches or more. The periods when the grunion lay their eggs are called grunion runs, since the grunion come ashore in masses.

Spawning occurs on the second, third, and fourth nights following the peak tides of the 14-day lunar cycle, when the tides that follow will be lower than those of the night before. Females begin to spawn each night just after the tide turns, so the eggs will not be washed away. The female drills into the damp sand with her tail until she is buried up to the pectoral fins, which anchor her in place. Then she lays her eggs into the hole she has dug, while the male curls himself around her and releases his milt, which fertilizes the eggs.

The newly laid eggs, buried a few inches below the surface, are buried farther beneath the sand by the action of the tides and waves. Grunion develop within a protective membrane that surrounds the fertilized eggs, and are ready to hatch in about nine days, but must wait until the tide is high enough to reach the eggs. Ten to fourteen days after fertilization the next series of very high tides occurs and an enzyme activated by the agitation of the surf causes the eggs to hatch; the newborn grunion are then carried out to sea by the waves. In about a year the grunion are mature and ready to spawn.

Young females lay about 1,000 eggs at each spawning, while older females can lay up to 3,000 eggs; older females may spawn as many as eight times during the season, thus spawning as many as 24,000 eggs in one season. Spawning runs last about three hours. The name grunion probably comes from the Spanish word grunon, meaning "one who grunts," since female grunion have been heard to make faint sounds like squeaks or grunts just after spawning.

Grunion may be collected by hand in March, June, July, and August by persons having a valid California fishing permit; no permit is required for those under 16. The Cabrillo Marine Museum sponsors a Grunion Run Program several nights a month during the spawning season, which includes a bonfire on the beach and discussion while waiting for the grunion run.

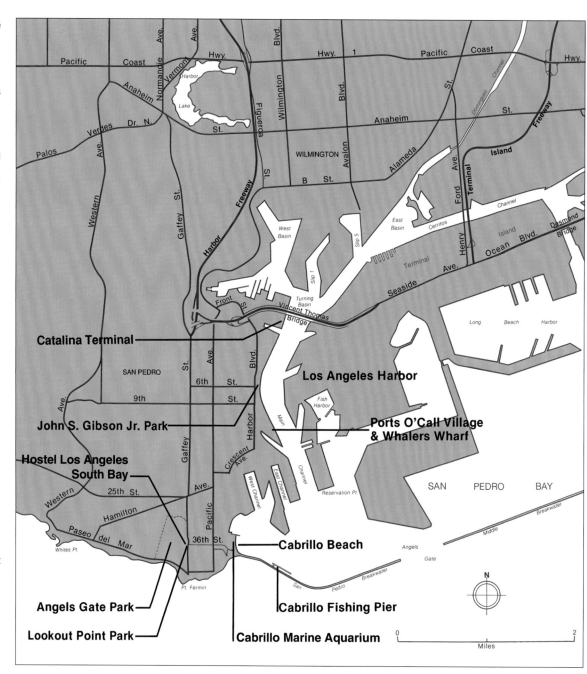

Los Angeles County

SAN PEDRO / LOS ANGELES HARBOR

NAME	LOCATION	Entrance/Parking Fee	Parking	Restrooms	Lifeguard	Campground	Showers	Firepits	Stairs to Beach	Path to Beach	Bike Path	Hiking Trail	Facilities for Disabled	Boating Facilities	Fishing	Equestrian Trail	Sandy Beach	Dunes	Rocky Shore	Upland from Beach	Stream Corridor	Bluff	Wetland
Angels Gate Park	930 Paseo del Mar, San Pedro		•	•									•									•	
Hostel Los Angeles South Bay	3601 South Gaffey St., Bldg. 613, San Pedro	•	•	•			•															•	
Lookout Point Park	E. side Gaffey St., between 34th and 36th streets, San Pedro		•																			•	
Cabrillo Beach	40th St. and Stephen M. White Dr., San Pedro	•	•	•	•		•	•					•	•	•		•		•				
Cabrillo Fishing Pier	E. of Cabrillo Beach, inside breakwater, San Pedro	•	•		•										•								
Cabrillo Marine Aquarium	3720 Stephen M. White Dr., San Pedro	•	•	•									•										
Ports O'Call Village and Whalers Wharf	Berth 77, L.A. Harbor, San Pedro		•	•											•							•	
John S. Gibson, Jr. Park	6th St. and Sampson Way, San Pedro		•										•									•	
Los Angeles Harbor	W. end of San Pedro Bay		•	•										•	•								
Catalina Terminal	Berths 94-95, L.A. Harbor, San Pedro	•	•	•										•	•								

ANGELS GATE PARK: View of the ocean and of Pt. Fermin. Facilities include wheelchair-accessible restrooms, a play area with basketball courts, and a historic site with a pagoda housing the Friendship Bell given by the Republic of Korea in 1976 to commemorate the U.S. Bicentennial and honor Korean War veterans; for park information, call: (310) 548-7705. Also within the park is the Fort MacArthur Military Museum, open Tues., Th., Sat., and Sun. 12-5 PM; (310) 548-2631. The Angels Gate Cultural Center, open Tues.-Sun. 11 AM-4 PM, features exhibits and offers classes; call: (310) 519-0936. The Marine Mammal Care Center at Fort MacArthur treats injured seals and sea lions, which can be viewed by the public daily during daylight hours; call: (310) 548-5677.

HOSTEL LOS ANGELES SOUTH BAY: Located within Angels Gate Park. Provides 60 beds, fully equipped kitchen, laundry facilities, and library; open daily 8 AM-noon and 4-11 PM. Call: (310) 831-8109.

LOOKOUT POINT PARK: Small overlook area with a view of San Pedro Bay; coin-operated telescopes.

CABRILLO BEACH: Stillwater beach inside the San Pedro Breakwater, and surf beach on the ocean; diving, surfing, and windsurfing. Facilities include picnic tables, grills, wheelchair-accessible restrooms, showers, dressing rooms, and playground; fee parking. Beach wheelchairs available from information booth at the Cabrillo Marine Aquarium. Open sunrise-10:00 PM. On the harbor side is a public boat launching ramp, four-lanes, concrete, open 24 hours. For information, call: (310) 832-1179 or (323) 913-7390.

CABRILLO FISHING PIER: 1,000-foot-long municipal fishing pier; connected to the Cabrillo Beach breakwater. Parking fee. Pier open daily. Call: (310) 832-1179 or (323) 913-7390.

CABRILLO MARINE AQUARIUM: Features interpretive displays, 35 aquaria containing marine plants and animals with touch tank, an auditorium, laboratories, classrooms, a gift shop, and an exhibition hall. Facility is wheelchair-accessible. Open 12 PM-5 PM Tues.-Fri., 10 AM-5 PM Sat.-Sun. The museum also sponsors guided tidepool tours at the Point Fermin Marine Life Refuge, whale watches on half-day boat trips, Channel Island boat trips, and grunion run programs. Call: (310) 548-7562. For information on whale-watching trips, call: (310) 548-8397.

PORTS O'CALL VILLAGE: Faces the main channel of the harbor. Ports O'Call Village is a facsimile of an old California seaport, with restaurants and shops; open 11 AM-7 PM daily. For harbor cruises, call: (888) 908-8800. For information on whale-watching cruises, call: (310) 831-1073 or (888) 908-8800. Whalers Wharf simulates a 19th-century New England seaport with shops and cobbled streets; harbor and dinner cruises available. Adjacent boardwalk offers freshly caught seafood cooked to order.

JOHN S. GIBSON, JR. PARK: Small grassy park overlooking L.A. Harbor. Los Angeles Maritime Museum, in the remodeled ferry building at the south end of the park (Berth 84 at the foot of 6th St.), contains a pictorial history of the Los Angeles Harbor area, marine artifacts, and ship models, and affords a view of the harbor from the promenade deck. The museum is wheelchair accessible. Open Tues.-Sun. 10 AM-5 PM. Free parking. Call: (310) 548-7618.

LOS ANGELES HARBOR: One of the world's largest artificial harbors, with 7,500 acres of land and water, and 35 miles of waterfront. The harbor has three districts—San Pedro, Wilmington, and Terminal Island; the Vincent Thomas Bridge joins the San Pedro and Terminal Island districts. Nation's largest commercial fishing fleet and canning center and the West Coast's most heavily used cruise port, with cargo terminals, shipyards, marinas, and slips. Facilities include dry storage, hoists, cranes, fuel dock, fishing licenses, bait and tackle, and marine supplies.

CATALINA TERMINAL: Take the Harbor Freeway south to the Harbor Blvd. exit and follow signs to the terminal; it is located at Berth 95 under the west end of the Vincent Thomas Bridge. Parking fee. Catalina Express runs daily ferries to Avalon and Two Harbors; 75-minute trip to Avalon and 90-minute trip to Two Harbors; for information, call: (310) 519-1212 or 1-800-481-3470. Catalina Classic Cruises operates two-hour trips to and from Avalon during the summer, leaving at 9:45 AM; for information, call: 1-800-641-1004. Island Express helicopters depart for Avalon daily; call: (310) 510-2525.

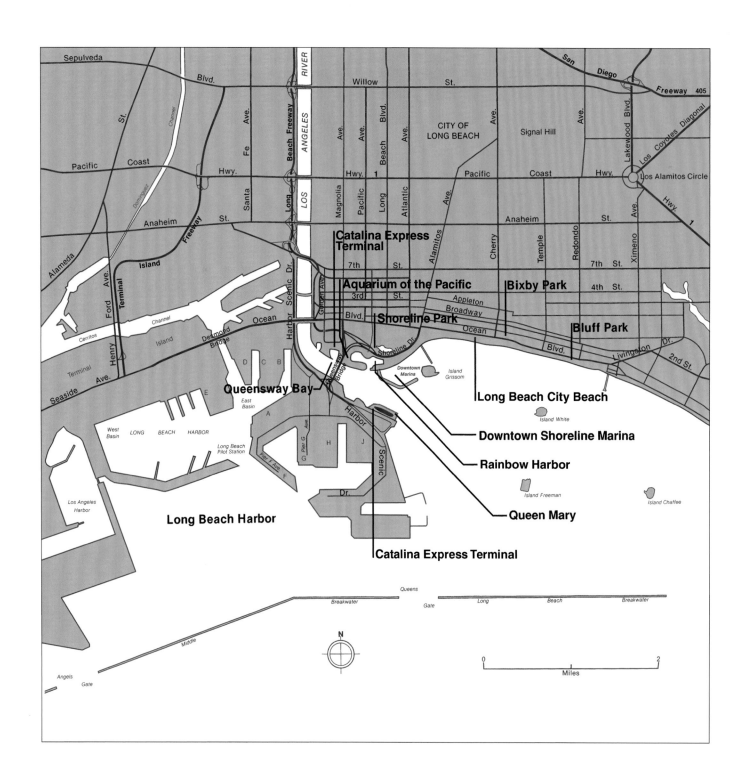

Sepulveda

Blvd.

St.

Channel

Pacific   Coast   Hwy.

Anaheim   St.

Alameda

Ford   Ave.

Terminal

Santa   Fe   Ave.

Beach Freeway

Long Beach Freeway

LOS ANGELES RIVER

Willow   St.

Ave.

Ave.

Beach   Blvd.

Ave.

CITY OF
LONG BEACH

Ave.

Signal Hill

Ave.

San   Diego   Freeway   405

Lakewood   Blvd.

Los Coyotes Diagonal

Pacific   Coast   Hwy.

Los Alamitos Circle

Hwy.   1

Hwy.   1

Magnolia

Pacific

Long

Atlantic

Alamitos   Ave.

Anaheim   St.

Cherry   Ave.

Temple   Ave.

Redondo   Ave.

7th   St.

Ximeno   Ave.

Catalina Express Terminal

7th   St.

Scenic Dr.

Golden Ave.

Harbor

Island   Terminal

Cerritos   Channel

Desmond Bridge

Island

Aquarium of the Pacific

3rd   St.

Blvd.

Ocean   Blvd.

Shoreline Park

Appleton

Broadway

Ocean   Blvd.

Bixby Park   4th   St.

Bluff Park

Livingston   Dr.

2nd St.

Shoreline Dr.

D   C   B

Queensway Bridge

Downtown
Marina

Island
Grissom

Queensway Bay

East
Basin

A

E

Seaside   Ave.

Henry Ford Ave.

Terminal

West
Basin

LONG   BEACH   HARBOR

Long Beach
Pilot Station

Pier G Ave.

G

H

J

Harbor   Scenic   Dr.

Long Beach City Beach

Island White

Downtown Shoreline Marina

Rainbow Harbor

Island Freeman

Island Chaffee

Los Angeles
Harbor

Pier F Ave.

F

Long Beach Harbor

Queen Mary

Catalina Express Terminal

Queens   Breakwater

Gate

Long   Beach   Breakwater

N

Middle

0   Miles   2

Angels   Gate

Los Angeles County

CITY OF LONG BEACH

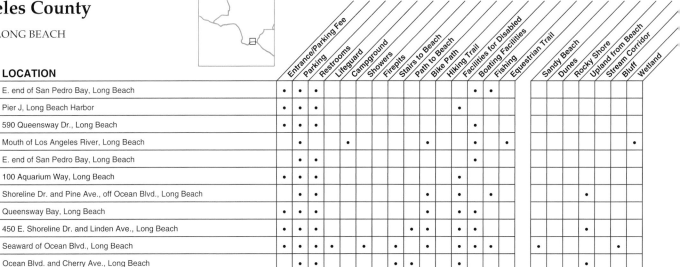

NAME	LOCATION	Entrance/Parking Fee	Parking	Restrooms	Lifeguard	Campground	Showers	Firepits	Stairs to Beach	Path to Beach	Bike Path	Hiking Trail	Facilities for Disabled	Boating Facilities	Fishing	Equestrian Trail	Sandy Beach	Dunes	Rocky Shore	Upland from Beach	Stream Corridor	Bluff	Wetland
Long Beach Harbor	E. end of San Pedro Bay, Long Beach	•	•	•										•	•								
Queen Mary	Pier J, Long Beach Harbor	•	•	•									•										
South Shore Public Boat Launch	590 Queensway Dr., Long Beach	•	•	•										•									
Queensway Bay	Mouth of Los Angeles River, Long Beach		•			•					•					•							•
Rainbow Harbor	E. end of San Pedro Bay, Long Beach		•	•										•									
Aquarium of the Pacific	100 Aquarium Way, Long Beach	•	•	•									•										
Shoreline Park	Shoreline Dr. and Pine Ave., off Ocean Blvd., Long Beach		•	•							•		•		•							•	
Terminals for Catalina Island Boat Service	Queensway Bay, Long Beach	•	•	•							•			•	•								
Downtown Shoreline Marina	450 E. Shoreline Dr. and Linden Ave., Long Beach	•	•	•						•	•		•	•								•	
Long Beach City Beach	Seaward of Ocean Blvd., Long Beach	•	•	•	•		•	•			•		•	•	•		•					•	
Bixby Park	Ocean Blvd. and Cherry Ave., Long Beach		•	•					•	•			•									•	
Bluff Park	Ocean Blvd. between 20th Pl. and Redondo Ave., Long Beach		•						•	•												•	

LONG BEACH HARBOR: Artificial harbor in the eastern part of San Pedro Bay with extensive cargo-handling facilities, slips, supplies, and repair yards. Charterboats and boat cruises; fishing licenses, bait, and tackle; hotels and restaurants along Queensway Bay.

QUEEN MARY: The *Queen Mary*, a one-time luxury liner, is the largest passenger ship ever built, weighing over 80,000 tons and carrying a crew of 1,200. The *Queen Mary* was launched in 1934 at Clydebank, Scotland and retired in 1964; now permanently berthed in Long Beach as a tourist attraction and a hotel, with restaurants and shops. The ship is wheelchair accessible; guided tours are available. Open daily 10 AM-6 PM. Entrance fee. Call: (562) 435-3511.

SOUTH SHORE PUBLIC BOAT LAUNCH: Small launch, open 24 hours, year round. Launch fee. Call: (562) 570-3203.

QUEENSWAY BAY: Boating and water-skiing area located at the mouth of the Los Angeles River. At 101 Golden Ave. is a fully equipped, 80-site R.V. campground; camping fee. Call: (562) 435-4646. The LARIO Trail, which runs along the east bank of the river in Long Beach, is an equestrian trail and bicycle path that extends 20 miles inland from the Golden Shore Marine Preserve to central Los Angeles.

RAINBOW HARBOR: Where downtown meets the shoreline; attractions include the Aquarium of the Pacific, Pine Avenue Pier, Shoreline Park, sport fishing charters, boat rentals, scenic cruises, fishing licenses and tackle, restaurants and shopping along the harbor's promenade, and shuttle boats and buses to other shoreline attractions. For information, call: (562) 8636.

AQUARIUM OF THE PACIFIC: Features more than 12,000 marine animals from three major regions of the Pacific Ocean; includes Shark Lagoon, interactive displays, and temporary exhibits. Open daily 9 AM-6 PM; entrance and parking fee. Call: (562) 590-3100.

SHORELINE PARK: Grassy park surrounding Rainbow Harbor, with fishing platforms, a working lighthouse, bicycle and pedestrian paths, picnic areas, parking, and wheelchair-accessible restrooms.

TERMINALS FOR CATALINA ISLAND BOAT SERVICE: The *Catalina Explorer* offers ferry service from the Pine Ave. Pier in Shoreline Park; call (877) 432-6276 or (562) 951-5801. Catalina Express boats depart from 1046 Queen's Hwy. next to the *Queen Mary* and from 320 Golden Shore at Catalina Landing on the downtown side of Queensway Bay; take the Long Beach Freeway south almost to the end, then the first downtown exit, then the Golden Shore exit. Call: 1-800-481-3470.

DOWNTOWN SHORELINE MARINA: The Downtown Shoreline Marina, with about 1,700 slips, and the nearby Rainbow Harbor are both part of the City of Long Beach Marina system, which also includes Alamitos Bay Marina to the south. Facilities include bicycle and transient public docks, pedestrian paths, parking, wheelchair-accessible restrooms, fuel dock, and pump-out stations. The Downtown Shoreline Marina Headquarters is located at the end of the eastern jetty. Call: (562) 570-4950. Shoreline Village is a tourist attraction near the marina with shops, restaurants, and a carousel. Rainbow Lagoon Park, located inland of the marina and Shoreline Dr. has a small seawater lagoon ringed by a pedestrian walkway; a stairway and an elevator lead up to the Promenade and an observation deck; the Promenade provides direct pedestrian access between Ocean Blvd. and Rainbow Harbor.

LONG BEACH CITY BEACH: Wide, sandy beach seaward of Ocean Blvd., which runs for several miles from 1st Place to 72nd Place; includes Belmont Shore and the Alamitos Peninsula. The section of the beach west of the Belmont Veterans Memorial Pier lies below a bluff, and can be reached via stairways at the ends of 2nd, 3rd, 5th, 8th, 9th, 10th, and 14th places, from the parking lot at the end of Junipero Ave., and from stairways at the ends of Molino, Orizaba, and Coronado avenues and 36th Place. Facilities include sand volleyball courts, concessions, lifeguard service, a paved bicycle path, wheelchair-accessible restrooms, outdoor showers, launch ramps for small boats available at Claremont Ave. and Granada Ave., and several pay parking lots located on the beach. Lots of sand, but very little surf, as the beach is inside the harbor breakwater.

BIXBY PARK: 10-acre park located at the intersection of Ocean Blvd. and Junipero Ave. with picnic tables, benches, a playground, a croquet lawn, shuffleboard courts, skateboarding area, wheelchair-accessible restrooms, and a bandshell with frequent concerts. Recreation Center used by senior citizens for social events, dances, and card-playing. Street parking is available.

BLUFF PARK: Linear, grassy park overlooking the beach runs 1 mile along the bluff on the south side of Ocean Boulevard. Street parking. Long Beach Museum of Art is at 2300 E. Ocean Blvd. at the west end of Bluff Park; open Tues.-Sun. 11 AM-5 PM; call: (562) 439-2119.

Bayshore Walk

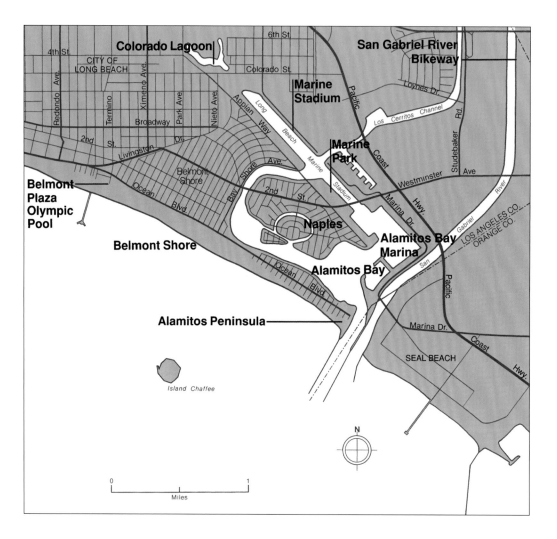

Los Angeles County

LONG BEACH / ALAMITOS BAY

NAME	LOCATION	Entrance/Parking Fee	Parking	Restrooms	Lifeguard	Campground	Showers	Firepits	Stairs to Beach	Path to Beach	Bike Path	Hiking Trail	Facilities for Disabled	Boating Facilities	Fishing	Equestrian Trail	Sandy Beach	Dunes	Rocky Shore	Upland from Beach	Stream Corridor	Bluff	Wetland
Belmont Shore	S. of Ocean Blvd., between 39th Pl. and 54th Pl., Long Beach	•	•	•	•					•			•	•			•						
Belmont Plaza Olympic Pool	Ocean Blvd. near 39th Pl., Long Beach	•	•	•	•		•												•				
Alamitos Peninsula	Off Ocean Blvd., 54th Pl. to 72nd Pl., Long Beach	•	•	•	•					•				•			•						
Alamitos Bay	W. of the San Gabriel River, Long Beach			•	•									•			•						
Alamitos Bay Marina	N.E. end of Alamitos Bay, Long Beach		•	•										•									
Naples	Off Pacific Coast Hwy., along 2nd St., in Alamitos Bay, Long Beach								•								•						
Marine Park	N. of Davies Bridge, Long Beach	•	•	•	•								•				•						
Marine Stadium	Appian Way and 2nd St., Long Beach	•	•	•									•	•			•					•	
Colorado Lagoon	Appian Way and 4th St., Long Beach	•	•	•	•			•	•								•					•	
San Gabriel River Bikeway	E. side San Gabriel River, Long Beach										•										•		

BELMONT SHORE: Part of Long Beach City Beach, extending from the Belmont Veterans Memorial Pier to the Alamitos Peninsula. The wide, sandy beach provides lifeguard service, sand volleyball, shore fishing, and metered parking; a paved bike path runs along the beach from the downtown area south to 54th Place. Two surfsail/catamaran launching ramps are located along Ocean Blvd. at Granada Ave. and at Claremont Ave.; fee lots for trailer storage.

The 1,620-foot-long, T-shaped Belmont Veterans Memorial Pier, located at the foot of 39th Place, is open dawn to dusk, and provides restrooms, a snack bar, and a bait shop. Metered parking is available at Termino Ave.; bike racks are on the beach near the pier.

BELMONT PLAZA OLYMPIC POOL: Indoor pool designed for Olympic-class swimming and diving events; used in training U.S. athletes for international events; also used for recreational swimming and diving. Metered parking. Call: (562) 570-1805.

ALAMITOS PENINSULA: Extends from 54th Place to the entrance channel of Alamitos Bay. The sandy beach on the ocean side is part of Long Beach City Beach; landscaped street ends between 54th Place and 69th Place lead to a wooden pedestrian boardwalk that parallels the ocean beach. Bayshore Beach and Playground at Bayshore Ave. and 54th Place has calm waters, children's playground, snack bar, sailing classes and basketball and handball courts. Bay Shore Walk is a public path that runs along the water from 55th to 65th Place. Bayshore Ave. is the southern continuation of the bike path that runs along Long Beach City Beach, and is closed to cars 9 AM-5 PM daily during the summer. Concerts are held at a small park at 72nd Place, at the south end of the peninsula; fee parking. Fishing from the jetty.

ALAMITOS BAY: Contains several marinas with berthing capacity for 2,000 boats; facilities include marine supplies, boat hoists, hand boat launch, restaurants, shops, yacht clubs, and rowing and sailing centers. Water-skiing is permitted only in Marine Stadium. Alamitos Bay Beach is a narrow, sandy swimming beach south of 2nd St. on the bay side;

restrooms, lifeguard service, and summer kayak rentals are provided. The Marine Bureau office is on the east side of the entrance channel, 205 Marina Dr., Long Beach 90803; (562) 570-3215.

ALAMITOS BAY MARINA: Provides public berths, supplies, and services; facilities include launch ramps, hoists, and fuel dock. Slips are available; temporary berthing for visitors on a first-come, first-served basis for up to 15 days in any month. Mooring controlled by the Marine Bureau, whose office is on the east side of the entrance channel, 205 Marina Dr., Long Beach 90803; (562) 570-3215.

Events include an annual Christmas parade of lighted boats and floats. Alamitos Bay Landing, a restaurant and shopping complex, is located on the east side of the entrance channel near the small beach town of Seal Beach.

NAPLES: Naples is composed of three islands in Alamitos Bay separated by canals. Public walkways run along each canal and encircle most of the islands. Along Sorrento Dr. are several narrow public walkways leading to the water's edge. Overlook Park at Naples Plaza and Vista del Golfo has benches, a small grassy area, and a view of the bay. Public parking is available on the streets.

MARINE PARK: Located on Naples at 5839 Appian Way, at the southwest end of Marine Stadium. Features a narrow sandy beach and calm bay waters known as Mothers Beach with a marked swimming area and grassy areas. Facilities include lifeguard service, sand volleyball, a small dock, paved paths, and wheelchair-accessible restrooms. Metered parking.

MARINE STADIUM: Built in the 1920s and used for the 1932 and 1984 Olympics. The Stadium is a long, narrow, rectangular body of sea water connected to Alamitos Bay; still used for rowing competitions and commercial water sport events. The City has successfully implemented a habitat restoration project at End Beach, on the northern end of Marine Stadium.

Marine Stadium's northwest side has a boat launching ramp and a narrow sandy beach. The Marine Stadium launch ramp is open all year, 8 AM-sunset (unless race events are scheduled). Fee for parking; wheelchair-accessible restrooms. At the southeast end of Marine Stadium, at 6201 E. Second St., is the Davies Launch Ramp, a 12-lane, concrete facility. Open 24 hours year-round; fee for parking. Scheduled events include the International Sea Festival, Girl and Boy Scout mariners, and university and other rowing races. Call: (562) 570-3203.

COLORADO LAGOON: 40 acres of water and land; the tidal lagoon receives sea water from Marine Stadium. Features a sandy beach, floating causeways, lifeguard, playgrounds, picnic areas with barbecue grills, restrooms, and metered parking.

Nearby Recreation Park, just north of the lagoon, contains a large golf course and offers picnicking in a eucalyptus grove. The community center offers programs and classes. Marina Vista Park, located south of the lagoon between Nieto and Havana avenues, provides picnic areas and playing fields.

SAN GABRIEL RIVER BIKEWAY: A paved bike path follows the river's east bank from the mouth inland to El Dorado Regional Park, then continues 30 miles to Azusa.

Santa Catalina Island

The largest of the Southern Channel Islands, Catalina Island is 21 miles long, 8 miles wide at its widest point, and has a resident population of about 3,200. The island was visited in 1542 by Juan Rodríguez Cabrillo, and again in 1602 by Sebastián Vizcaíno. It's now a popular tourist spot with mountains, canyons, and beaches; herds of bison roam free in the interior. Catalina is 22 miles off the coast of Los Angeles, and can be reached by air or sea.

To get around on the island, bicycles or golf carts can be rented, and taxis and buses can be hired. Tours include cruises, bus tours, and inland motor tours, and visit such sites as the Casino in Avalon, the Undersea Gardens (in a glass-bottomed or semi-submersible boat) just offshore from Avalon, and El Rancho Escondido in the interior, where pure-bred Arabian horses are raised and trained. For information on guided horseback riding trips, contact Catalina Stables: (310) 510-0478. For hotel reservations and more information about transportation and tours, contact the Avalon Chamber of Commerce and Visitor's Information Center on the Pleasure Pier, P.O. Box 217, Avalon 90704, (310) 510-1520.

The Santa Catalina Island Conservancy, a non-profit organization, owns 88 percent of the Island's 76 square miles and manages it as a preserve. Permits are required for hiking (no cost) and bicycling (fee) into the interior and are available at the Conservancy office in Avalon, 125 Claressa Avenue, (310) 510-2595; at Two Harbors Visitor Services at the foot of the pier, (310) 510-4205; through Two Harbors Enterprises, (310) 510-0303; and at the Airport in the Sky, (310) 510-0143.

Fresh water is available only at certain spots; hikers are advised to carry water. Backpackers must camp in designated campgrounds; fees are charged and permits and reservations required. Campers under 18 must be accompanied by an adult. Pets are not allowed in any campground or in the interior. (See specific campground entries for more information on particular campgrounds.) Skin and scuba diving is permitted; licenses, rentals, and air refills are available at Avalon and at Two Harbors. Mooring fees are not required for private boats when landing ashore for less than two hours; transient moorings are available on a first-come, first-served basis.

Santa Catalina Island, Avalon Bay

Transportation to Catalina:

CATALINA CLASSIC CRUISES: Operates two-hour trips during the summer, leaving San Pedro at 9:45 AM and returning from Avalon at 6 PM. Call: 1-800-641-1004.

CATALINA PASSENGER SERVICE: The *Catalina Flyer*, a fast catamaran, departs Newport Harbor from the Balboa Pavilion daily at 9 AM and returns from Avalon at 4:30 PM; limited service December thru mid-March. Additional fee for bicycles. For reservations, call: (949) 673-5245.

CATALINA EXPRESS: Runs commuter boats year-round from San Pedro, Long Beach and Dana Point to Avalon and Two Harbors. Reservations required; call: (310) 519-1212 or 1-800-481-3470.

CATALINA EXPLORER: Operates ferries from Long Beach and Dana Point to Avalon and Two Harbors; limited service during winter. Call: toll-free (877) 432-6276, or (310) 510-9404 (Avalon), (562) 951-5801 (Long Beach), (949) 492-5308 (Dana Point).

ISLAND EXPRESS HELICOPTER SERVICE: Helicopters fly daily to Avalon from Long Beach and from the Catalina Terminal in San Pedro. For information, call: (310) 510-2525.

AIRPORT IN THE SKY: Located 25 minutes from Avalon by bus service; private planes land here. Open daily. For information, call: (310) 510-0143.

Los Angeles County

SANTA CATALINA ISLAND EAST

NAME	LOCATION	Entrance/Parking Fee	Parking	Restrooms	Lifeguard	Campground	Showers	Firepits	Stairs to Beach	Path to Beach	Bike Path	Hiking Trail	Facilities for Disabled	Boating Facilities	Fishing	Equestrian Trail	Sandy Beach	Dunes	Rocky Shore	Upland from Beach	Stream Corridor	Bluff	Wetland
Hamilton Cove	1 mi. N. of Avalon Bay		•						•	•			•				•		•				
Descanso Beach	N. of Avalon Bay, Avalon	•	•				•						•				•						
Avalon Bay	E. end of the island, on the leeward side													•	•		•						
Crescent Beach	Avalon Bay, Avalon			•	•		•						•				•						
Hermit Gulch Campground	Avalon Canyon Rd., 1 mi. inland from Avalon	•	•			•	•	•				•	•			'				•			
Toyon Junction	On Old Stage Rd., 3.9 mi. from Avalon											•			•					•			
Black Jack Campground	9.4 mi. N.W. of Avalon	•	•			•	•	•				•			•					•			

HAMILTON COVE: Sandy cove adjacent to condominium complex. Amenities include restrooms, picnic area, and pier. Accessible from Avalon by traveling to guardhouse, then walking down steep hill via Playa Azul. Also accessible by kayak or sailboat.
Transit: Ferry to Avalon; walk, bike, or take taxi one mile.

DESCANSO BEACH: The private beach club is open to the public May-Sept. for a small daily fee. Facilities include showers and changing rooms, snack bar and restaurant, kayak and snorkeling rentals, gift shop. Call: (310) 510-7410.
Transit: Ferry to Avalon.

AVALON BAY: Avalon Harbor is within the bay. There are no berths; moor to buoys. For mooring assignments, check in with Harbor Patrol boat in front of harbor; call: (310) 510-0535 for mooring availability. Pleasure Pier in the south part of the bay has concessions, equipment rentals, and boat hoist. Diving supplies are sold and rented on the pier. Also available are marine supplies, bait and tackle, and fishing licenses. A fuel dock is located adjacent to the Casino, open daily; call: (310) 510-0046; sewage pumpout station nearby. Water-skiing and diving are permitted beyond the breakwater.

In Avalon are golf and tennis courts, stables, bike rentals, and miniature golf. The Casino, at the west end of the bay, is a popular tourist attraction and contains a ballroom, theater, and museum.

The 38-acre Wrigley Memorial and Botanical Garden, at 1400 Avalon Canyon Rd., is 1.7 miles from downtown Avalon; walk or take a tram. The Memorial and Garden are open 8 AM-5 PM year-round; small fee. The Memorial, completed in 1934, is an imposing structure of stone, marble, and tile. The Botanical Garden below features cactus, succulents, and other native Catalina plants. The Memorial and Garden commemorate William Wrigley, Jr., who purchased control of the Santa Catalina Island Company in 1919 and spent the rest of his life improving Catalina.
Transit: Ferry to Avalon.

CRESCENT BEACH: Sandy, surf-free beach. Wheelchair-accessible restroom and outdoor shower available.
Transit: Ferry to Avalon.

HERMIT GULCH CAMPGROUND: One and a half miles inland from Avalon, 46 campsites with picnic tables and barbecue grills. Showers are coin operated; wheelchair-accessible restrooms. Overnight fee; group rates available; no pets. Reservations recommended; hiking trail permits and maps are available from ranger. Small camp store open daily. Call: (310) 510-8368.
Transit: Ferry to Avalon; hike 1.5 miles inland or take a tram.

TOYON JUNCTION: Picnic area; water available.
Transit: Ferry to Avalon, hike 3.9 miles.

BLACK JACK CAMPGROUND: The campground is located at 1500' elevation; water and picnic tables. Overnight fee; group rates are available. Permit and reservations required. Call: (310) 510-2800 or 510-0303.

Transit: Ferry to Avalon, hike or take the island's Safari Bus 7.9 miles to Black Jack Junction, then hike 1.5 miles on the trail to the campground.

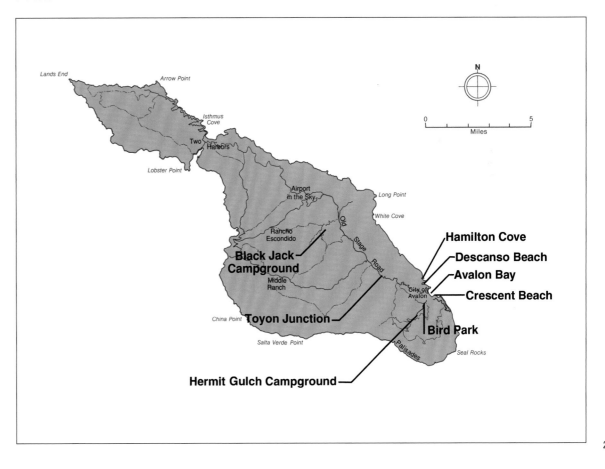

Little Fisherman's Cove

Santa Catalina Island, Little Fisherman's Cove

Little Fisherman's Cove Campground

Los Angeles County

SANTA CATALINA ISLAND WEST

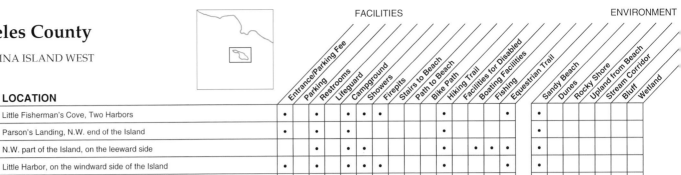

NAME	LOCATION	Entrance/Parking Fee	Parking	Restrooms	Lifeguard	Campground	Showers	Firepits	Stairs to Beach	Path to Beach	Bike Path	Hiking Trail	Facilities for Disabled	Boating Facilities	Fishing	Equestrian Trail	Sandy Beach	Dunes	Rocky Shore	Upland from Beach	Stream Corridor	Bluff	Wetland
Two Harbors Campground	Little Fisherman's Cove, Two Harbors	•		•		•	•	•							•		•						
Parson's Landing Campground	Parson's Landing, N.W. end of the Island	•		•		•									•		•						
Two Harbors (Isthmus)	N.W. part of the Island, on the leeward side			•		•	•						•	•	•	•		•						
Little Harbor Campground	Little Harbor, on the windward side of the Island	•		•		•	•	•							•		•						
Ben Weston Beach	W. end of Middle Canyon Trail, windward side of the Island														•		•						

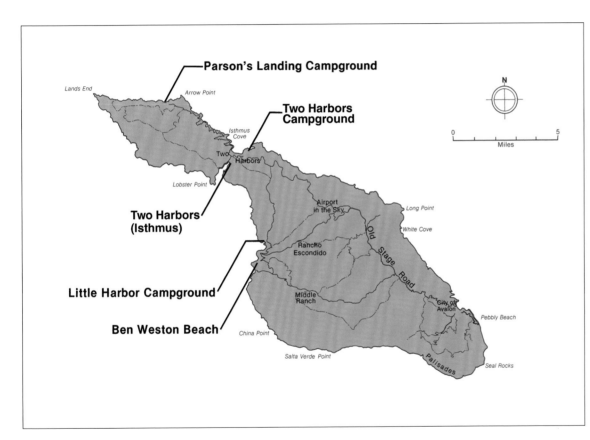

Parson's Landing Campground

Lands End — Arrow Point

Two Harbors Campground

Isthmus Cove

Two Harbors

Two Harbors (Isthmus)

Lobster Point

Airport in the Sky — Long Point

White Cove

Little Harbor Campground

Rancho Escondido

Old Stage Road

Middle Ranch

City of Avalon

Pebbly Beach

Ben Weston Beach

China Point

Salta Verde Point

Palisades

Seal Rocks

N

0 — 5 Miles

TWO HARBORS CAMPGROUND: 43 campsites near the water's edge. Amenities include tables, outdoor cold showers, fire rings, barbecues, firewood for sale, and fresh water. Ranger on duty. Overnight fee; permit and reservations required. For information, call: (310) 510-0303. Restaurant, bar, snack bar, dive shop, general store, and camping equipment rentals located nearby in Two Harbors.
Transit: Ferry to Two Harbors.

PARSON'S LANDING CAMPGROUND: Primitive campground with 8 sites located on the beach. Picnic tables; ranger on duty. Overnight fee; permit and reservations required. Campers receive fresh water and firewood after making reservations; call Two Harbors Enterprises: (310) 510-0303.
Transit: Ferry to Two Harbors; hike 7.0 miles to Parson's Landing or take shore boat or Harbor Patrol boat to Emerald Bay, then hike 2.5 miles.

TWO HARBORS (ISTHMUS): 12 miles by water from Avalon; mooring assignments are on a first-come, first-servedd basis. Two Harbors Enterprises leases moorings for the entire island (except the Avalon area); for information, call: (310) 510-4253.

Facilities at Two Harbors include a fuel dock, dinghy dock, boat and motor rentals, shore boat service, public restrooms, shower, and laundry. There is also a restaurant, bar, snack bar, dive shop, and general store. Fishing licenses and bait and tackle are available. Isthmus Cove is a popular water-skiing area. For information, call: (310) 510-4205. The USC Wrigley Marine Science Center, 1½ miles from the Two Harbors pier, holds public open house from 2-4 PM on Sat. during summer; call (310) 510-0811.
Transit: Ferry to Two Harbors.

LITTLE HARBOR CAMPGROUND: The campground is located in a sheltered cove on the windward side of the Island. A wide, sandy beach offers swimming and surfing. Picnic tables, fire rings, barbecues. Overnight fee; permit and reservations are required. For information, call: (310) 510-2800.
Transit: Ferry to Two Harbors, hike 6.8 miles or take the island's Safari Bus.

BEN WESTON BEACH: Beach and picnic area; day use only.
Transit: Ferry to Two Harbors, hike 11.6 miles, or take Safari Bus to Little Harbor campground, hike 4.8 miles.

Laguna Beach

Orange County

Orange County is well known for its excellent surfing conditions at Huntington Beach, the yacht harbor at Newport Beach, and the artists' community at Laguna Beach. Since the first American settlement in the late 1800s, the Orange County coast has been a resort area for nearby residents due to its year-round warm weather, high cliffs, wide beaches, and sandy coves. This pleasant environment provides opportunities for swimming, sunbathing, boating, fishing, diving, and surfing.

From Seal Beach to Newport Beach, the broad sandy beaches are backed by low-lying plains and wetlands. Several of these wetlands have been established as nature preserves, such as Bolsa Chica Ecological Reserve and Upper Newport Bay Ecological Reserve.

In the late 1800s, a wharf was built at Newport, which transformed the quiet little village into a bustling community and initiated an era of shipping and trade. Newport Pier has since been re-built, and many of the original buildings still exist and have been renovated as part of Cannery Village, a cluster of shops.

At about the same time the first wharf was built, the first commercial grove of orange trees was planted. Most of the orange trees are now gone, due to the rapid development of homes and businesses in the county. The harbors, however, are still flourishing. Huntington Harbor, Newport Harbor, and Dana Point Harbor are all popular destinations for private sailboats and yachts, and Newport and Dana Point harbors also have commercial fishing boat rentals and whale-watching trips.

Newport Harbor was one of the earliest summer resorts in southern California; the Balboa Pavilion, built in 1905 and designed to be a "magnificent pleasure pavilion," was the terminus of a streetcar line that originated in Los Angeles. The pavilion was used by fashionable guests as a boathouse, and later as a gambling casino and dancehall; today it remains as a prominent Victorian landmark, housing a restaurant, gift shop, and banquet room.

South of Newport Beach, uplifted marine terraces form cliffs at the water's edge, providing numerous coves for swimming and diving. Laguna Beach comprises a cluster of these coves, with the town centered near the longest beach, called Main Beach. In the early 1900s the rugged terrain and dramatic headlands drew painters to the area who established Laguna Beach as a center for artists. The city now has shops, galleries, and an art festival each summer that lasts several weeks.

Located inland from San Clemente, San Juan Capistrano, established in 1776, was the first Spanish mission in Orange County. The Gabrielino and Juaneño Indians were the original inhabitants of the Orange County coast, and while most of the Indians perished from exposure to European diseases, a band of Juaneño Indians still lives in the area. Archaeological remains of the Juaneño culture have been found in Crystal Cove State Park.

The California Coastal Commission's "Orange County Beach Access Map" lists the information in this chapter in a brochure-sized, portable guide. It also includes detailed maps and color photographs. To order, call: 1-800-COAST 4U or see the Coastal Commission's website at www.coastal.ca.gov/publiced/OC-beach-access-map.html.

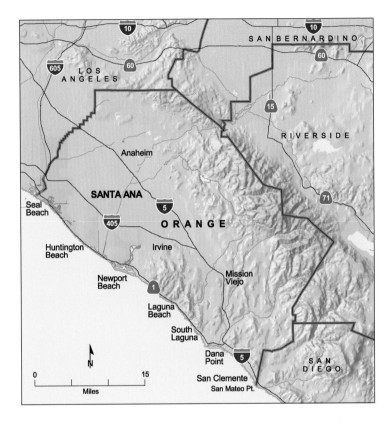

For more information on coastal towns in Orange County, write or call the appropriate Chamber of Commerce: Huntington Beach Chamber of Commerce, 2100 Main St. #200, Huntington Beach 92648, (714) 536-8888; Newport Beach Conference and Visitors Bureau, 3300 W.Coast Hwy., Newport Beach 92663, (949) 722-1611 or 1-800-94-COAST; Laguna Beach Chamber of Commerce, 357 Glenneyre St., Laguna Beach 92651, (949) 494-1018; Dana Point Chamber of Commerce, 24681 La Plaza, Suite 115, Dana Point 92629, (949) 496-1555; or San Clemente Chamber of Commerce, 1100 N. El Camino Real, San Clemente 92672, (949) 492-1131.

The Orange County Transportation Authority provides transit service to many of the beaches and parks in this guide. Transit lines link inland points to the beach and also run along Pacific Coast Highway. For information, call: (714) 636-7433, TDD: (714) 636-4327 or see www.octa.net for route maps and service information. For the Laguna Beach area, call Laguna Beach Municipal Transit: (949) 497-0746.

Seal Beach

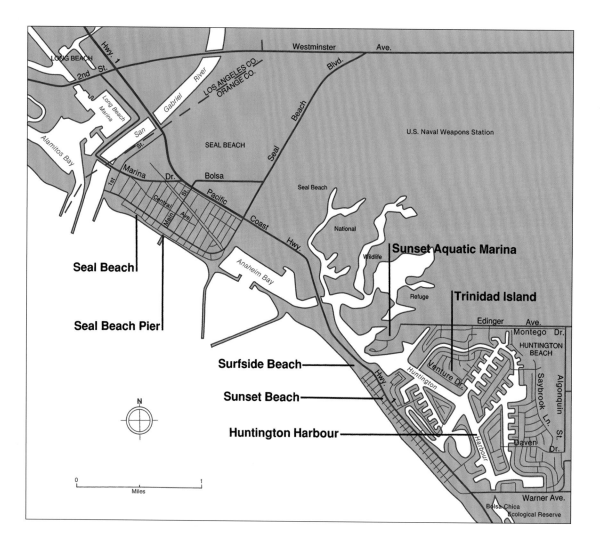

Orange County

SEAL BEACH / HUNTINGTON HARBOR

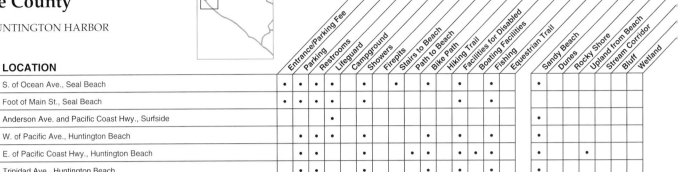

NAME	LOCATION	Entrance/Parking Fee	Parking	Restrooms	Lifeguard	Campground	Showers	Firepits	Stairs to Beach	Path to Beach	Bike Path	Hiking Trail	Facilities for Disabled	Boating Facilities	Fishing	Equestrian Trail	Sandy Beach	Dunes	Rocky Shore	Upland from Beach	Stream Corridor	Bluff	Wetland
Seal Beach	S. of Ocean Ave., Seal Beach	•	•	•	•		•		•		•		•		•		•						
Seal Beach Pier	Foot of Main St., Seal Beach	•	•	•	•		•				•		•										
Surfside Beach	Anderson Ave. and Pacific Coast Hwy., Surfside				•												•						
Sunset Beach	W. of Pacific Ave., Huntington Beach		•	•	•		•				•		•		•		•						
Huntington Harbour	E. of Pacific Coast Hwy., Huntington Beach		•	•			•				•	•	•	•	•							•	
Trinidad Island	Trinidad Ave., Huntington Beach		•	•			•					•	•		•		•						
Sunset Aquatic Marina	2901 Edinger Ave., Huntington Beach		•	•			•						•	•								•	

SEAL BEACH: Long, sandy beach south of the mouth of the San Gabriel River. There is a paved path from Seal Beach Pier south along the beach, then inland to Pacific Coast Highway. Street ends between 1st St. and Electric Ave. lead to the beach; beach parking at 1st , 8th, and 10th streets with bicycle racks, wheelchair-accessible restrooms, and dressing rooms. Bike path starting at the water follows the south bank of the river inland for several miles. Aquatic sports and swimming regulations are seasonal and marked by signs. Popular surfing and wind surfing spot. Electric Ave., which starts at the south end of the beach, has a long grass meridian with benches, trees, and a paved pathway. For information, call: (562) 430-2613.

SEAL BEACH PIER: A restaurant and fish-cleaning counter are located at the end of the pier. Parking lots are at the ends of 8th and 10th streets. The pier adjoins a grassy park with benches at the foot of Main Street. Playground at base of pier in sand. Restrooms are located under pier and outdoor showers are located next to it; both are accessible by steep ramps.

SURFSIDE BEACH: Pedestrian and bicycle access only through the Anderson St. gate of the private Surfside community. No public parking. Poles for volleyball nets. For information, call: (562) 430-2613.

SUNSET BEACH: Street ends from Anderson Ave. to Warner Ave. lead to a sandy beach. A grassy linear park one block east runs the length of the beach; it has a bike path, wheelchair-accessible restrooms and showers, beach wheelchair, and street parking. The beach area has volleyball nets; surfing is allowed. For information, call: (949) 509-6683.

HUNTINGTON HARBOUR: Some public docks and facilities are available; water entry from Anaheim Bay. At Peter's Landing, inland from Anderson Ave., some overnight slips are available for rental, and there is a pump-out station, restrooms, showers, and laundry; for information, call: (714) 840-1387. Motels, grocery stores, and shops are near the marina off Pacific Coast Highway. Free boat launch ramp is at the southeast end of the harbor at Warner Ave.

A fishing dock and metered parking are located nearby at the Earl D. Percy Marine Park. Public dock also at Huntington Yacht Club. Seabridge Park, near entrance of harbor, has a sandy beach, playground, grassy areas, picnic tables, free parking, wheelchair-accessible restrooms, and an outdoor shower. Small public beaches are at the intersection of Davenport Dr. and Edgewater Lane and at Humboldt Dr. and Mandalay Circle. Small parks are found on most of the islands, some with tennis courts and play areas.

TRINIDAD ISLAND: Small island, mostly residential, has a waterfront park at the corner of Trinidad and Sagamore lanes. Park provides playground, sandy beach, outdoor shower, and wheelchair-accessible restrooms, and leads to a greenbelt that crosses the island to Venture Drive. A bike path and walkway parallels Venture Dr. and Typhoon Lane along the harbor, and meanders through the center of the island. Small public fishing dock off Venture Dr. walkway, open 5 AM-10 PM. Respect private property; do not trespass.

SUNSET AQUATIC MARINA: A public small boat harbor at the end of Edinger Avenue. Ocean access is from Anaheim Bay. Grassy areas with pathways and picnic tables overlook the bay; wheelchair-accessible picnic areas and restrooms. There are 300 privately owned boat slips; eight rented guest slips are available. Concrete seven-lane boat ramp is open 24 hours with a fee for use; register at office. Dry storage, haul-out, and repair facilities. Call: (714) 846-0179. The park is adjacent to Seal Beach National Wildlife Refuge, 1,200 acres of salt marshes that are within the boundaries of the U.S. Naval Weapons Station; no public access.

Seal Beach Pier

Huntington Beach

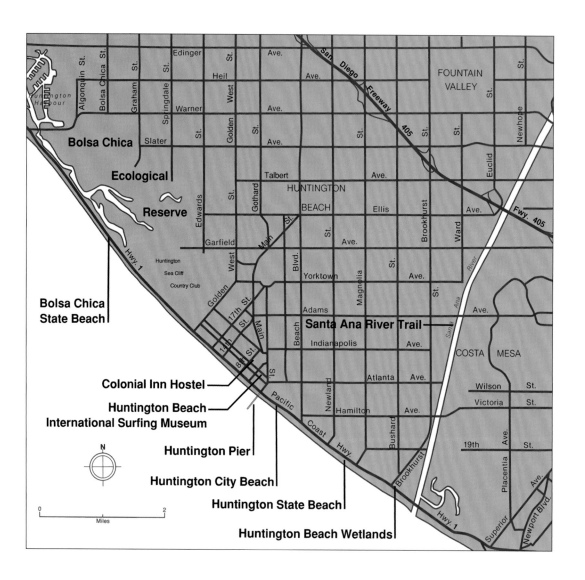

Orange County

HUNTINGTON BEACH

NAME	LOCATION	Entrance/Parking Fee	Parking	Restrooms	Lifeguard	Campground	Showers	Firepits	Stairs to Beach	Path to Beach	Bike Path	Hiking Trail	Facilities for Disabled	Boating Facilities	Fishing	Equestrian Trail	Sandy Beach	Dunes	Rocky Shore	Upland from Beach	Stream Corridor	Bluff	Wetland
Bolsa Chica Ecological Reserve	E. of Pacific Coast Hwy., S. of Warner Ave., Huntington Beach	•	•																				•
Bolsa Chica State Beach	W. of Pacific Coast Hwy, Warner Ave. to Huntington Pier, Huntington Beach	•	•	•	•	•	•	•	•	•	•		•		•		•					•	
Colonial Inn Hostel	421 8th St., Huntington Beach			•	•		•													•			
Huntington Beach International Surfing Museum	411 Olive Ave., Huntington Beach																			•			
Huntington Beach Pier	Main St. and Pacific Coast Hwy., Huntington Beach	•	•	•									•		•								
Huntington City Beach	W. of Pacific Coast Hwy., Main St. to Beach Blvd., Huntington Beach	•	•	•	•		•	•	•	•			•		•		•						
Huntington State Beach	W. of Pacific Coast Hwy., Beach Blvd. to Santa Ana River, Huntington Beach	•	•	•	•		•	•					•		•		•						
Huntington Beach Wetlands (Talbert Marsh)	Pacific Coast Hwy. and Brookhurst St., Huntington Beach		•										•		•								•
Santa Ana River Trail	Along Santa Ana River, Huntington Beach										•	•				•					•		

BOLSA CHICA ECOLOGICAL RESERVE: 300-acre wetland reserve, owned by the state, where thousands of birds can be seen, including three endangered species: Belding's savannah sparrow, clapper rail, and California least tern. A parking lot and interpretive center are at Warner Ave. near Pacific Coast Hwy. (where shore fishing is allowed); call: (714) 846-1114. Another lot is on Pacific Coast Hwy. across from the main Bolsa Chica Beach entrance, and features a loop trail with interpretive signs and benches for wildlife viewing. Tours of the reserve on first Saturday of each month; call: (714) 840-1575.

BOLSA CHICA STATE BEACH: The northern half of the six-mile-long beach provides fee parking lots, lifeguards, picnic areas, firepits, food concessions, cold showers, wheelchair-accessible restrooms, dressing rooms, and beach wheelchair. Entrance to beach parking is on Pacific Coast Hwy. between Warner Ave. and Golden West St.; open 6 AM-10 PM all year. Enroute campsites are available for self-contained R.V.'s; overnight fee; reservations required; call 1-800-444-7275. Paved multi-use path extends the length of the beach and connects to the Santa Ana River Trail in the south. Call: (714) 846-3460.

Blufftop Park, located on the low bluffs at the narrow southern end of the beach, has picnic areas and wheelchair-accessible vista points. There is pedestrian access to the beach area below from Pacific Coast Hwy. at Golden West, 20th, 17th, 14th, 11th, and 9th streets. A ramp at 9th St. runs from the highway onto the sand; wheelchair-accessible pit toilet. Beach is not usable during high tide. At low tide clamming and fishing are popular; good surfing. Grunion runs occur between March and August; call: (714) 848-1566.

COLONIAL INN HOSTEL: Located four blocks from the beach on 8th Street. Facilities include 48 beds, showers, fully equipped kitchen, washer/dryer, T.V. room, and a locked storage shed for bicycles. Open daily 8 AM-11 PM. Overnight fee; open all year. Call: (714) 536-3315.

HUNTINGTON BEACH INTERNATIONAL SURFING MUSEUM: Features exhibits and a collection of surfing memorabilia, including antique surfboards and photos. Located at 5th St. and Olive Ave; open noon-5 PM daily. Call: (714) 960-3483.

HUNTINGTON BEACH PIER: The 1,850-foot-long pier has snack bars, several stores including a bait and tackle shop, and a restaurant. It is floodlit at night for fishing and surfing. Wheelchair-accessible restroom. Fee parking at 5th St. and Pacific Coast Highway.

HUNTINGTON CITY BEACH: Site of international surfing competitions. Facilities include food and surf equipment concessions, sand volleyball courts, fire rings, outdoor showers, beach wheelchairs, and restrooms. Fee parking lots at First St., Huntington St., and Beach Boulevard. Stairways to beach between First St. and Beach Boulevard. Camping allowed November through the weekend before Easter; overnight fee. Entrance is off Pacific Coast Hwy. opposite First Street. For information, call: (714) 536-5280, or write: City of Huntington Beach, 2000 Main Street, 5th floor, Huntington Beach 92648.

HUNTINGTON STATE BEACH: A two-mile-long sandy beach. Facilities include picnic areas, poles for volleyball nets, food concessions, outdoor cold showers, wheelchair-accessible restrooms, a two-mile-long multi-use path, beach wheelchair, and six wheelchair access ramps that reach almost to the water. Fee parking lots and pedestrian access at Beach Blvd., Newland St., Magnolia St., and Brookhurst St.; limited street parking nearby. A five-acre least tern (an endangered bird species) preserve is between Brookhurst St. and the Santa Ana River jetty. Rangers and lifeguards are on duty all year. Call: (714) 536-1454.

HUNTINGTON BEACH WETLANDS (TALBERT MARSH): Restored 25-acre salt marsh located west of the mouth of the Santa Ana River. Parking available along Brookhurst St.; a short, wheelchair accessible path with interpretive signs runs along the marsh, connecting to the Santa Ana River Trail. Call: (714) 963-2123.

SANTA ANA RIVER TRAIL: This paved multi-use trail, which joins the Bolsa Chica State Beach path, runs along the west side of the Santa Ana River from the mouth to a bridge, where it crosses over to the east side and follows the river inland to Gypsum Canyon Bridge in Yorba Linda. An unpaved riding and hiking trail along the west side starts at Adams Street. A paved bikeway runs along the east side of the river as well.

Bolsa Chica State Beach

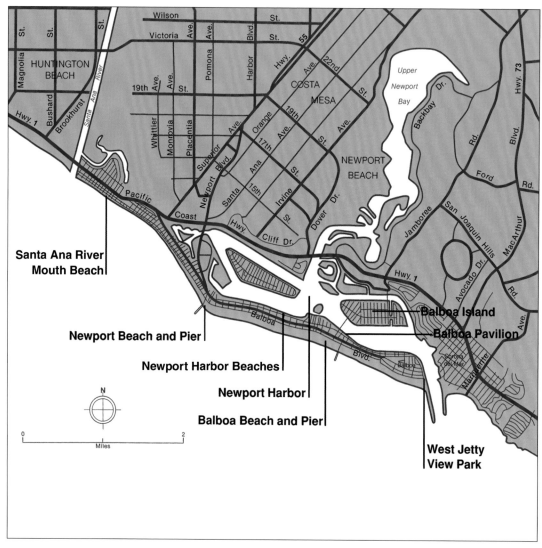

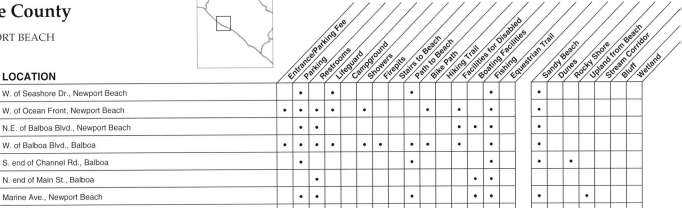

NAME	LOCATION	Entrance/Parking Fee	Parking	Restrooms	Lifeguard	Campground	Showers	Firepits	Stairs to Beach	Path to Beach	Bike Path	Hiking Trail	Facilities for Disabled	Boating Facilities	Fishing	Equestrian Trail	Sandy Beach	Dunes	Rocky Shore	Upland from Beach	Stream Corridor	Bluff	Wetland
Santa Ana River Mouth Beach	W. of Seashore Dr., Newport Beach		•	•					•						•		•						
Newport Beach and Pier	W. of Ocean Front, Newport Beach	•	•	•	•		•			•	•				•		•						
Newport Harbor	N.E. of Balboa Blvd., Newport Beach		•	•										•	•		•						
Balboa Beach and Pier	W. of Balboa Blvd., Balboa	•	•	•	•		•	•		•					•		•						
West Jetty View Park	S. end of Channel Rd., Balboa		•						•						•		•			•			
Balboa Pavilion	N. end of Main St., Balboa		•											•	•								
Balboa Island	Marine Ave., Newport Beach		•	•					•					•	•		•				•		
Newport Harbor Beaches	Street ends along Newport Harbor, Newport Beach		•	•	•		•										•						

SANTA ANA RIVER MOUTH BEACH: Sandy strip that runs along both sides of the river mouth. Street ends between Summit St. and 61st St. lead to the beach. Seasonal lifeguard; metered parking nearby. Lifeguard station: (949) 499-3312.

NEWPORT BEACH AND PIER: The long, sandy beach is narrow at the north end and wide at the southern end. Popular surfing spots just north of the pier and at small jetties between 18th and 56th streets. Surfing regulated by season and conditions; flags designating conditions fly at each lifeguard station. Volleyball nets on the beach; beach wheelchairs at lifeguard station. Outdoor showers and wheelchair-accessible restrooms are located at the base of the pier. The pier is a center for morning surfing and fishing activities; fish-cleaning sinks at end of pier. The Newport Dory Fishing Fleet returns around 9:30 AM on the north side of the pier to sell catch on the beach. Full-service restaurant on pier. Tackle shops and the Cannery Village shopping area are nearby. A paved path called the Boardwalk runs along beach from 36th St. to F Street. For weather and surf conditions, call: (949) 673-3371.

West Newport Park, a grassy strip one block inland from the beach between Seashore Dr. and Pacific Coast Hwy., stretches from Olive St. south to 56th Street. Facilities include wheelchair-accessible restrooms, outdoor shower, picnic tables, playground, and tennis and racquetball courts; metered parking.

NEWPORT HARBOR: Large yachting harbor with 1,230 residential piers, 2,219 commercial slips and side ties, and over 50 guest slips and 1,221 bay moorings for rent. Facilities include dry storage, gas docks and service stations, shipyards, engine and hull maintenance, marine supplies, and sewage pump-outs. Harbormaster: 1901 Bayside Dr., Corona del Mar 92625; or call: (949) 723-1002.

There are numerous small parks, vistas, and walkways around the harbor. Channel Place Park, on W. Balboa Blvd. at 44th St., has covered picnic tables with grills, wheelchair-accessible restrooms, and a playground adjacent to a small sandy beach. Newport Island Park, at 38th St. and Marcus Ave., provides a basketball court, a playground, and a small sandy beach; Lido Park, at Lafayette Ave. and 31st St., is a small harbor viewing area with benches, trees, and a paved path. Las Arenas Park, located on Balboa Blvd. between 15th and 16th streets, has covered picnic tables, grills, basketball and tennis courts, and a playground. The Newport Harbor Nautical Museum is located onboard a floating riverboat at 151 E. Pacific Coast Hwy.; open Tues.–Sunday. Call: (949) 675-8915.

BALBOA BEACH AND PIER: Most street ends lead to the sandy ocean beach, which provides outdoor showers and wheelchair-accessible restrooms. A bicycle path/walkway continues from Newport Beach along the beach; equipment rentals are available near the Balboa Pier. Restaurant at end of pier. Beach wheelchair available. Adjacent to the pier is Pier Plaza and Peninsula Park, with a bandstand, baseball diamonds, a playground, picnic tables, and barbecue grills. For weather and surf conditions, call: (949) 673-3371.

WEST JETTY VIEW PARK: Located between the West Jetty and Balboa Beach. The park overlooks the Wedge, a popular but dangerous bodysurfing spot.

BALBOA PAVILION: The large pavilion is the hub of Newport Harbor. A restaurant, gift shop, and banquet room operate in the building. It is also the terminal for Catalina Island tours, harbor cruises, whale-watching trips, and charter boats; for information, call: (949) 673-5245. Other facilities include boat and motor rentals, fishing licenses, and bait and tackle sales. A boardwalk runs from the pavilion to the Newport Harbor Yacht Club along the bay. The Fun Zone is a small permanent carnival with rides. Many shops, food concessions, and restaurants are located near the Pavilion.

BALBOA ISLAND: Small island in Newport Harbor. A bayfront boardwalk circles the island with access to periodic small beaches and boat slips. Marine Ave. has shops and restaurants. There is a bridge to the island on the north side; an auto ferry connects the island to Balboa Peninsula on the south side. Hours of operation—summer months: 24 hours; rest of year: until midnight Sun.-Thurs. and until 2 AM Fri. and Sat.

NEWPORT HARBOR BEACHES: Beaches facing the bay at the foot of both 18th St. and 10th St. on the Balboa Peninsula are popular with families due to small waves. Lifeguard, shower and restrooms at 18th St. Beach. Stairs lead to a beach at the end of M St. in a residential area; showers. Small public beach at the end of Montero St. and another near the Yacht Club near Buena Vista Boulevard. Limited metered and some free street parking. Most sandy beaches along north and south sides of Lido Isle are open to the public via several street-end public walkways; parks adjacent to walkways are private; pedestrian and bicycle access; very limited parking.

Balboa Pavilion, Newport Bay

Pelican Point, Crystal Cove State Park

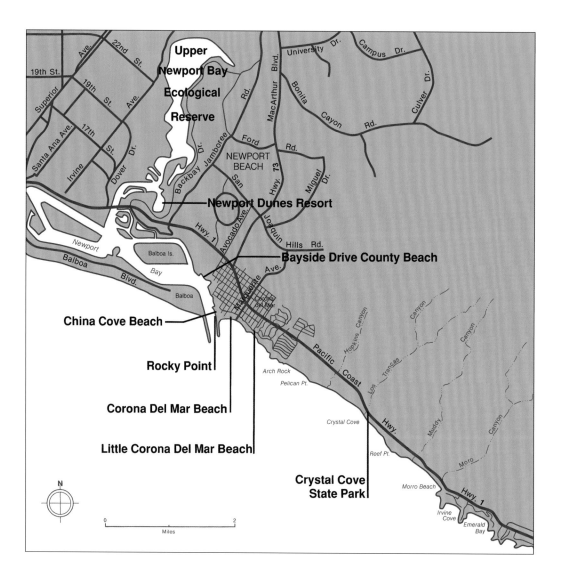

Orange County

UPPER NEWPORT BAY TO CRYSTAL COVE STATE PARK

FACILITIES · ENVIRONMENT

NAME	LOCATION	Entrance/Parking Fee	Parking	Restrooms	Lifeguard	Campground	Showers	Firepits	Stairs to Beach	Path to Beach	Bike Path	Hiking Trail	Facilities for Disabled	Boating Facilities	Fishing	Equestrian Trail	Sandy Beach	Dunes	Rocky Shore	Upland from Beach	Stream Corridor	Bluff	Wetland
Newport Dunes Resort	Bayside Dr., Newport Beach	•	•	•		•	•	•					•	•	•		•			•			•
Upper Newport Bay Ecological Reserve	W. of Backbay Dr., Newport Beach		•	•			•				•	•	•	•	•	•	•			•		•	•
Bayside Drive County Beach	Adjacent to 1901 Bayside Dr., Corona del Mar		•	•							•		•				•						
China Cove Beach	Along Harbor Channel, W. of Shell St., Corona del Mar								•	•							•						
Rocky Point	Along Harbor Channel, S. of China Cove Beach, Corona del Mar								•	•									•				
Corona Del Mar State Beach	Ocean Blvd. and Iris Ave., Corona del Mar	•	•	•	•		•	•	•	•			•		•		•		•			•	
Little Corona Del Mar Beach	Ocean Blvd. and Poppy Ave., Corona del Mar		•	•						•					•		•		•			•	
Crystal Cove State Park	W. of Pacific Coast Hwy., between Arch Rock and Irvine Cove	•	•	•	•	•	•	•		•	•	•	•		•	•	•		•			•	

NEWPORT DUNES RESORT: A 15-acre lagoon with swimming, boating, and camping facilities. Day-use fee. Resort includes picnic tables, playground, volleyball and basketball courts, swimming pool, outdoor showers, dressing rooms, and beach fire rings; entire facility is wheelchair accessible. Cabanas and other beach equipment can be rented. Eight-lane boat ramp open 24 hours; boat launch fee. Slips and dry storage are available as well as paddleboard, kayak, sea cycle, wind surfboard, and sailboat rentals. Restaurant open daily. Camping facilities include 406 R.V. spaces with full hookups; tent camping at R.V. sites. Reservations are required; fee. For information, call: (714) 729-3863. Reservations: 1-800-288-0770.

UPPER NEWPORT BAY ECOLOGICAL RESERVE: 752-acre estuary, intertidal mudflat, and salt marsh wildlife reserve; wintering area for migratory shorebirds and waterfowl. A regional park with signed entrances at Bayview Ave. and University Ave. includes hiking, biking, and equestrian trails; parking. A paved bike path follows Backbay Dr. and provides overlooks. Path continues around bay and joins trail along Irvine Avenue. Remain on established trails and roads. Galaxy Park, at southwest end of reserve, provides a paved blufftop path and overlooks the bay. Farther south, west of Polaris Dr., is North Star Beach, with a small-craft launch and kayak rentals, wheelchair-accessible restrooms, showers, and free parking.

The Ecological Reserve is surrounded by the Upper Newport Bay Nature Preserve. An interpretive center is at 2301 University Drive at the corner of Irvine Ave.; call: (714) 973-6820. Newport Bay Naturalists and Friends offers guided bird-watching tours the second Saturday of October through March; no reservations required; call: (714) 973-6826 or (949) 640-6746. The San Joaquin Freshwater Marsh Reserve, just east of the ecological reserve at Campus Dr. near Jamboree Rd., protects 202 acres of marsh above Newport Bay; part of the University of California Natural Reserve System. Unpaved trail from Upper Newport Bay along west side of San Diego Creek passes near edge of Reserve. Stay on trail. For information on permits and group tours, call: (949) 824-6031.

BAYSIDE DRIVE COUNTY BEACH: Sandy beach adjacent to the Orange County Sheriff/Harbor Patrol Bureau at 1901 Bayside Drive. Parking lot; wheelchair-accessible restrooms and showers; volleyball net; jetty for fishing.

CHINA COVE BEACH: A pair of small sandy coves along the harbor channel; volleyball net. Pedestrian access from Shell and Cove streets, and from two stairways at Ocean Blvd.: one at foot of Fernleaf and one from Lookout Point blufftop park at foot of Goldenrod.

ROCKY POINT: Small sandy cove along harbor channel; formerly called Pirates Cove for its many small wave-formed caves. Access is by stairs or footpaths down the bluff from Ocean Blvd. near the foot of Heliotrope Avenue. Paths also lead from China Cove Beach to the east and over the rocks from Corona del Mar State Beach to the west.

CORONA DEL MAR STATE BEACH: Popular large sandy beach just east of the eastern jetty at the entrance to Newport Harbor. Fee parking lot at the beach; free parking on the bluff. Beach facilities include picnic tables, volleyball nets, outdoor showers, restrooms, dressing rooms, beach wheelchair, food and equipment concessions, and lifeguard service; rock jetty area used by surfers. A concrete path with overlooks and benches leads from Inspiration Point, a grassy area at the foot of Orchid Ave., to the southeast end of the beach. The beach can also be reached from the foot of Jasmine St. at Lookout Point, a grassy park with benches and views of beaches and harbor entrance; park is on the blufftop along Ocean Ave. between Iris and Goldenrod avenues. Stairs and paths lead to Corona Del Mar, Rocky Point, and China Cove beaches. For information, call: (949) 644-3151 or (949) 644-3047.

LITTLE CORONA DEL MAR BEACH: Sandy cove with rocky tidepools and reefs. A walkway leads down to the beach from the foot of Poppy Avenue. Call: (949) 640-3151 or (949) 644-3047. Newport Marine Life Refuge is offshore; tours offered; call: (949) 644-3038 for reservations.

Crystal Cove State Park includes Pelican Point, Los Trancos, Reef Point, and El Moro Canyon.

CRYSTAL COVE STATE PARK: Approximately 2,700 acres in the San Joaquin Hills, formerly part of the Irvine Ranch, comprising four park units. 3.25 miles of rocky coves and sandy beaches, grassy terraces, and wooded canyons; offshore is an underwater park and the Irvine Coast Marine Life Refuge. Visitors can dive, sport fish, swim, and surf; tidepools are plentiful. Park is open from 6 AM–sunset; day-use fee. Parking at each unit; no dogs or horses on beach. For information on diving and fishing, or to make campsite reservations, call: (949) 494-3539 or 1-800-444-7275. Mailing address: Orange Coast District, 3030 Avenida del Presidente, San Clemente 92672.

Pelican Point is the northernmost entrance to Crystal Cove State Park and has paved blufftop trails, overlooks, and steep beach access ramps. Outdoor showers, wheelchair-accessible restrooms, and dressing rooms are available; diving access.

A large parking lot is located at Los Trancos on the east side of Pacific Coast Hwy. with dressing rooms and wheelchair-accessible restrooms. Paved path on south end by restrooms leads to access tunnel under the highway, then past cottages in the Historic District to a sandy beach with tidepools and a visitor center; beach wheelchair; handicapped parking only below highway. Pathway can also be reached by crossing highway at the crosswalk.

Reef Point has two steep, paved ramps and a stairway to a sandy beach with tidepools. Diving access; hiking trail and bike path, outdoor shower, dressing rooms, and wheelchair-accessible restrooms are available. Pit toilets on beach at bottom of ramps.

The El Moro Canyon unit provides access to inland campsites and 18 miles of hiking, bicycling, and equestrian trails. Water is available only at ranger stations. Three environmental and equestrian campgrounds with picnic tables and a total of 32 sites; no fires or smoking allowed; no direct access to the beach. Visitor center on east side of highway has wheelchair-accessible restrooms.

The adjacent undeveloped Laguna Coast Wilderness Park extends west of Laguna Canyon Rd. between Crystal Cove and Aliso and Wood Canyons. The park can be entered on special public access days, including weekends; call ahead: (949) 494-9352.

Pocket Beach, Laguna Beach

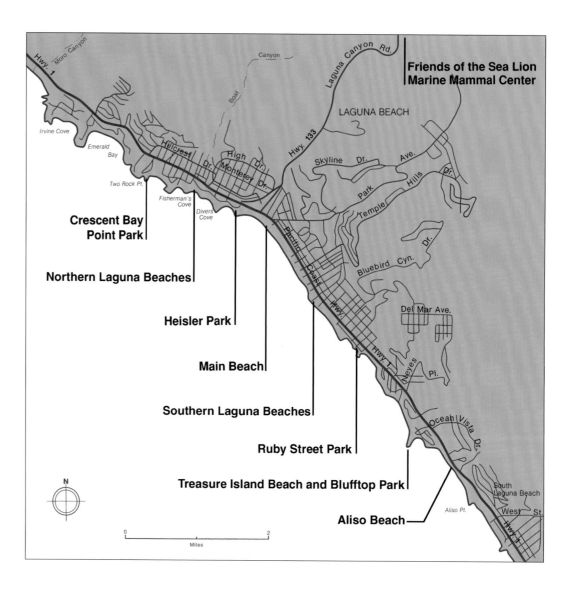

Friends of the Sea Lion
Marine Mammal Center

LAGUNA BEACH

Hwy. 1

Moro Canyon

Canyon

Laguna Canyon Rd.

Hwy. 133

Irvine Cove

Emerald
Bay

Hillcrest Dr.

High Dr.

Monterey Dr.

Skyline Dr.

Ave.

Hills Dr.

Two Rock Pt.

Park

Temple

Fisherman's
Cove

Divers
Cove

Pacific Coast Hwy

Bluebird Cyn.

Dr.

Crescent Bay
Point Park

Del Mar Ave.

Northern Laguna Beaches

Heisler Park

Main Beach

Hwy. 1

Neyes Pl.

Southern Laguna Beaches

Ocean Vista Dr.

Ruby Street Park

Treasure Island Beach and Blufftop Park

South
Laguna Beach

Aliso Beach

Aliso Pt.

West St.

Hwy. 1

N

0   2
Miles

Orange County

LAGUNA BEACH

NAME	LOCATION	Entrance/Parking Fee	Parking	Restrooms	Lifeguard	Campground	Showers	Firepits	Stairs to Beach	Path to Beach	Bike Path	Hiking Trail	Facilities for Disabled	Boating Facilities	Fishing	Equestrian Trail	Sandy Beach	Dunes	Rocky Shore	Upland from Beach	Stream Corridor	Bluff	Wetland
Crescent Bay Point Park	W. end of Crescent Bay Dr., Laguna Beach	•											•									•	
Northern Laguna Beaches	W. of Cliff Dr., Laguna Beach	•	•	•		•		•	•				•				•		•				
Heisler Park	W. of Cliff Dr., 400 block, Laguna Beach	•	•			•		•	•				•									•	
Friends of the Sea Lion Marine Mammal Center	20612 Laguna Canyon Rd., Laguna Beach	•	•																	•			
Main Beach	W. end of Broadway, W. of Pacific Coast Hwy., Laguna Beach	•	•	•		•							•				•						
Southern Laguna Beaches	Street ends W. of Pacific Coast Hwy., Laguna Beach	•		•				•	•						•		•		•				
Ruby Street Park	W. end of Ruby St., Laguna Beach																	•				•	
Treasure Island Beach and Blufftop Park	30800 block of Pacific Coast Hwy., South Laguna	•	•	•	•		•	•	•	•			•		•		•					•	
Aliso Beach	31000 block of Pacific Coast Hwy., South Laguna	•	•	•	•		•	•		•			•		•		•		•				

CRESCENT BAY POINT PARK: View of Laguna Beach and Seal Rock from viewing area along lawn and paved walkway; wheelchair accessible. Sea lions haul out on offshore rocks.

NORTHERN LAGUNA BEACHES: Several small, rocky coves north of Laguna's main beach can be reached by walkways off Cliff Dr.: Crescent Bay, which provides restrooms, volleyball net, and lifeguard, is accessible from a stairway off Circle Way and from a steep ramp off Barranca St.; Shaw's Cove from the west end of Fairview St.; Fisherman's Cove and Divers Cove from walkways 20 yards apart in the 600 block of Cliff Dr.; Picnic Beach from a ramp at the west end of Myrtle St. in Heisler Park; and Rockpile Beach from stairs at the west end of Jasmine St. in Heisler Park. Picnic Beach has a lifeguard and showers. Signs at all beach and cove accessways explain fishing regulations. The Glenn E. Vedder Ecological Reserve extends offshore from Divers Cove to the north end of Main Beach. No shell or vegetation gathering or fishing within reserve borders, which roughly parallel Heisler Park. Good tidepools and wildlife viewing.

HEISLER PARK: Grassy area on the bluff above Main, Picnic, and Rockpile beaches; great views. Picnic tables, shuffleboard, and lawn bowling available. Wheelchair accessible restrooms, outdoor showers, and stairs and pathways leading to nearby beaches. Metered parking on adjoining streets.

FRIENDS OF THE SEA LION MARINE MAMMAL CENTER: A facility that rehabilitates sea lions and seals that wash up on the beach. Public can visit and observe any animals being treated. Small gift shop; free parking in front of the center. Located on Laguna Canyon Rd. 2.5 miles from the beach. Open 10 AM-4 PM daily. Call: (949) 494-3050.

MAIN BEACH: Sandy beach and grassy area just south of Heisler Park. A frontage park where Laguna Canyon Rd. meets Pacific Coast Hwy. includes a boardwalk, benches and tables, a playground, and basketball and volleyball courts. Restrooms at north and south end of beach; restrooms at south end are wheelchair-accessible. Beach wheelchairs at lifeguard headquarters at north end. Outdoor showers at both ends. Access to Main Beach also from the ends of Legion St. (through mall) and Sleepy Hollow. Laguna Beach Marine Life Refuge extends from Crescent Bay Point to Camel Point in southern Laguna. Call: (949) 497-0706.

SOUTHERN LAGUNA BEACHES: The following street ends, from north to south starting just south of the main beach, have paths and/or stairs to small sandy and rocky beaches: Cleo St., St. Ann's Dr., Thalia St. (with wheelchair-accessible viewing platform and bench), Anita St., Oak St. (with wheelchair-accessible viewing platform and bench), Brooks St. (popular surfing spot), Cress St., Mountain Rd., Bluebird Canyon Dr., Agate St., Pearl St., Diamond St., and Moss Street. Woods Cove can be reached from a stairway at the 1900 block of Ocean Way between Diamond and Pearl. Sandy strip containing Victoria and Lagunita beaches can be reached from the foot of Sunset Terrace and the foot of Dumond Dr., both off Victoria Drive. Limited on-street parking; lifeguards. Good surfing and diving. Respect private property.

RUBY STREET PARK: Small park overlooking the ocean; no beach access. Features picnic table, benches, and shade trees.

TREASURE ISLAND BEACH AND BLUFFTOP PARK: A grassy park on the blufftop provides benches, picnic tables, wheelchair-accessible restrooms, showers, and paved and dirt paths along the bluff. Stairs and two ramps lead to the beach and tidepools. Fee parking lot at Wesley Dr.; path leads past hotel from lot to the beach and park. Pedestrian access also from Montage Resort Drive.

ALISO BEACH: County park features a sandy beach popular for boogie boarding, with lifeguard service, picnic tables, grills, firepits, play area, outdoor showers, concessions, and snack bar. Beach wheelchairs available from lifeguard headquarters; call: (949) 499-3312. Park extends onto the east side of Pacific Coast Hwy.; beach and inland sides are connected by a pedestrian tunnel; fee parking. Call: (949) 661-7013 ext. 21.

Main Beach, Laguna Beach

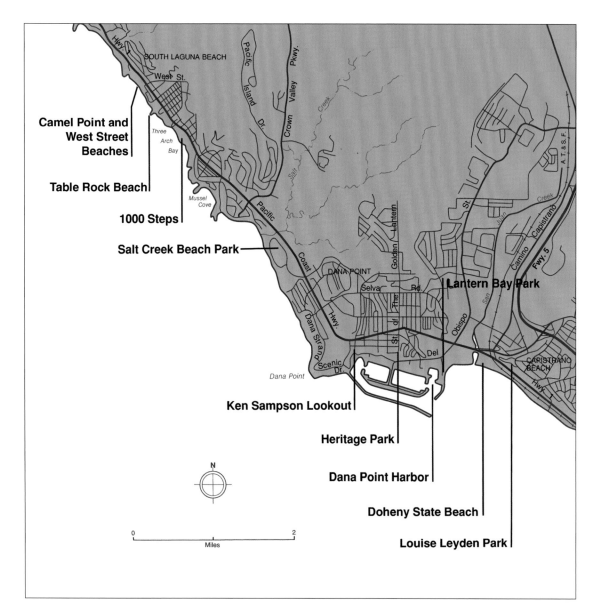

Camel Point and West Street Beaches

Table Rock Beach

1000 Steps

Salt Creek Beach Park

Lantern Bay Park

Ken Sampson Lookout

Heritage Park

Dana Point Harbor

Doheny State Beach

Louise Leyden Park

SOUTH LAGUNA BEACH

West St.

Three Arch Bay

Mussel Cove

Pacific

DANA POINT

Selva

Dana Point

Scenic Dr.

Dana Strand Hwy

Pacific Coast

Pacific Island Dr.

Crown Valley Pkwy.

Creek

Salt

Creek

Golden Lantern

Lantern

St. of The

Del

Obispo

San Jun

Camino Capistrano

A.T.&S.F.

Fwy. 5

CAPISTRANO BEACH

Hwy. 1

N

0 ——— 2
Miles

1000 Steps

254

Orange County

SOUTH LAGUNA/DANA POINT

FACILITIES   ENVIRONMENT

NAME	LOCATION	Entrance/Parking Fee	Parking	Restrooms	Lifeguard	Campground	Showers	Firepits	Stairs to Beach	Path to Beach	Bike Path	Hiking Trail	Facilities for Disabled	Boating Facilities	Fishing	Equestrian Trail	Sandy Beach	Dunes	Rocky Shore	Upland from Beach	Stream Corridor	Bluff	Wetland
Camel Point and West Street Beaches	Between ends of Camel Pt. Dr. and West St., South Laguna		•	•					•						•		•						
Table Rock Beach	End of Table Rock Dr., South Laguna								•						•		•						
1000 Steps	Pacific Coast Hwy., opposite 9th Ave., South Laguna		•	•	•		•								•		•						
Salt Creek Beach Park	W. of Pacific Coast Hwy. off Ritz Carlton Dr., South Laguna	•	•	•	•				•	•	•	•	•		•		•		•	•		•	
Three Arch Cove Beach	W. of Pacific Coast Hwy., N. of Salt Creek Beach Park, South Laguna								•						•		•						
Ken Sampson Lookout	S. end of St. of the Blue Lantern, Dana Point		•																	•		•	
Heritage Park	N. of Dana Point Harbor Dr., Dana Point		•										•							•		•	
Lantern Bay Park	End of St. of the Golden Lantern, Dana Point		•	•						•										•		•	
Dana Point Harbor	Del Obispo and Island Way, Dana Point		•	•	•						•		•	•	•								
Doheny State Beach	Del Obispo and Pacific Coast Hwy., Dana Point	•	•	•	•	•	•	•		•		•		•	•		•		•				
Louise Leyden Park	End of Villa Verde, off Camino Capistrano, Capistrano Beach		•																			•	

CAMEL POINT AND WEST STREET BEACHES: A walkway to a middle section of Aliso Beach County Park begins at the end of Camel Pt. Dr.; enter through pedestrian gate; respect private property. Two walkways south of Camel Pt. Dr. marked by signs among houses also lead down from the 31300 block of Pacific Coast Highway. A stairway at the end of West St. and another stairway at the Coast Royale condominium building 200 feet north of West St. also lead to these sandy beaches. South Laguna Marine Life Refuge extends south to Three Arch Bay. Lifeguard, pit toilets, volleyball net. Limited street parking.

TABLE ROCK BEACH: A stairway at the end of Table Rock Dr., two blocks south of West St., leads to a sandy cove at the southern end of Aliso Beach County Park. Limited street parking.

1000 STEPS BEACH: A long, steep stairway on Pacific Coast Hwy. at the foot of 9th Ave. leads down the bluff to the beach; lifeguard, restrooms, shower. Limited street parking.

SALT CREEK BEACH PARK: The long, sandy beach is a popular surfing spot. Both upland and beach facilities can be reached from Ritz Carlton Dr. off Pacific Coast Highway. Parking and wheelchair-accessible restrooms are available at a large metered parking lot, open 5 AM-midnight, on Ritz Carlton Drive. From the west side of the lot, a paved path leads to the beach and to the grassy seven-acre Bluff Park, which has benches, barbecue grills, picnic tables, and wheelchair-accessible restrooms.

Just south of Ritz Carlton Dr., a pedestrian gate, open 24 hours, leads through the hotel grounds to a paved path along the blufftop above Salt Creek beach; the path, which skirts the southern perimeter of the grounds, has benches and overlooks providing views of gray whale migration routes, Catalina Island, and the coastline. The Niguel Marine Life Refuge lies offshore. At its northern end, the path joins a paved multi-use trail leading north along the shoreline past Bluff Park. Stairways and paths lead from the multi-use trail to the beach, changing rooms, outdoor showers, restrooms, snack bar, and a path along the seawall. Paths and trails lead north and east through a golf course and into Salt Creek Regional Park, featuring several miles of trails for hiking, biking, and horseback riding.

The southern entrance to Salt Creek Beach Park is off Selva Rd. and features picnic tables, barbecue grills, restrooms, and path and stairs to Dana Strands Beach. The parking lot is free and open 5 AM-midnight. For information, call: (949) 661-7013 ext. 21.

Inland of the Salt Creek Corridor (north on Crown Valley Pkwy.) are two regional parks, Aliso and Wood Canyons and Laguna Niguel, all featuring hiking, biking, and riding trails. Aliso and Wood Canyons is a wilderness portion of the Laguna Greenbelt, and includes scenic canyons, woodlands, and native grasslands; call: (949) 831-2790. Laguna Niguel features a 44-acre recreational lake; call: (949) 831-2791.

THREE ARCH COVE BEACH: A long walk from the north end of Salt Creek Beach Park, past the private Monarch Beach community, leads over shoreline rocks and large boulders to a small sandy cove. Public access across the Monarch Beach shoreline must be on wet sand at low tide only. Tidepools are plentiful. Respect private property.

KEN SAMPSON LOOKOUT: View of Dana Point Harbor with sheltered benches. Other lookout points and benches are at the south end of St. of the Amber Lantern and off St. of the Violet Lantern.

HERITAGE PARK: Grassy blufftop park with benches and paved wheelchair-accessible paths overlooking Dana Point Harbor. A stairway leads down to harbor. Parking is at the foot of St. of the Golden Lantern at El Camino Capistrano (take Del Prado Ave. east from Pacific Coast Highway).

LANTERN BAY PARK: Grassy park on bluff above Dana Point Harbor just east of Heritage Park. Facilities include a basketball court, par course, picnic areas with grills, restrooms, and path with overlooks; views of harbor. Open 6 AM-10 PM. Main park entrance is off St. of the Park Lantern; stairs lead into park from St. of the Golden Lantern.

DANA POINT HARBOR: Once a major port for square-rigged ships plying the hides trade, the harbor, named for author Richard Henry Dana, now features extensive boating facilities and services. There are grassy areas with picnic tables and walkways throughout the area, and the small stillwater "Baby Beach" off Ensenada Place; near the beach is a wheelchair-accessible fishing platform. Restaurants and small shops line a boardwalk along the marina; restrooms are wheel-

chair-accessible. The Dana Point Marine Life Refuge lies offshore at the north end of the harbor.

Boating facilities include 42 guest slips, 15-lane launch ramp, dry storage, hoist, engine and hull maintenance, fuel, and marine supplies. For information on slip rentals and information, call the Harbor Patrol: (949) 248-2222. For information on boat charters, sport fishing and seasonal whale-watching trips, call: (949) 496-5794. For diving, breakwater fishing, sailing lessons, bicycle rentals, and trailer space, call: (949) 496-6177.

The Ocean Institute, at the harbor's west end at 24200 Dana Pt. Harbor Dr., offers classes and educational programs, including trips aboard the "Floating Lab" research vessel. Most programs are open to groups on weekdays, and to individuals and families on weekends. The historic square-rigged sailing ship, the Pilgrim, is anchored at the Institute, and can be toured year round on Sundays 10 AM-2:30 PM. The public can also take cruises aboard the tallship, Spirit of Dana Point. The Institute also has some exhibits and a gift shop and is open to the public 10 AM-3 PM on weekends. For free schedules, reservations, or information, call: (949) 496-2274.

DOHENY STATE BEACH: Campground and day-use area; fee for use. Day-use area has .75 mile of beachfront and a five-acre lawn with picnic tables, group picnic areas, fire rings, changing rooms, and food and supply concessions. Surfing is allowed in designated areas. Beach wheelchairs; boardwalk at north end from picnic area onto sand; wheelchair-accessible fire ring at south end. The offshore area is an underwater park for divers. Campground area has 120 campsites with fire rings, tables and drinking water, and a trailer sanitation station. Four sites are wheelchair accessible; some sites for hikers and bicyclists. For camping information and reservations, call 1-800-444-7275; for group picnic reservations, call: (949) 496-3617. A Visitor Center, includes exhibits and a gift shop; call: (949) 496-2704. For park information, call: (949) 496-6172 or 492-0802. The San Juan Creek Bikeway, which runs along the west side of the creek, begins just north of Doheny State Beach and continues inland along the creek past San Juan Capistrano.

LOUISE LEYDEN PARK: An overlook with a walkway and benches; grassy areas. View of Doheny State Beach and Dana Point Harbor.

Linda Lane City Park

San Clemente City Beach

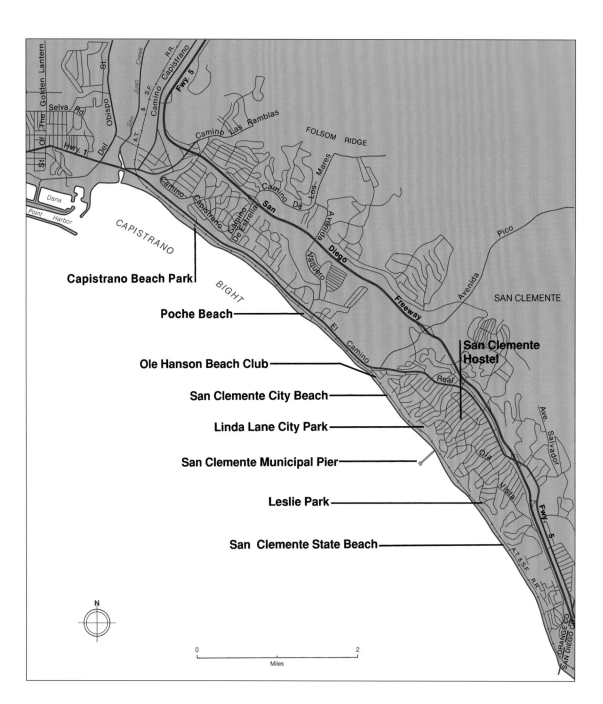

Orange County

SAN CLEMENTE

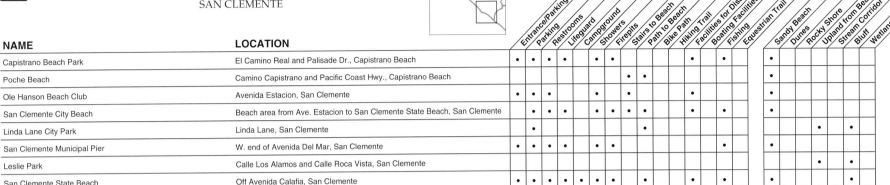

NAME	LOCATION	Entrance/Parking Fee	Parking	Restrooms	Lifeguard	Campground	Showers	Firepits	Stairs to Beach	Path to Beach	Bike Path	Hiking Trail	Facilities for Disabled	Boating Facilities	Fishing	Equestrian Trail	Sandy Beach	Dunes	Rocky Shore	Upland from Beach	Stream Corridor	Bluff	Wetland
Capistrano Beach Park	El Camino Real and Palisade Dr., Capistrano Beach	•	•	•	•		•	•					•		•		•						
Poche Beach	Camino Capistrano and Pacific Coast Hwy., Capistrano Beach								•	•							•						
Ole Hanson Beach Club	Avenida Estacion, San Clemente	•	•	•			•			•			•				•						
San Clemente City Beach	Beach area from Ave. Estacion to San Clemente State Beach, San Clemente		•	•	•		•	•	•	•			•		•		•						
Linda Lane City Park	Linda Lane, San Clemente		•							•										•		•	
San Clemente Municipal Pier	W. end of Avenida Del Mar, San Clemente	•	•	•	•		•	•					•				•						
Leslie Park	Calle Los Alamos and Calle Roca Vista, San Clemente																			•		•	
San Clemente State Beach	Off Avenida Calafia, San Clemente	•	•	•	•	•	•		•				•		•		•					•	

CAPISTRANO BEACH PARK: Sandy beach with metered parking lot. Restrooms are wheelchair-accessible. Outdoor showers, basketball and volleyball courts, firepits, food and equipment concessions, picnic tables, and viewing platforms are available. Call: (949) 661-7013 ext. 21.

POCHE BEACH: A stairway on the inland side of Pacific Coast Hwy. just south of the end of Camino Capistrano leads to an underpass beneath the highway and the railroad tracks; follow the path to a sandy beach. Volleyball court. Call: (949) 661-7013 ext. 21.

OLE HANSON BEACH CLUB: Facilities include a snack bar at North Beach, playground, and heated six-lane public pool and wading pool with dressing room and showers; fee. Overlooks the beach. There are wheelchair-accessible restrooms, free parking, and stairs and a wheelchair-accessible ramp to beach. Located upstairs from pool, the banquet hall is available for weddings and parties. Open year-round. For information, call: (949) 361-8207.

SAN CLEMENTE CITY BEACH: The beach extends south for two miles from the Ole Hanson Beach Club. Most beach access requires crossing railroad tracks, and most paths or stairways have limited on-street parking. Avenida Pico, which ends at Ole Hanson Beach Club, leads directly to North Beach; metered parking lot, snack stands, body board rentals, beach wheelchairs, picnic areas, volleyball, swings, firepits and grills, outdoor showers, wheelchair-accessible restrooms, lifeguard. The west ends of Dije Ct., Buena Vista at W. El Portal, and W. Escalones at W. Mariposa each have paths and/or stairs that lead to the beach north of the pier. At Calle Los Alamos near Calle Lasuen an unmaintained dirt path down a steep slope leads to the beach; the path is hard to find. Avenida Esplanade at Paseo del Cristobal leads to a pedestrian railroad overpass and stairway; lifeguard, restrooms, showers, picnic facilities, metered parking available. Surfing is allowed, depending on the crowds and the season. For weather or surf conditions, call: (949) 492-1011. For general information, call: (949) 361-8219.

LINDA LANE CITY PARK: An upland park that slopes down to the beach, located just north of the municipal pier; playground, parking, and picnic area. A storm drain tunnel under the railroad tracks provides access to the beach.

SAN CLEMENTE MUNICIPAL PIER: A popular fishing pier with food concessions, fish-cleaning sinks, and a bait and tackle shop, restrooms, and lifeguard tower. The beach area has picnic facilities, fire rings, outdoor showers, and wheelchair-accessible restrooms. Avenida Del Mar has metered parking, and leads to the beach and the pier. AMTRAK stops right at the pier. There is one underpass to the beach; all other access is across railroad tracks. Full-service restaurant at the base of the pier. For information, call: (949) 361-8219.

LESLIE PARK: A small park on a hill overlooking the beaches, accessible only by climbing up the hillside. Limited on-street parking. No beach access.

SAN CLEMENTE STATE BEACH: Provides 85 tent and 72 full hookup camping sites on the bluff at the southern end of the park, with one group site for up to 50 people. Three tent and three hookup campsites are wheelchair-accessible, as are restrooms and showers. Two steep trails lead to the beach. Day-use and camping fee. Beach wheelchair available at entrance kiosk. The city of San Clemente operates the day-use area at the foot of Avenida Calafia as Calafia Beach Park, which provides access to the narrow sandy beach via a controlled pedestrian crossing at the railroad tracks. Facilities include metered parking, picnic areas, snack bar, wheelchair-accessible restrooms, and outdoor shower; lifeguard. At the north end of Plaza a la Playa street is a tunnel that crosses under railroad tracks and leads to the beach. Surfing is allowed in designated areas. For information, call: (949) 492-3156 or (949) 492-0802.

San Clemente Municipal Pier

Nicholson Point Park

San Diego County

San Diego County was first inhabited by humans more than 10,000 years ago, when Native Americans migrated west from the Colorado River area and formed several permanent settlements near the bay. In 1542, the European explorer Juan Rodríguez Cabrillo spent six days in San Diego Bay, landed on Point Loma, and named the area San Miguel. In 1602, Sebastián Vizcaíno, a merchant navigator who was charting the California coast for the Spanish government, visited the same spot and renamed it San Diego. Today, thousands of people visit Point Loma annually to watch the migration of the California gray whale to its winter breeding grounds in the Gulf of California.

Maritime interests have since taken advantage of the shelter provided by the landlocked San Diego Bay. The bay has a long history as the home port for a large commercial fishing fleet. During World War I the U.S. Navy established headquarters for the 11th Naval District in the bay; today, the facilities in San Diego Bay constitute the largest naval base on the west coast. In addition to the commercial and military shipping activity, a number of sportfishing charters operate out of the bay, and numerous marinas and launching facilities provide berthing space and services for a large portion of Southern California's boating population.

San Diego County's 76 miles of shoreline include virtually every kind of coastal landscape. The north county coast consists primarily of a series of long, sandy beaches backed by steep eroded bluffs ranging in elevation from a few feet to more than 100 feet above sea level. The La Jolla coastline south includes a number of rocky headlands, coves, and points whose beaches are transformed into tidepools during low tide periods. Some of the largest and best remaining examples of coastal wetlands in Southern California are located in San Diego; examples include Los Peñasquitos Marsh in the Torrey Pines State Reserve, and the Tijuana River Estuary near the U.S.-Mexico border.

As in all of Southern California, the mild Mediterranean climate has had a major influence on the county's development. Many coastal communities were originally established as agricultural centers. Pacific Beach, a residential community located only eight miles north of downtown San Diego, was considered a farming district as recently as the late 1930s. Principal crops throughout the county include citrus fruits, avocados, flowers, and vegetables. Along the coast, major agricultural areas are located on the terraces and floodplains north of Del Mar; large farms also operate inland of San Diego Bay, near Chula Vista.

The mild air and water temperatures, combined with a short rainy season and diverse coastal topography, make San Diego County's beaches popular for sunbathing, swimming, surfing, surf and rock fishing, diving, and tidepool exploring. Beachgoers should check tide tables; due to a lack of sand, certain beaches are limited in size at higher tides. The San Diego coast also supports a large recreational boating community. Marinas and launching facilities from Oceanside to Chula Vista provide opportunities for thousands of people to enjoy boating activities.

The variety of aquatic recreation activities available in San Diego County is demonstrated dramatically at Mission Bay. The bay, previously known as False Bay because ships would mistake its entrance for the entrance to San Diego Bay to the south, originally was a salt marsh and estuary system at the mouth of the San Diego River. In 1946, construction began to convert the wetland into a boating and water recreation complex. Today, Mission Bay contains dock facilities and slips for more than 1,900 boats, as well as sheltered swimming bays, an aquatic park, water-skiing areas, picnic areas, campgrounds, playgrounds, and numerous fishing spots.

For more information on San Diego County, write or call the San Diego Convention and Visitors Bureau, 401 B St. Suite 1400, San Diego, CA 92101-4237; (619) 236-1212; go to the Visitor's Center at the intersection of First Ave. and F St. in downtown San Diego or 7966 Herschel Avenue in La Jolla; or see www.sandiego.org. More detailed information can also be obtained from local Chamber of Commerce offices. Public transit in San Diego County consists of trains, trolleys, and buses. For coastal transit information, call the San Diego Metropolitan Transit System at 1-800-266-6883 or 1-888-722-4889 TTY/TDD in northern San Diego County or at (619) 233-3004 or (619) 234-5005 TTY/TDD in southern San Diego County, or visit the Transit Store, 102 Broadway, for maps, tickets, timetables, and information, or see www.sdcommute.com. The Coaster train connects Oceanside with downtown San Diego, with stops at intermediate coastal points; call: 1-800-262-7837 or (619) 234-5005 TTY, or see www.gonctd.com. Most transit vehicles are wheelchair-accessible. Bus stops served by accessible buses are marked with the wheelchair symbol. Call the transit numbers for specific information on Dial-A-Ride service.

Coastal Industry

The various physical characteristics and resources of California's coastal environment provide an opportunity for the development of diverse industrial operations.

Geologic formations along portions of the coast and beneath the continental shelf offshore contain petroleum resources that have been extracted since the turn of the century; other coastal resources that have been extracted commercially include sand and kelp. As the use of coastal resources has increased and expanded, shoreside facilities have been developed to refine and process the raw materials into marketable products. For example, many petroleum refineries are located in the Los Angeles basin between El Segundo and Long Beach, and a number of kelp processing plants operate within the Port of San Diego.

Electric power plants and sewage treatment facilities have also been located along the coast in order to take advantage of the large volumes of water provided by the ocean. Power plants require copious amounts of water to cool circulating steam, while sewage treatment plants use the ocean as a location for the disposal of treated effluent.

Commercial shipping and the trades that support it constitute a major portion of California's coastal industry. California's ports have served as the principal gateways for the import and export of raw materials and finished goods such as petroleum, ores, timber, agricultural products, and automobiles. Support industries at many ports include tanker terminals, shipbuilding and repair facilities, lumber mills, truck and railroad terminals, and seafood processing plants and canneries.

Although the operations at various industrial facilities may differ significantly, many coastal dependent or coastal related facilities employ heavy equipment, high voltage power lines, or potentially hazardous substances in their operation. Consequently, many of these facilities are closed to the public. To promote public safety and prevent interference with the functions of coastal industries, please respect fences, warning signs, and other barriers installed by the operators of the facilities when using beaches near these facilities.

San Onofre Nuclear Generating Station

San Diego County

CAMP PENDLETON

NAME	LOCATION	Entrance/Parking Fee	Parking	Restrooms	Lifeguard	Campground	Showers	Firepits	Stairs to Beach	Path to Beach	Bike Path	Hiking Trail	Facilities for Disabled	Boating Facilities	Fishing	Equestrian Trail	Sandy Beach	Dunes	Rocky Shore	Upland from Beach	Stream Corridor	Bluff	Wetland
San Onofre State Beach (North)	S.W. of I-5 at Basilone Rd., San Onofre	•	•	•	•	•		•		•		•			•		•		•	•	•	•	
San Onofre State Beach (South)	S.W. of I-5, 2.5 mi. S. of Basilone Rd. off-ramp, San Onofre	•	•	•	•	•	•	•		•		•	•		•		•					•	
Bike Path	Along I-5 from San Clemente to S. entrance of Camp Pendleton						•													•			
Las Flores Viewpoint	W. of I-5, 5 mi. S. of San Onofre		•																	•			
Camp Pendleton Beach Access	W. of I-5 at Las Pulgas Rd., Camp Pendleton	•	•	•		•							•				•				•		
Aliso Creek Roadside Rest	W. of I-5, 6 mi. N. of Oceanside	•	•	•															•				

SAN ONOFRE STATE BEACH NORTH: Wide, sandy beach below sandstone bluffs. North of San Onofre Nuclear Generating Station, the state beach includes the well-known surfing spots, Trestles Beach to the north and Surf Beach to the south, as well as a wetland area at the mouth of San Mateo Creek. Trestles Beach can be reached at two points. A paved trail leads to the beach from a parking lot at Cristianitos Rd. and El Camino Real near the southern boundary of the city of San Clemente; the parking lot is on the east side of the highway next to a fast food restaurant and is open 6 AM-10 PM (8 PM winter months). Trestles Beach can also be reached from the north end of Basilone Rd. just west of the highway; a drop-off point provides access to an unpaved trail through the blufftop San Mateo Creek Preserve. No parking permitted. Surf Beach is reached by continuing south on Basilone Rd.; entrance to beach and parking lot is west of Basilone Road. San Mateo Campground, a unit of San Onofre State Beach, includes 160 developed campsites, some with hookups, east of the highway on Cristianitos Road; for information, call: (949) 362-2531. An unpaved trail leads from the campground to the paved Trestles Beach trail at Cristianitos Road. Restrooms at Surf Beach are wheelchair accessible.

SAN ONOFRE STATE BEACH SOUTH: Called Bluffs Beach. A campground along the abandoned highway on the bluffs above the beach has 221 tent and trailer spaces and a group camp for up to 50 people; facilities include wheelchair-accessible restrooms, cold outdoor showers, and two trailer sanitation stations (no individual hookups). 26 primitive tent spaces are available at the Echo Arch hike-in camp, located on a terrace between the blufftop and the beach; closed during the winter rainy season. Enroute campsites are also available. Day-use and camping fees are charged. There are several hiking trails from the campgrounds to the beach, with panoramic views of the coast and whale-watching spots. Dogs on leashes are allowed on Trail 1 and Trail 6. For information, call: (949) 492-0802.

BIKE PATH: Paved path runs from Ave. del Presidente in San Clemente and crosses to the west side of the highway at Cristianitos Road. The path follows Basilone Rd. through San Onofre State Beach Campground and continues down the coast.

LAS FLORES VIEWPOINT: Parking area with a view of the adjacent bluffs and the beach.

CAMP PENDLETON BEACH ACCESS: Public access to Las Pulgas (Red) Beach is for surf fishing and self-contained R.V. camping only; no hookups or facilities. No swimming or surfing. To apply for a fishing or camping permit, go to the Game Wardens Office, Building 25155. For fishing, a valid state fishing license with an ocean enhancement stamp is required. For camping, a valid registration for vehicle and any trailer is required. There is a fee for permits. For more information, call: (760) 725-3360.

ALISO CREEK ROADSIDE REST: Landscaped area with picnic tables, a dog run, and a map display showing local points of interest.

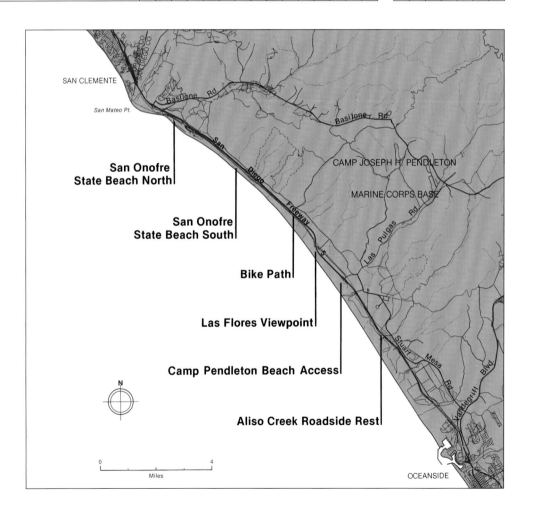

Oceanside Harbor

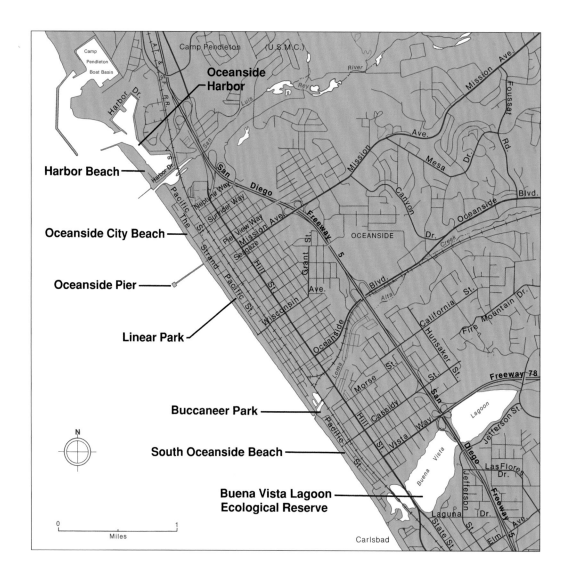

Oceanside Harbor

Harbor Beach

Oceanside City Beach

Oceanside Pier

Linear Park

Buccaneer Park

South Oceanside Beach

Buena Vista Lagoon
Ecological Reserve

NAME	LOCATION	FACILITIES																ENVIRONMENT						
		Entrance/Parking Fee	Parking	Restrooms	Lifeguard	Campground	Showers	Firepits	Stairs to Beach	Path to Beach	Bike Path	Hiking Trail	Facilities for Disabled	Boating Facilities	Fishing	Equestrian Trail	Sandy Beach	Dunes	Rocky Shore	Upland from Beach	Stream Corridor	Bluff	Wetland
Oceanside Harbor	Along Harbor Dr., Oceanside	•	•	•									•	•	•								
Harbor Beach	Corner of Harbor Dr. South and Pacific St., Oceanside	•	•		•	•		•					•		•		•						
Oceanside City Beach	Along The Strand, between San Luis Rey River and Witherby St., Oceanside	•	•	•	•		•	•	•				•		•		•					•	
Oceanside Pier	W. end of Pier View Way, Oceanside	•	•	•					•				•		•								
California Surf Museum	223 North Coast Hwy., Oceanside				•															•			
Linear Park	Along Pacific St., between Sportsfisher Dr. and Wisconsin Ave., Oceanside								•				•							•		•	
Buccaneer Park	N.E. corner of Pacific and Morse streets, Oceanside	•	•				•						•							•			•
South Oceanside Beach	W. of Pacific St., between Morse and Eaton streets, Oceanside			•					•				•				•	•					
Buena Vista Lagoon Ecological Reserve	E. and W. of I-5, 2.5 mi. S. of downtown Oceanside	•	•										•	•						•			•

OCEANSIDE HARBOR: Small-craft harbor with berthing for about 960 vessels. Transients arrange berthing in person at the Harbor District office on the east side of the harbor, opposite the channel entrance, or by pre-paid reservations. Harbormaster: 1540 Harbor Dr. North, Oceanside 92054; (760) 435-4000. Harbor facilities and services include shops, restaurants, fuel dock, fishing pier, marine supplies, haul-out yard, charter sportfishing boats, fishing licenses, bait and tackle, lockers, and sailing instruction. A paved path runs along the harbor. Boat launching ramp on the west side of the harbor, with four concrete lanes, open 24 hours; launching is free. Picnic tables, most restrooms, and the fishing pier are wheelchair accessible. Whale-watching trips available November to March; (760) 722-2133. Some free parking is available.

HARBOR BEACH: Wide, sandy beach; surf and rock fishing. Facilities include wheelchair-accessible picnic areas, fire rings, playground equipment, and campsites for self-contained R.V.'s; some free parking available. Lifeguards on duty only during the summer. For information, call the Harbor District: (760) 435-4000.

OCEANSIDE CITY BEACH: Popular beach for swimming, surfing, and surf fishing; ramps or stairways located at several street ends between Breakwater Way and Witherby St. and in the 1400 block of Pacific St. provide beach access. Facilities include covered picnic tables, barbecue grills, outdoor showers, and wheelchair-accessible restrooms; lifeguard service is provided. A landscaped concrete walkway with benches runs along The Strand from Breakwater Way to the Oceanside Pier. Public parking lots at ends of Breakwater Way, Windward Way, Pier View Way, and Wisconsin, Hayes, and Forster streets; fee at some lots. Tyson Street Park, at the west end of Tyson St. along The Strand, is a grassy park with benches, a playground, and wheelchair-accessible restrooms; can also be reached via stairs from Pacific Street. For information, call: (760) 435-5041.

OCEANSIDE PIER: 1,954-foot-long municipal pier, lighted and patrolled at night, with bait and tackle shop, restaurant, lifeguard station, and wheelchair-accessible restrooms. Bait and tackle shop: (760) 966-1406. No license required to fish from pier. A motorized, wheelchair-accessible trolley runs from the pier gate to the end. Strand Pier Plaza at the base of the pier has an amphitheater and a community center with meeting rooms and gym facilities; the center can be rented for special events, and the gym facilities are open to the public for a small fee. Fee parking. Call: (760) 966-1406.

CALIFORNIA SURF MUSEUM: Displays of surfing equipment and photographs; gift shop. Open Thursday-Monday 10 AM-4 PM; free admission. For information, call: (760) 721-6876.

LINEAR PARK: Landscaped concrete walkway along the blufftop with benches and view platforms; views of the shoreline. Stairs lead down to The Strand at Civic Center Drive, Pier View Way, Mission Avenue, and Tyson Street.

BUCCANEER PARK: Provides parking, covered picnic tables, outdoor shower, wheelchair-accessible restrooms, playground, athletic field, basketball and volleyball courts, and concession stand (summer only). A concrete path along the bank of Loma Alta Marsh, a small lagoon, leads beneath the railroad trestle to Hill St.; a variety of bird species can be seen from the path.

SOUTH OCEANSIDE BEACH: Swimming, surfing, and surf fishing. Includes Buccaneer Beach, accessible via stairs in the 1500 block of Pacific St., north of Morse Street. Public stairs are also adjacent to 1639 Pacific St., and at the end of Cassidy Street. Lifeguards in summer at Buccaneer Beach.

BUENA VISTA LAGOON ECOLOGICAL RESERVE: The natural reserve protects wetland vegetation and bird habitat. Plant and animal life can be viewed from many points around the lagoon shoreline; fishing is permitted from the shore. Parking is available along Jefferson St. between I-5 and Route 78; duck feeding area on Jefferson St. south of Route 78 with paved parking lot, benches, and views.
The Buena Vista Audubon Nature Center, at 2202 S. Hill St. in Oceanside on the north side of the bridge, provides nature exhibits, wheelchair-accessible restrooms, and a short trail along the lagoon shoreline. Open Tuesday-Saturday 10 AM-4 PM; call for information on group tours and monthly lectures: (760) 439-2473.

Oceanside City Beach, Along the Strand

Amtrak Pacific Surfliner

Amtrak's passenger train, the Pacific Surfliner, offers almost hourly departures between Los Angeles and San Diego between 6 AM and 10 PM. (Some of the Pacific Surfliner trains also provide service as far north as San Luis Obispo, following the route of the Amtrak Coast Starlight; see p. 178.) For about one hour of this two hour and forty minute trip, the Pacific Surfliner travels along the Orange and San Diego County shorelines, providing unobstructed views of the coast between Dana Point in Orange County and the Del Mar area of San Diego County. Modern trains offer laptop computer outlets, food service, and bicycle racks.

About 25 minutes out of the Santa Ana station going south (7 minutes out of San Clemente going north), the Pacific Surfliner meets the coast at Capistrano Beach, located just south of Dana Point. The Dana Point Harbor breakwater is visible just before the train turns south to parallel the shoreline. Dana Point was named after the author Richard Henry Dana, who vividly described this area of the coast in his book *Two Years Before the Mast,* written in 1840.

The Pacific Surfliner arrives at San Clemente about five minutes after reaching the coast (15 minutes out of Oceanside going north). Four trains stop at San Clemente Beach and Municipal Pier every day; the 1,100-foot-long pier is a popular fishing spot. Five minutes after leaving the San Clemente Pier (about 10 minutes out of Oceanside going north), the Pacific Surfliner passes the San Onofre Nuclear Generating Station. The power plant consists of two pressurized water reactor units. The nuclear reactors for Units 2 and 3 have been operating since the early 1980s, and are contained in the

170-foot-high dome-shaped structures. When operating at full capacity, the power plant provides electrical power for more than 2.7 million people, and more than 1.6 million gallons of ocean water are used per minute to condense reactor-produced steam after the steam turns the generating turbines.

After passing San Onofre going south (just out of Oceanside going north), the Pacific Surfliner begins its run through the Camp Pendleton Marine Corps Base. The base extends along 17 miles of shoreline, and is a major center for amphibious landing operations and other advanced training programs; occasionally, train passengers may see military exercises taking place on either side of the train.

The Pacific Surfliner passes the Oceanside Harbor and arrives at the Oceanside station immediately after leaving Camp Pendleton (about 20 minutes out of Del Mar going north). The city of Oceanside was incorporated in 1888, during one of Southern California's railroad booms. By the 1920s, Oceanside was known as a beach resort and an agricultural produce distribution center.

Between Oceanside and Del Mar, the Pacific Surfliner travels along the mid-San Diego County coast and passes through the towns of Carlsbad, Encinitas, and Solana Beach. This area has long been one of the country's major centers for the production of cut flowers, shrubs, bulbs, and other ornamental plants. The coastal landscape here is characterized by narrow beaches backed by high coastal bluffs, except where natural drainage areas meet the ocean. Some of the best examples of California's remaining coastal wetlands are found in these drainage basins; the Pacific Surfliner passes along or across

a total of five coastal marshes and lagoons between Oceanside and Del Mar.

About 20 minutes out of Oceanside (30 minutes out of San Diego going north), the Pacific Surfliner passes the Del Mar Race Track and San Diego County Fairground complex, located on the north side of the San Dieguito River estuary; the race track and fairgrounds are visible from the inland side of the train. The Pacific Surfliner continues for a short distance to the Del Mar station; for the first 1-½ miles south of the Del Mar Station, the train travels immediately adjacent to the coast, less than 100 feet from the edges of the sea cliffs.

About three minutes out of Del Mar (27 minutes out of San Diego going north), the Pacific Surfliner travels inland along the Los Peñasquitos Marsh Natural Preserve, a part of Torrey Pines State Park. The marsh contains valuable habitat for rare and endangered bird species such as the California least tern, and is also a nesting and feeding place for waterfowl and other migratory birds.

Between the coast at Los Peñasquitos Marsh and the downtown San Diego station, the Pacific Surfliner follows an inland route. Minutes before reaching the station, the train passes inland of Mission Bay, the largest aquatic park on the West Coast. The downtown station is only four blocks from the bay edge, and is located near the Broadway Pier, the Maritime Museum, and other bayside points of interest. The historic Gaslamp District and other attractions are nearby.

For scheduling, fares, or other information on the Pacific Surfliner, call Amtrak at 1-800-872-7245.

Amtrak Pacific Surfliner

San Diego County

CARLSBAD

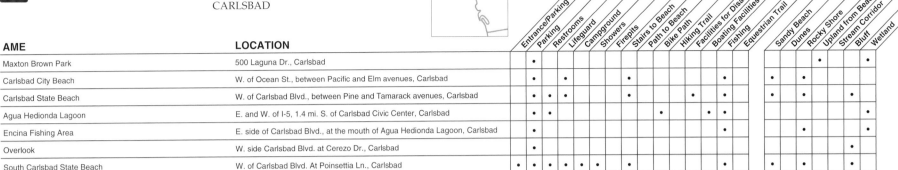

AME	LOCATION	Entrance/Parking Fee	Parking	Restrooms	Lifeguard	Campground	Showers	Firepits	Stairs to Beach	Path to Beach	Bike Path	Hiking Trail	Facilities for Disabled	Boating Facilities	Fishing	Equestrian Trail	Sandy Beach	Dunes	Rocky Shore	Upland from Beach	Stream Corridor	Bluff	Wetland
Maxton Brown Park	500 Laguna Dr., Carlsbad		•																	•			•
Carlsbad City Beach	W. of Ocean St., between Pacific and Elm avenues, Carlsbad		•	•					•						•		•		•				
Carlsbad State Beach	W. of Carlsbad Blvd., between Pine and Tamarack avenues, Carlsbad	•	•	•					•		•		•		•		•		•			•	
Agua Hedionda Lagoon	E. and W. of I-5, 1.4 mi. S. of Carlsbad Civic Center, Carlsbad	•	•							•				•	•								•
Encina Fishing Area	E. side of Carlsbad Blvd., at the mouth of Agua Hedionda Lagoon, Carlsbad	•													•				•				•
Overlook	W. side Carlsbad Blvd. at Cerezo Dr., Carlsbad	•																		•			
South Carlsbad State Beach	W. of Carlsbad Blvd. At Poinsettia Ln., Carlsbad	•	•	•	•	•	•		•						•		•	•				•	

MAXTON BROWN PARK: On the shore of Buena Vista Lagoon. A paved path leads to an overlook with benches, picnic table, and views of the lagoon. Call: (760) 602-7513.

CARLSBAD CITY BEACH: Access to the beach is from stairways at the north end of Ocean St., at the ends of Christiansen Way, Grand Ave., and Elm Ave., and from the parking lot on Ocean St. at Carlsbad Boulevard. Popular beach for swimming, surfing, and surf fishing; surfing hours restricted May-October. Lifeguards during the summer only.

CARLSBAD STATE BEACH: Sandy and rocky beach backed by bluffs, with benches, picnic tables, grassy areas, and overlooks along the blufftop. Stairways lead from view platforms to a paved pedestrian-only path that runs along the seawall. Additional stairways to the beach and the seawall path are at the ends of Sycamore, Maple, Cherry, Hemlock, and Tamarack avenues. Wheelchair-accessible restrooms. Popular beach for swimming, surfing, diving, and rock and surf fishing; lifeguards during summer only. Call: (760) 438-3143.

AGUA HEDIONDA LAGOON: Calm water within the lagoon makes it a popular swimming, fishing, and water-skiing spot. From Cove Dr. a paved walkway adjacent to a condominium development leads to views of the lagoon and a trail along a small sandy beach. A paved blufftop pedestrian and bicycle path and vista point off Park Dr. west of Neblina Dr. provide views of the lagoon. Several informal dirt paths lead to and along the shoreline from Adams St. east of Hoover St., and from the south end of Hoover Street. The Snug Harbor Marina, at 4215 Harrison St., provides a launch ramp, a loading dock, boat and motor sales, equipment rentals, snack bar, picnic area, and restrooms; fees. Call: (760) 434-3089. The Agua Hedionda Lagoon Foundation's new Discovery Center is scheduled to open Sept. 2003 at 1580 Cannon Rd., Carlsbad.

ENCINA FISHING AREA: Provides fishing access to the western basin of Agua Hedionda Lagoon. Private property; permission to pass revocable by owner. No overnight parking.

OVERLOOK: A small blufftop parking area provides views of the coast. The sandy pocket beach below can be reached only by scrambling down the steeply eroded cliffs.

SOUTH CARLSBAD STATE BEACH: Facilities include 214 campsites along the bluffs, hot showers, restrooms, laundromat, grocery and bait store, beach equipment rentals, and a trailer sanitation station. Access from the blufftop to the beach is provided by stairways; the beach is noted for swimming, surfing, surf fishing, and diving. The north end of the beach is also accessible by trails criss-crossing the bluffs and dunes west of the highway. Limited day-use parking within the park; roadside parking is available along Carlsbad Blvd. at San Marcos Creek. Day-use and camping fees. Park information: (760) 438-3143. Camping reservations: 1-800-444-7275.

To the east is Batiquitos Lagoon Ecological Reserve, an important migratory waterfowl habitat. A trail leads along the north shore of the lagoon, with four parking lots along Batiquitos Drive. A fifth parking lot is located at the Nature Center at the end of Gabbiano Lane; for information, contact the Batiquitos Lagoon Foundation, P.O. Box 130491, Carlsbad 92013, (760) 845-3501.

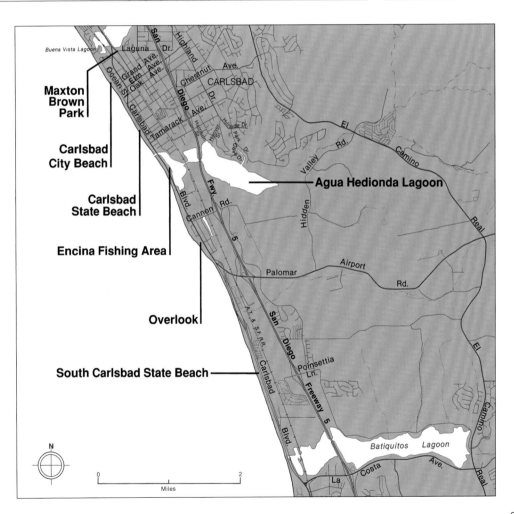

Surfing

Surfing is the sport of riding a wave as it breaks along the shoreline. This sport has been practiced for hundreds of years; Captain James Cook observed natives surfing when he visited the Sandwich (Hawaiian) Islands in 1778. Surfing became known on the California coast during the early 1900s. However, the sport was seen as an activity for the truly hardy, and caught on slowly; a major reason for this was that early surfboards were 14- to 18-foot-long redwood planks that weighed up to 150 lbs.

With the advent of lightweight balsa boards in the 1940s and polyurethane boards in the 1950s, the popularity of surfing increased significantly. A modern surfboard is five to nine feet long and may weigh as little as eight pounds. Today, there are also a number of variations to the sport; kayaks, canoes, small sailboats, inflated rubber mats, and bodyboards are used to ride waves. However, using a surfboard and body surfing (riding waves without the aid of any floating object) are still the most popular forms of surfing.

Regardless of the particular type of surfing one does, the basic principles are the same. The surfer heads toward the shore just ahead of an incoming, unbroken wave. As the wave moves forward and catches up to the surfer, the surfboard (or body, boat, raft, etc.) starts to slide down the face of the wave. At this point the surfer has caught the wave and now must stay just ahead of the breaking portion of the wave to maintain the ride; this is done by shifting body weight to maneuver the board in the appropriate direction.

The most highly regarded surfing spots are usually locations where waves either approach the shore at an angle or are refracted around a point, jetty, or other protruding landform. In these locations incoming waves reach the breaking point in stages. The portion of the wave closest to the beach or in the shallowest water breaks first; the remaining portions of the wave break in succession as they move toward the shore.

Surfers nearest the breaking portion of a wave have the right-of-way; don't try to catch a wave in front of someone. In crowded surfing areas, watch out for unattended boards; try to hold on to your own board at all times to protect yourself and others from injury.

Most public beach areas where surfing is popular have designated surfing areas or times; look for signs that indicate where and when surfing is allowed, or ask a lifeguard. At most beaches in Southern California, a black ball flag is raised at the lifeguard tower to signal surfers to get out of the water. Body surfing is usually permitted in any swimming area. Rip currents are common in many surfing areas; if caught in a rip current, swim or paddle parallel to the shore to escape the current.

There are a number of popular surfing beaches in California. In San Diego County, San Onofre and Windansea beaches are frequented by surfers. Other well-known Southern California surfing areas include Newport and Huntington beaches in Orange County, Malibu and Leo Carrillo beaches in Los Angeles County, and the Rincon area in Ventura County. Excellent surfing conditions also exist farther north; Steamer Lane is a popular surfing spot in Santa Cruz County.

San Diego County

ENCINITAS

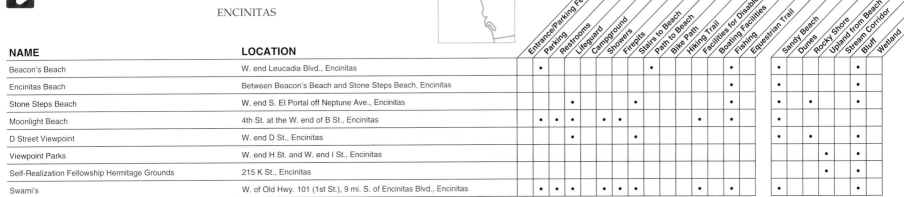

NAME	LOCATION	Entrance/Parking Fee	Parking	Restrooms	Lifeguard	Campground	Showers	Firepits	Stairs to Beach	Path to Beach	Bike Path	Hiking Trail	Facilities for Disabled	Boating Facilities	Fishing	Equestrian Trail	Sandy Beach	Dunes	Rocky Shore	Upland from Beach	Stream Corridor	Bluff	Wetland
Beacon's Beach	W. end Leucadia Blvd., Encinitas		•						•						•		•					•	
Encinitas Beach	Between Beacon's Beach and Stone Steps Beach, Encinitas														•		•					•	
Stone Steps Beach	W. end S. El Portal off Neptune Ave., Encinitas				•				•						•		•		•			•	
Moonlight Beach	4th St. at the W. end of B St., Encinitas	•	•	•			•	•					•		•		•					•	
D Street Viewpoint	W. end D St., Encinitas				•												•					•	
Viewpoint Parks	W. end H St. and W. end I St., Encinitas																			•		•	
Self-Realization Fellowship Hermitage Grounds	215 K St., Encinitas																			•		•	
Swami's	W. of Old Hwy. 101 (1st St.), 9 mi. S. of Encinitas Blvd., Encinitas	•	•	•			•	•					•		•		•					•	

BEACON'S BEACH: City-managed park, formerly known as Leucadia State Beach. Parking is available; a wide dirt path with railings leads to a sandy, bluff-backed beach. Good surfing, swimming, surf fishing, and skin diving. Parking also available at the end of Grandview Street. Call: (760) 633-2740 or 633-2880 (recorded message).

ENCINITAS BEACH: Narrow, sandy beach popular for swimming and surf fishing. No access from the bluffs to the beach; to reach this area, walk south along the shore from Beacon's Beach or north from Stone Steps. Call: (760) 633-2740 or 633-2880.

STONE STEPS BEACH: A long, partially stone stairway leads to an extremely narrow cobble beach used for swimming, surfing, and surf fishing. Lifeguard service. Bench at top of stairway; on-street parking is available along Neptune Avenue. For information, call: (760) 633-2740 or 633-2880.

MOONLIGHT BEACH: A sandy beach popular for surfing, swimming, and surf fishing. Facilities include volleyball and tennis courts, equipment rentals, a snack bar, picnic tables, fire rings, outdoor showers, and wheelchair-accessible restrooms. Parking lots are located at 4th and C streets, and at the end of C St.; parking for vehicles carrying wheelchairs is available at 4th and B streets. Seasonal lifeguard service. For information, call: (760) 633-2740 or 633-2880.

D STREET VIEWPOINT: Features a bench, shoreline views, and a long, wooden stairway that leads down to a narrow cobble beach. Lifeguards; on-street parking only.

VIEWPOINT PARKS: Small day-use parks at the ends of H and I streets provide grassy areas with picnic tables, paved paths, benches, and views of the coast. No beach access. On-street parking.

SELF-REALIZATION FELLOWSHIP HERMITAGE GROUNDS: Meditation area; panoramic views of the coastline from the gardens. Open Tuesday-Saturday 9 AM-5 PM; Sunday 11 AM-5 PM. Commercial photography and bathing attire are prohibited. For information, call: (760) 753-2888.

SWAMI'S: Small blufftop park with stairs leading to a narrow sandy beach that is known as an excellent surfing spot; surf fishing, scuba diving, and swimming are also popular. Lifeguard service is available in summer only. The blufftop area has a grassy picnic area with barbecue grills, restrooms, and parking. Restrooms and picnic area are wheelchair accessible. An outdoor shower is located near the base of the stairway. For information, call: (760) 633-2740 or 633-2880.

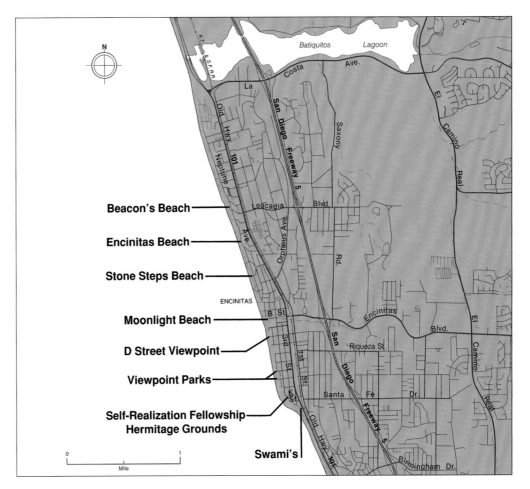

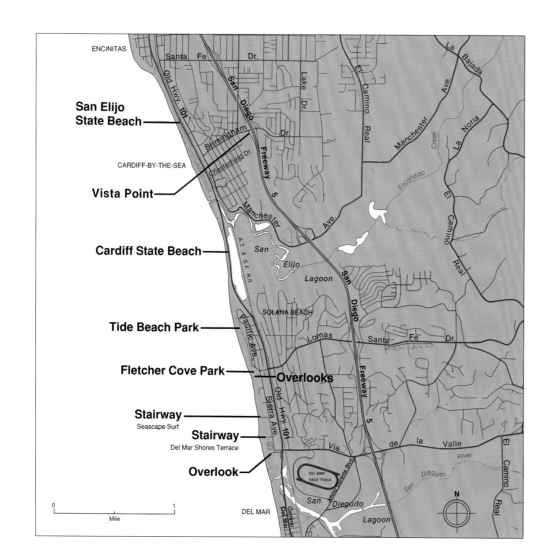

San Elijo State Beach

Vista Point

Cardiff State Beach

Tide Beach Park

Fletcher Cove Park — **Overlooks**

Stairway
Seascape Surf

Stairway
Del Mar Shores Terrace

Overlook

ENCINITAS

CARDIFF-BY-THE-SEA

SOLANA BEACH

DEL MAR

N

0   1
    Mile

Tide Beach County Park

Cardiff State Beach

San Diego County

CARDIFF/SOLANA BEACH

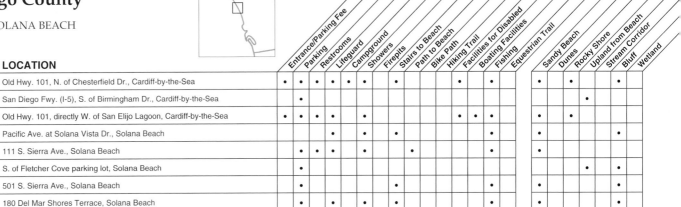

NAME	LOCATION	Entrance/Parking Fee	Parking	Restrooms	Lifeguard	Campground	Showers	Firepits	Stairs to Beach	Path to Beach	Bike Path	Hiking Trail	Facilities for Disabled	Boating Facilities	Fishing	Equestrian Trail	Sandy Beach	Dunes	Rocky Shore	Upland from Beach	Stream Corridor	Bluff	Wetland
San Elijo State Beach	Old Hwy. 101, N. of Chesterfield Dr., Cardiff-by-the-Sea	•	•	•	•	•	•		•				•		•		•		•			•	
Vista Point	San Diego Fwy. (I-5), S. of Birmingham Dr., Cardiff-by-the-Sea		•																	•			
Cardiff State Beach	Old Hwy. 101, directly W. of San Elijo Lagoon, Cardiff-by-the-Sea	•	•	•	•		•						•	•	•		•		•				
Tide Beach Park	Pacific Ave. at Solana Vista Dr., Solana Beach			•	•		•		•						•		•					•	
Fletcher Cove Park	111 S. Sierra Ave., Solana Beach		•	•	•					•					•		•						
Overlooks	S. of Fletcher Cove parking lot, Solana Beach		•																•			•	
Stairway to Beach (Seascape Surf)	501 S. Sierra Ave., Solana Beach		•				•							•			•					•	
Stairway to Beach (Del Mar Shores Terrace)	180 Del Mar Shores Terrace, Solana Beach		•		•		•							•			•					•	
Overlook	End of Border Ave., Del Mar		•																	•			

SAN ELIJO STATE BEACH: Features the southernmost developed campground in the State Beach Park System. The campground, located atop a bluff overlooking the beach, has 171 sites that can accommodate trailers up to 35 feet long. Each site has a picnic table, fire ring, and grill. Other facilities include hot showers, a grocery store, a bait shop, beach equipment rentals, and wheelchair-accessible restrooms; laundry facilities are available during the summer and most holidays. The Campfire Center offers weekly nighttime programs including films, lectures, sing-alongs, and slide shows. A tidepool exhibit is at the park headquarters; there is also a native plant garden. Stairways lead from the blufftop to a cobble beach, which is popular for surf fishing, swimming, skin diving, and surfing; outdoor shower and wheelchair-accessible restrooms. Day-use area open sunrise to sunset. Fees for camping and for day-use parking within the park; some parking is available along the highway. Call: (760) 753-5091. For camping reservations, call:1-800-444-7275.

VISTA POINT: Accessible from the southbound lanes of San Diego Freeway (I-5). Overlooks San Elijo Lagoon and the Pacific Ocean.

CARDIFF STATE BEACH: Sandy beach popular for surfing, swimming, and surf fishing; swimming is allowed on the northern half of the beach, surfing on the southern half. On the north end of the beach is a landing and launch zone for carry-on and soft-bottom boats; the launch area is closed in summer. Tidepools are at the south end. The beach is patrolled by lifeguards year-round, but lifeguards are in the towers summer only. Wheelchair-accessible restrooms are available. Fee for parking. Parking lot hours 7 AM-sunset; the beach is open 6 AM-11 PM.

TIDE BEACH PARK: A stairway down the bluff leads to a sandy beach; surfing, skin diving, swimming, and surf fishing. Lifeguards are on duty during the summer.

FLETCHER COVE PARK: A ramp provides access to a sandy beach; popular activities include diving, swimming, surfing, surf fishing, and catching grunion. Park facilities include basketball and shuffleboard courts, sand volleyball courts, picnic tables, and restrooms. Lifeguard headquarters are located in the park; lifeguard service year-round. The Community Center Building, located at 133 Pacific Ave., can be rented for private functions; call: (760) 793-2564 or 793-2571. Beach parking lot is open 6 AM-6 PM winter, and 6 AM-10 PM summer; beach is open 24 hours.

OVERLOOKS: A paved trail in the Fletcher Cove parking lot leads up the bluff to the Las Brisas and Surfsong viewpoints, which afford panoramic views of the ocean and the coastline. Along the trail are benches and a landscaped garden.

STAIRWAY TO BEACH (SEASCAPE SURF): A paved sidewalk leads to a wooden stairway to a sandy beach popular for surfing, skin diving, surf fishing, swimming, and catching grunion. There are also sand volleyball courts. Parking in lots at 335 and 550 S. Sierra Avenue.

STAIRWAY TO BEACH (DEL MAR SHORES TERRACE): A stairway at the end of a paved walkway leads to a sandy beach, popular for surf fishing, catching grunion, swimming, and surfing. Sand volleyball courts; lifeguards in summer. Parking lots at 721 and 733 S. Sierra Avenue.

OVERLOOK: There is an overlook at the terminus of Border Avenue.

San Elijo State Beach

269

Torrey Pines State Reserve

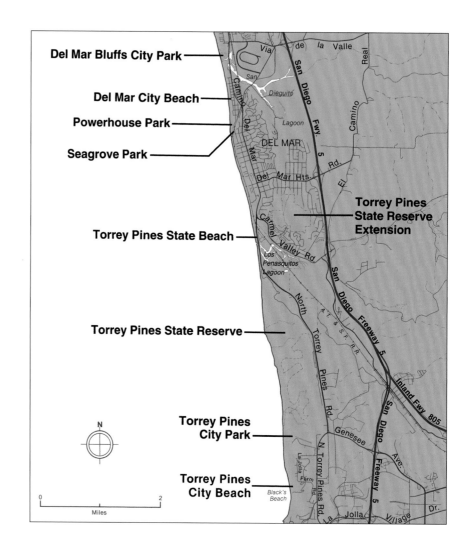

Del Mar Bluffs City Park

Del Mar City Beach

Powerhouse Park

Seagrove Park

DEL MAR

Torrey Pines State Beach

Torrey Pines State Reserve

Torrey Pines City Park

Torrey Pines City Beach

Torrey Pines State Reserve Extension

Via de la Valle

San Dieguito

Lagoon

San Diego Fwy

Camino Real

Camino

Del Mar Hts

El Camino Rd.

Carmel Valley Rd.

Los Penasquitos Lagoon

San Diego Freeway 5

North Torrey Pines Rd.

A T & S F R R

Inland Fwy 805

Genesee Ave.

San Diego Freeway 5

La Jolla Farm

N. Torrey Pines Rd.

Black's Beach

La Jolla Village Dr.

N

0   Miles   2

San Diego County

DEL MAR / TORREY PINES

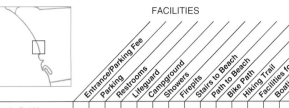

NAME	LOCATION	Entrance/Parking Fee	Parking	Restrooms	Lifeguard	Campground	Showers	Firepits	Stairs to Beach	Path to Beach	Bike Path	Hiking Trail	Facilities for Disabled	Boating Facilities	Fishing	Equestrian Trail	Sandy Beach	Dunes	Rocky Shore	Upland from Beach	Stream Corridor	Bluff	Wetland
Del Mar Bluffs City Park	W. of Camino Del Mar, just N. of the San Dieguito River mouth, Del Mar														•		•				•	•	
Del Mar City Beach	W. of Camino Del Mar, from 29th St. to Torrey Pines State Beach, Del Mar	•	•		•		•								•		•						
Powerhouse Park	Coast Blvd. between 17th and 15th streets, Del Mar	•	•				•			•										•			
Seagrove Park	Ocean Ave. at 15th St., Del Mar	•	•										•							•			
Torrey Pines State Beach	McGonigle Rd., off Carmel Valley Rd., San Diego	•	•	•	•								•		•		•						
Torrey Pines State Reserve Extension	W. of I-5 off Del Mar Hts. Rd. at Mar Scenic Dr., San Diego		•									•								•		•	
Torrey Pines State Reserve	W. of N. Torrey Pines Rd., 2 mi. N. of Genessee Ave., San Diego	•	•	•								•	•							•		•	•
Torrey Pines City Park	W. of N. Torrey Pines Rd., at the end of Torrey Pines Scenic Dr., San Diego		•	•					•	•										•		•	
Torrey Pines City Beach (Black's Beach)	La Jolla Farms Rd. at Blackgold Rd., San Diego									•					•		•					•	

DEL MAR BLUFFS CITY PARK: Sandy beach at the mouth of the San Dieguito River; swimming and surf fishing. On-street parking only.

DEL MAR CITY BEACH: Wide, sandy beach; good swimming and surfing; other activities include catching grunion and surf fishing. Beach access is at street ends from 29th to 18th streets. Off-street fee parking adjacent to the Amtrak station on Coast Blvd. near 15th St.; also some on-street metered parking. Lifeguard service is provided year round from the main station at 17th St., summers only at 20th and 25th streets. Outdoor showers at 17th St., 20th St., and 25th St. stations. Dogs on beach subject to seasonal restrictions; call: (858) 755-1556.

POWERHOUSE PARK: Located at the south end of Del Mar City Beach. Grassy park with picnic areas, playground, and paved path to beach; outdoor shower. Fee parking lot adjacent to the Amtrak station.

SEAGROVE PARK: Grassy area with paved, wheelchair-accessible path and benches on the bluffs overlooking the ocean. Metered parking is available at L'Auberge restaurant.

TORREY PINES STATE BEACH: Wide, sandy beach backed by steep sandstone bluffs. Extends from 6th St. near Del Mar Heights Rd. south to Torrey Pines City Beach. Beach access and main parking lot are at the North Beach area adjacent to Los Peñasquitos Lagoon. Popular activities include picnicking, swimming, surfing, surf fishing, clamming, and skin diving. Beach is open 8 AM-11 PM; lifeguard service in summer and on spring weekends, with jeep patrols throughout the year. No dogs. Restrooms are wheelchair accessible. Fee parking lot closes at sunset; some off-street parking available. This beach supports snowy plovers, a threatened species, and there may be some restrictions on public access in this area to protect the habitat. For information, call: (619) 755-2063.

TORREY PINES STATE RESERVE EXTENSION: Numerous trails through the Extension offer spectacular views of Los Peñasquitos Lagoon, the main reserve, and the ocean. Rare and endangered plant and animal life, such as black-shouldered kites and the snowy plovers, may also be observed. Parking is available at various street ends off Del Mar Heights Road. A hiking trail and an undeveloped scenic overlook are located off Mar Scenic Drive. Within the Torrey Point subdivision at the end of Nogales Dr., a gated, paved pedestrian path leads to a public overlook that provides views of the reserve extension and Los Peñasquitos Lagoon. Please respect private property. Call: (858) 755-2063.

TORREY PINES STATE RESERVE: The reserve is situated on a series of steep bluffs interspersed with deep ravines; it is the only natural continental habitat for the world's rarest pine tree, the Torrey pine. The reserve contains two natural preserves: the Torrey Pines Natural Preserve has the finest stands of trees; the Los Peñasquitos Marsh Natural Preserve, part of one of the few remaining salt marsh and lagoon areas in Southern California, is the habitat of a number of rare and endangered bird species, such as the least tern and the light-footed clapper rail, and is an important feeding and nesting place for migratory waterfowl and shorebirds.

A network of trails leading to the beach and to several viewpoints, provides opportunities for observing plant and animal life and viewing the adjacent coastline. A visitor center with a small museum offers interpretive programs, trail maps, and species lists; guided walks are given on weekends and holidays at 10:00 AM and 2:00 PM. Interpretive displays are located at various points within the reserve. The reserve accommodates a limited number of people in order to protect the natural resources; visitors may be asked to come back at a later time or date if the reserve is full; it is full most weekend afternoons. Day-use fee. For information, call: (858) 755-2063.

TORREY PINES CITY PARK: The park is located on the sandstone bluffs overlooking the ocean. Steep paths and stairs lead to the beach from the north and south ends of the parking lot. A radio-controlled model aircraft field and a glider port are located within the park. Hang-gliding off the bluffs is also popular. Pit toilets.

TORREY PINES CITY BEACH (BLACK'S BEACH): Popular sandy beach backed by highly eroded, hazardous bluffs. The beach is noted for good swimming and surf fishing. Access is via a very steep pedestrian path at the south end of Blackgold Road. The path is the property of the University of California; permission to use is revocable by the owner. Very limited on-street parking, restricted to two hours.

Torrey Pines State Reserve

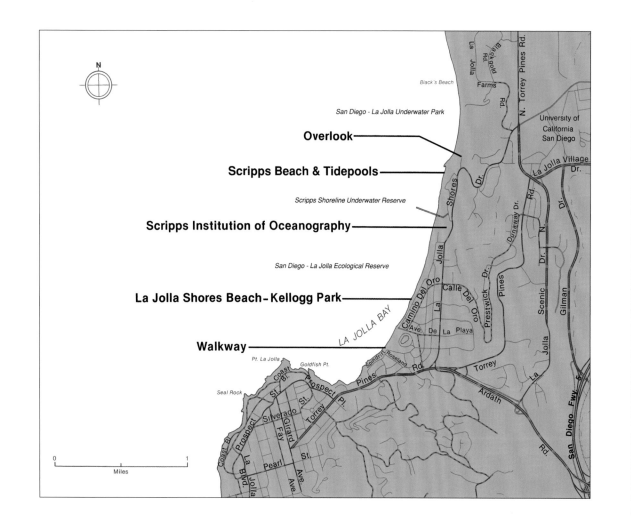

Overlook

Scripps Beach & Tidepools

Scripps Shoreline Underwater Reserve

Scripps Institution of Oceanography

San Diego - La Jolla Ecological Reserve

La Jolla Shores Beach - Kellogg Park

Walkway

San Diego - La Jolla Underwater Park

Black's Beach

University of California San Diego

LA JOLLA BAY

Pt. La Jolla

Goldfish Pt.

Seal Rock

N

0   1
         Miles

Scripps Beach

San Diego County

SCRIPPS/LA JOLLA

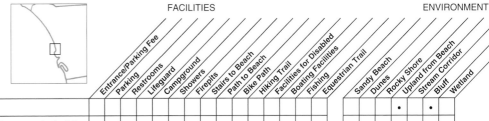

NAME	LOCATION	Entrance/Parking Fee	Parking	Restrooms	Lifeguard	Campground	Showers	Firepits	Stairs to Beach	Path to Beach	Bike Path	Hiking Trail	Facilities for Disabled	Boating Facilities	Fishing	Equestrian Trail	Sandy Beach	Dunes	Rocky Shore	Upland from Beach	Stream Corridor	Bluff	Wetland
Overlook	N. end of La Jolla Shores Lane, La Jolla																		•			•	
Scripps Beach and Tidepools	W. of La Jolla Shores Dr. at the Scripps Institution of Oceanography, La Jolla	•															•		•			•	
Scripps Institution of Oceanography	8602 La Jolla Shores Dr., La Jolla	•	•										•						•				
Underwater Reserves	From the S. end of the city of Del Mar to Goldfish Pt., La Jolla																		•				
La Jolla Shores Beach-Kellogg Park	Camino Del Oro at Calle Frescota, La Jolla	•	•	•			•	•					•	•	•		•						
Walkway to Beach	W. of Spindrift and Roseland drives, La Jolla								•	•							•						

OVERLOOK: A short path leads to a blufftop overlook that provides views of the offshore marine reserves. On-street parking only. Private property adjoins; do not trespass.

SCRIPPS BEACH AND TIDEPOOLS: Sandy beach and tidepools north of the 1,090-foot-long Ellen Browning Scripps Memorial Pier, a research facility of the Scripps Institution of Oceanography. Beach and tidepools are popular for observing intertidal marine life; part of the Scripps Shoreline-Underwater Reserve, protected by the California Department of Fish and Game. Please do not touch or disturb the marine life. On-street and metered parking.

SCRIPPS INSTITUTION OF OCEANOGRAPHY: Public facilities at the Institution include an aquarium and museum complex, open daily 9 AM-5 PM; partially wheelchair accessible. An artificial tidepool, located near the entrance of the aquarium, contains marine flora and fauna representative of a naturally occurring tidepool ecosystem. In the aquarium, tanks containing examples of local, deep-sea, and tropical marine life are on display. The museum contains several exhibits that demonstrate wave action and tide and current patterns. Admission is by donation. The Stephen Birch Aquarium-Museum is located across La Jolla Shores Dr. from the aquarium; call: (619) 534-FISH. Tours and educational programs are available; for information and registration, call: (858) 534-8665. Parking on street and in metered lot.

UNDERWATER RESERVES: The Scripps Shoreline-Underwater Reserve, the San Diego-La Jolla Underwater Park, and the San Diego-La Jolla Ecological Reserve are located along the coast from the southern limits of the city of Del Mar south to Goldfish Point in La Jolla. The reserves, comprising nearly 6,000 acres of tidal and submerged lands, were established to preserve the marine life of La Jolla Canyon. At low tide much of the shoreline is exposed, revealing numerous tidepools that support a diversity of marine plants and animals. Protected areas; do not touch or disturb the marine life.

LA JOLLA SHORES BEACH-KELLOGG PARK: Wide, sandy beach with good swimming conditions; also noted for surfing, surf fishing, and skin diving. Native American artifacts have been recovered by divers near the north end of the beach. Facilities include outdoor showers and wheelchair-accessible restrooms; lifeguard year-round. A launching area for hand-held boats is located at the end of Avenida de la Playa, at the south end of the beach. Kellogg Park, east of the beach promenade, has grassy areas, picnic tables, and firepits.

WALKWAY TO BEACH: A narrow concrete walkway and steps lead to a wide sandy beach at La Jolla Bay. Sunbathing and swimming only; no surfing allowed. On-street parking.

La Jolla Shores Beach

Cave, La Jolla

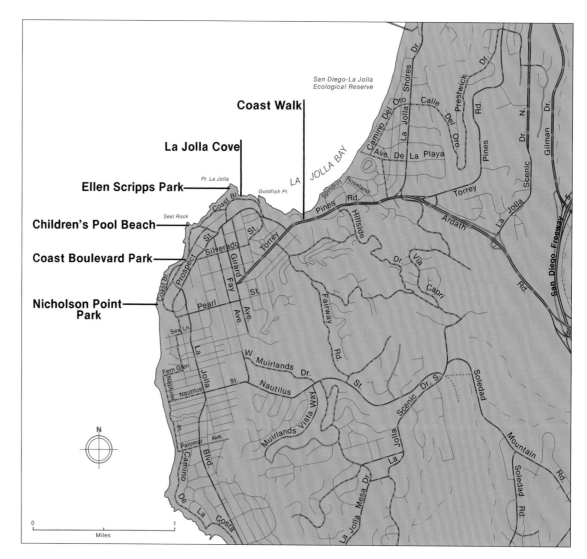

Coast Walk

La Jolla Cove

Ellen Scripps Park

Children's Pool Beach

Coast Boulevard Park

Nicholson Point
Park

San Diego-La Jolla
Ecological Reserve

LA JOLLA BAY

Pt. La Jolla

Goldfish Pt.

Seal Rock

N

0   1
    Miles

274

San Diego County

LA JOLLA BLUFFS

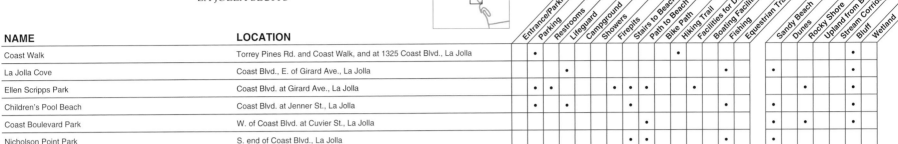

NAME	LOCATION	Entrance/Parking Fee	Parking	Restrooms	Lifeguard	Campground	Showers	Firepits	Stairs to Beach	Path to Beach	Bike Path	Hiking Trail	Facilities for Disabled	Boating Facilities	Fishing	Equestrian Trail	Sandy Beach	Dunes	Rocky Shore	Upland from Beach	Stream Corridor	Bluff	Wetland
Coast Walk	Torrey Pines Rd. and Coast Walk, and at 1325 Coast Blvd., La Jolla	•										•										•	
La Jolla Cove	Coast Blvd., E. of Girard Ave., La Jolla				•										•		•					•	
Ellen Scripps Park	Coast Blvd. at Girard Ave., La Jolla	•	•					•	•	•			•						•			•	
Children's Pool Beach	Coast Blvd. at Jenner St., La Jolla	•	•						•						•		•					•	
Coast Boulevard Park	W. of Coast Blvd. at Cuvier St., La Jolla									•							•		•			•	
Nicholson Point Park	S. end of Coast Blvd., La Jolla								•	•					•		•						

COAST WALK: A dirt path along the La Jolla bluffs provides a panoramic view of the ocean, beach, and caves along the shoreline. The walk can be entered at Torrey Pines Rd., just east of the Prospect St. intersection, or from the Goldfish Point area, adjacent to 1325 Coast Boulevard. The Sunny Jim is one of seven wave-created caves in La Jolla; a tunnel and stair lead from the Cave Store at 1325 Cave St.; open 9 AM-5 PM every day; admission fee.

Free public parking is available during weekends and non-business hours at the Bank of America lot at Fay Ave. and Kline St., and serves coastal accessways from Coast Walk south to Nicholson Point Park.

LA JOLLA COVE: Small cove with a sandy beach; very popular for diving, swimming, and surf fishing. Limited on-street parking.

ELLEN SCRIPPS PARK: A grassy picnic area on the blufftop includes picnic tables, firepits, and shuffleboard courts. Paved path to the bluff edge; path and stairs down the bluff lead to the rocky shore and Boomer Beach, well known for body surfing. Paths (but not restrooms) are wheelchair accessible.

CHILDREN'S POOL BEACH: Stairs lead to the sandy beach and to a paved path along the breakwater. No swimming; harbor seal haul-out site and rookery. On-street parking.

COAST BOULEVARD PARK: Shoreline park with benches and picnic tables; paths along Coast Blvd. lead to the sandy beach and rocky shore. Parking is available along nearby streets.

NICHOLSON POINT PARK: A hard-to-find public path between 100 and 202 Coast Blvd. leads to a stairway to the sandy beach; noted for swimming, diving, fishing, and body surfing.

La Jolla Caves, viewed from Coast Walk

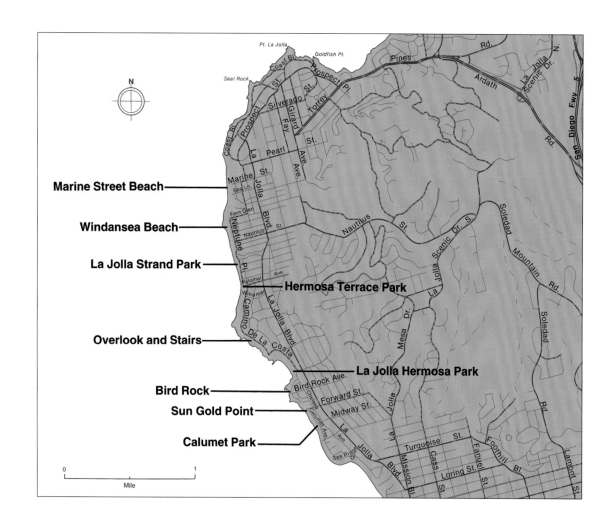

Marine Street Beach

Windansea Beach

La Jolla Strand Park

Hermosa Terrace Park

Overlook and Stairs

La Jolla Hermosa Park

Bird Rock

Sun Gold Point

Calumet Park

Windansea Beach

San Diego County

LA JOLLA

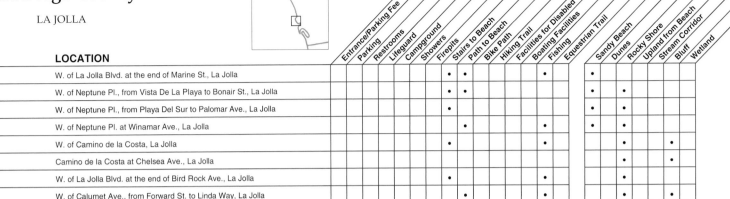

NAME	LOCATION	Entrance/Parking Fee	Parking	Restrooms	Lifeguard	Campground	Showers	Firepits	Stairs to Beach	Path to Beach	Bike Path	Hiking Trail	Facilities for Disabled	Boating Facilities	Fishing	Equestrian Trail	Sandy Beach	Dunes	Rocky Shore	Upland from Beach	Stream Corridor	Bluff	Wetland
Marine Street Beach	W. of La Jolla Blvd. at the end of Marine St., La Jolla								•	•					•		•						
Windansea Beach	W. of Neptune Pl., from Vista De La Playa to Bonair St., La Jolla								•	•							•		•				
La Jolla Strand Park	W. of Neptune Pl., from Playa Del Sur to Palomar Ave., La Jolla								•								•		•				
Hermosa Terrace Park	W. of Neptune Pl. at Winamar Ave., La Jolla									•					•		•		•				
Overlook and Stairs	W. of Camino de la Costa, La Jolla								•						•				•			•	
La Jolla Hermosa Park	Camino de la Costa at Chelsea Ave., La Jolla																		•			•	
Bird Rock	W. of La Jolla Blvd. at the end of Bird Rock Ave., La Jolla								•						•				•				
Sun Gold Point	W. of Calumet Ave., from Forward St. to Linda Way, La Jolla									•					•				•			•	
Calumet Park	W. of Calumet Ave., between Midway St. and Colima Ct., La Jolla	•																	•			•	

MARINE STREET BEACH: Sandy beach popular for swimming, surfing, skin diving, and surf fishing. Stairways to the beach are located at the ends of Marine St. and Sea Lane; a paved walk at the end of Vista del Mar also leads to the beach. Horseshoe Reef, a popular surfing spot, is just north of Marine Street. On-street parking only.

WINDANSEA BEACH: Also known as Neptune Park. One of the most popular surfing spots on the coast; swimming is also popular. The waves are sometimes too large for inexperienced surfers and swimmers. Paths lead to the beach from the end of Vista de la Playa and from Vista de la Playa at Fern Glen; beach is also accessible by stairways along Neptune Place south of Fern Glen. On-street parking only.

LA JOLLA STRAND PARK: This seasonally sandy beach is noted for swimming and surfing. The beach can be reached by stairways along the blufftop walkway that parallels Neptune Place and from street ends throughout the subdivision. On-street parking only.

HERMOSA TERRACE PARK: The small, seasonally sandy beach can be reached from a paved path just north of the end of Winamar Avenue. Swimming and surfing; fishing from the rocky shore south of Winamar. A smooth, easily climbed rock formation separates the beach from La Jolla Strand Park. On-street parking only.

OVERLOOK AND STAIRS: Viewpoint with bench and table off Camino de la Costa. Concrete stairway leads from the viewpoint to the rocky shoreline, which is a noted fishing spot. Overlooks at the ends of Cortez Pl. and Mira Monte Pl. off Camino de la Costa; informal paths lead down hazardous bluffs to the shoreline. On-street parking only.

LA JOLLA HERMOSA PARK: Small blufftop park with benches and picnic tables; stairs lead to overlooks of the coastline. The rocky beach below can be reached by traversing a steep eroded bluff face.

BIRD ROCK: This rocky point gets its name from the large guano-covered rock rising above the water just offshore. Tidepools are exposed at low tide. The area is also a popular fishing spot; surfing and skin diving here are recommended for experts only. Viewpoints at Moss Lane in the 5700 block of Dolphin Pl. and at the end of Bird Rock Ave.; a stairway at Bird Rock Ave. provides access to the riprap seawall below. Parking along adjacent streets.

SUN GOLD POINT: Viewpoints at the ends of Midway and Forward streets overlook a rocky beach with tidepools. At the end of Chelsea Pl. an informal path leads down the hazardous bluff face to the beach. At the end of Linda Way a path leads to benches and a stairway to a rocky cove, popular for fishing. On-street parking only.

CALUMET PARK: A landscaped picnic area with benches situated on the bluffs above a cobblestone beach. Views of the coast; foot path to beach at north end of park. Parking is available along Calumet Avenue.

Sun Gold Point

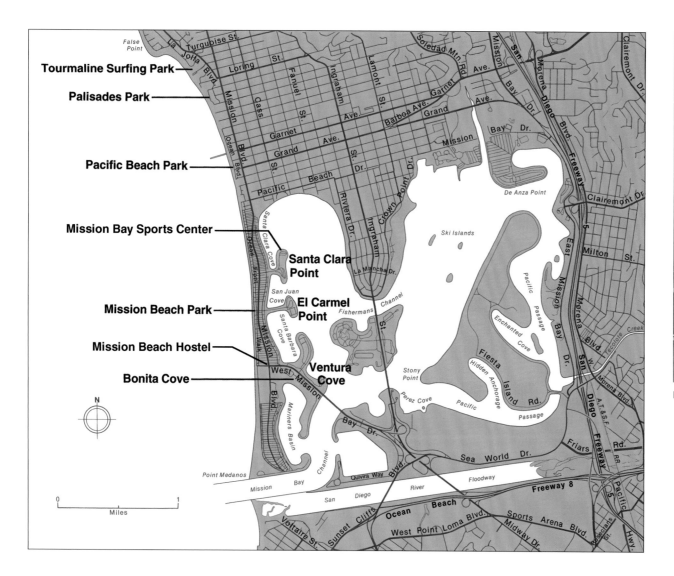

Tourmaline Surfing Park

Palisades Park

Pacific Beach Park

Mission Bay Sports Center

Santa Clara Point

El Carmel Point

Mission Beach Park

Mission Beach Hostel

Bonita Cove

Ventura Cove

False Point

Turquoise St.

La Jolla Blvd.

Loring St.

Cass St.

Fanuel St.

Ingraham St.

Lamont St.

Soledad Mtn. Rd.

Mission Bay Dr.

Morena Blvd.

Mission Bay Freeway

Clairemont Dr.

Balboa Ave.

Grand Ave.

Garner Ave.

Garnet Ave.

Mission

Bay Dr.

De Anza Point

Clairemont Dr.

5

Milton St.

Grand Ave.

Garnet Ave.

Beach Dr.

St. Dr.

Riviera Dr.

Crown Point Dr.

Ingraham St.

Ski Islands

Pacific Passage

Mission Bay Dr.

East

Morena Blvd.

Tecolote Creek

Pacific Beach Blvd.

Ocean Blvd.

Santa Clara Cove

La Mancha Dr.

Channel

Enchanted Cove

San Juan Cove

Fishermans

St.

Fiesta Island Rd.

Santa Barbara Cove

Mission Walk

Stony Point

Hidden Anchorage

W. Morena Blvd.

San Diego Freeway

A.T.&S.F. RR.

West Mission Blvd.

Mariners Basin

Perez Cove

Pacific Passage

Bay Dr.

Friars Rd.

Point Medanos

Channel

Quivira Way

Blvd.

Sea World Dr.

Floodway

Freeway 8

Pacific Hwy.

5

Rosecrans St.

Mission Bay

San Diego River

Voltaire St.

Sunset Cliffs

Ocean Beach

West Point Loma Blvd.

Sports Arena Blvd.

Midway Dr.

N

0   1
Miles

Mission Beach Park

San Diego County

PACIFIC BEACH/MISSION BEACH

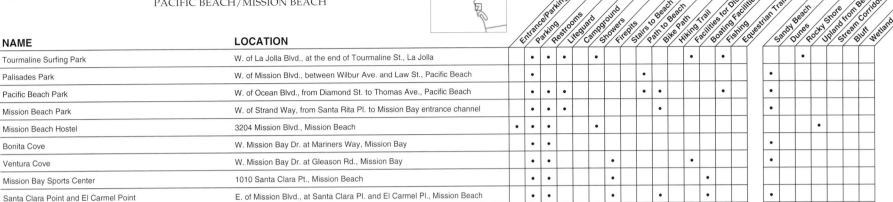

NAME	LOCATION	Entrance/Parking Fee	Parking	Restrooms	Lifeguard	Campground	Showers	Firepits	Stairs to Beach	Path to Beach	Bike Path	Hiking Trail	Facilities for Disabled	Boating Facilities	Fishing	Equestrian Trail	Sandy Beach	Dunes	Rocky Shore	Upland from Beach	Stream Corridor	Bluff	Wetland
Tourmaline Surfing Park	W. of La Jolla Blvd., at the end of Tourmaline St., La Jolla		•	•	•		•						•		•		•						
Palisades Park	W. of Mission Blvd., between Wilbur Ave. and Law St., Pacific Beach		•						•								•						
Pacific Beach Park	W. of Ocean Blvd., from Diamond St. to Thomas Ave., Pacific Beach		•	•	•				•	•					•		•						
Mission Beach Park	W. of Strand Way, from Santa Rita Pl. to Mission Bay entrance channel		•	•	•								•				•						
Mission Beach Hostel	3204 Mission Blvd., Mission Beach	•	•	•			•															•	
Bonita Cove	W. Mission Bay Dr. at Mariners Way, Mission Bay		•	•													•						
Ventura Cove	W. Mission Bay Dr. at Gleason Rd., Mission Bay		•	•				•					•				•						
Mission Bay Sports Center	1010 Santa Clara Pt., Mission Beach		•	•				•						•									
Santa Clara Point and El Carmel Point	E. of Mission Blvd., at Santa Clara Pl. and El Carmel Pl., Mission Beach		•	•				•					•	•			•						

TOURMALINE SURFING PARK: Rocky beach, popular for surfing, fishing, skin diving, and sea kayaking. Large waves break well offshore. Facilities include an outdoor shower, wheelchair-accessible restrooms, and picnic area.

PALISADES PARK: Paths lead from a grassy picnic area to a wide, sandy beach. On-street parking.

PACIFIC BEACH PARK: A landscaped picnic area overlooks the ocean; street ends provide access to the sandy beach, which is used for surfing and swimming. Separate paved paths for pedestrians and bicycles parallel the beach. Restrooms are available. Crystal Pier, located at the end of Garnet Ave., is open from 7 AM-sunset for walking and fishing; call: (858) 483-6983.

MISSION BEACH PARK: Sandy beach, with areas posted for surfing and body surfing. A paved promenade runs along the beach and is popular for walking, jogging, roller skating, and bicycling; skate and bike rentals are available along the boardwalk. A motorized beach wheelchair is available. There are several free parking lots in the area. Belmont Park, located at 3126 Mission Blvd. on the old amusement park site, has shops, restaurants, and family amusements. For information and hours for the park's carousel and the restored Giant Dipper, a 65-year-old roller coaster, call: (858) 488-1549. The historic Plunge, a public swimming pool, is also located here; call for hours: (858) 488-3110. Shops line the promenade south of W. Mission Bay Drive. Basketball courts and a play field are located at the south end of the beach.

MISSION BEACH HOSTEL: Also called International House at the 2nd Floor, offering dorm rooms, continental breakfast, kitchen, library, Internet access. Open 9-12 noon and 6-9 PM; call (858) 539-0043.

The following sites are within Mission Bay Park, the largest aquatic park on the west coast. Additional sites within the park are listed on the following page.

BONITA COVE: Used for swimming, picnicking, over-the-line softball, and volleyball. Paved path, large playground, and restroom along the cove; shops, restaurants, and recreation equipment rentals within walking distance. Transient vessel anchorage allowed in cove.

VENTURA COVE: Sandy beach, grassy picnic area with fire rings, paved path, and wheelchair-accessible restrooms; the very calm waters make it a popular swimming spot for small children. Other facilities include a hotel-marina complex. Bahia Point to the north has a grassy picnic area and a small sandy beach.

MISSION BAY SPORTS CENTER: Water sports equipment rental, sailing lessons, and children's summer camps are offered; for information, call: (858) 488-1004.

SANTA CLARA POINT AND EL CARMEL POINT: Facilities at Santa Clara Point include the city's boat house, a one-lane concrete boat launch, a dock, a water-ski landing and takeoff area, barbecue grills, a baseball field, tennis courts, and multi-purpose courts, and a recreation center. The Mission Bay Yacht Club (858-488-0501), a parking area, a restroom, and a sandy beach are located on El Carmel Point, about .4 mile south of Santa Clara Point. A sandy beach is located between the two points in San Juan Cove, and continues south through Santa Barbara Cove. Bayside Walk, a bicycle and pedestrian path, parallels the beach.

Pacific Beach

Sea World, Hubbs Research Center

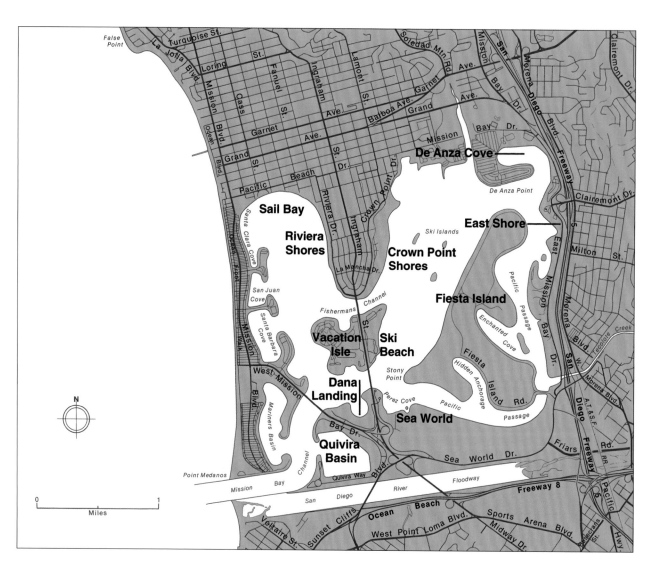

N

0   1
Miles

False Point

Turquoise St.

Loring   St.

Garnet   Ave.

Balboa Ave.

Grand   Ave.

Mission   Bay   Dr.

De Anza Cove

De Anza Point

East Shore

Clairemont Dr.

Milton   St.

Sail Bay

Riviera Shores

Ski Islands

Crown Point Shores

Pacific   Passage

Fiesta Island

San Juan Cove

Santa Barbara Cove

Fishermans   Channel

Vacation Isle   **Ski Beach**

Enchanted   Cove

Fiesta   Island   Rd.

Stony Point

Hidden   Anchorage

Dana Landing

Perez Cove

Pacific   Passage

Sea World

Mariners'   Basin

Quivira Basin

Point Medanos

Mission   Bay   Channel

Quivira   Way

Bay   Dr.

Sea   World   Dr.

Friars   Freeway

Mission   Bay   Floodway

San   Diego   River

Freeway 8

Voltaire St.

Sunset   Cliffs

Ocean   Beach

West   Point   Loma   Blvd.

Sports   Arena   Blvd.

Midway   Dr.

Pacific   Hwy.

280

San Diego County

MISSION BAY

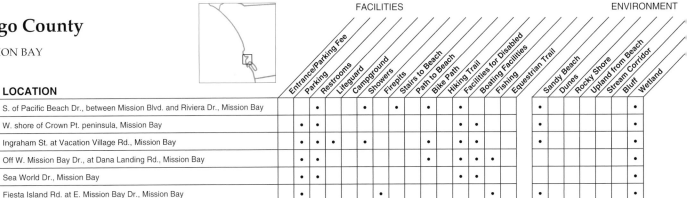

NAME	LOCATION	Entrance/Parking Fee	Parking	Restrooms	Lifeguard	Campground	Showers	Firepits	Stairs to Beach	Path to Beach	Bike Path	Hiking Trail	Facilities for Disabled	Boating Facilities	Fishing	Equestrian Trail	Sandy Beach	Dunes	Rocky Shore	Upland from Beach	Stream Corridor	Bluff	Wetland
Sail Bay and Riviera Shores	S. of Pacific Beach Dr., between Mission Blvd. and Riviera Dr., Mission Bay		•			•		•	•		•		•				•						•
Crown Point Shores	W. shore of Crown Pt. peninsula, Mission Bay	•	•									•	•				•						•
Vacation Isle and Ski Beach	Ingraham St. at Vacation Village Rd., Mission Bay	•	•	•		•				•							•						•
Dana Landing and Quivira Basin	Off W. Mission Bay Dr., at Dana Landing Rd., Mission Bay	•	•								•		•	•	•								•
Sea World	Sea World Dr., Mission Bay	•	•								•	•											•
Fiesta Island	Fiesta Island Rd. at E. Mission Bay Dr., Mission Bay	•				•							•			•						•
East Shore	Along E. Mission Bay Dr., Clairemont Dr. to Fiesta Island Rd., Mission Bay	•	•							•		•			•							•
De Anza Cove	N.E. corner of Mission Bay	•	•	•	•	•				•		•	•			•						•

Mission Bay Park encompasses the entire shoreline of Mission Bay; the following areas, except for Sea World, are all part of the park. For information, contact the Mission Bay Visitor Center: 2688 E. Mission Bay Dr., San Diego CA, 92109; (619) 276-8200; open daily 9 AM-5 PM.

SAIL BAY AND RIVIERA SHORES: The northwest portion of Mission Bay. The ends of Gresham, Fanuel, Everts, Dawes, and E. and W. Briarfield streets provide access to the sandy beach at Sail Bay. Fanuel Street Park at the end of Fanuel St. is a grassy park with a children's play area, parking, wheelchair-accessible restrooms, and an outdoor shower. Bayside Walk, a paved bicycle and pedestrian path, follows the bay shore from Sail Bay south to W. Mission Bay Drive. The Briarfield Cove Bridge provides pedestrian and bicycle access across Little Briarfield Cove. Riviera Shores is a sandy swimming beach with a water-ski takeoff and landing area. Access is provided by stairs at the west ends of Moorland, La Mancha, La Cima, and Edge Cliff drives, and by a path at 3862 Riviera Drive. Only on-street parking is available.

CROWN POINT SHORES: Sandy beach with picnic area, children's play area, physical fitness course, water-ski landing and takeoff areas, and public loading dock. Parking in three parking lots and along nearby streets. Just north of the beach is the Kendall-Frost Mission Bay Marsh Reserve, an ecological study area that hosts endangered clapper rails; no public access.

VACATION ISLE AND SKI BEACH: Vacation Isle is bisected by Ingraham St.; the west side has a hotel and golf course, boat rentals, a youth camp, and a model boat pond. North Cove Public Beach, on the northwest side, is a small sandy beach with picnic tables, a paved path, and wheelchair-accessible restrooms; South Cove is located by the boat basin on the southwest side. Ski Beach, on the east side of the island, is a noted water-skiing and boating spot, with a water-ski takeoff and landing area, a four-lane concrete boat launch, a dock, picnic areas, volleyball courts, wheelchair-accessible restrooms, outdoor showers, and a swimming beach.

DANA LANDING AND QUIVIRA BASIN: Dana Inn Hotel and Marina provides 142 boat slips; call: (619) 222-6440. The adjacent Dana Basin Boat Launching Facility is a 24-hour, 5-lane concrete boat ramp. The Dana Landing Market and Fuel Dock is open daily 6 AM-7 PM; call: (619) 226-2929. The 90-slip Dana Landing Marina is located across the basin; call: (619) 224-2513. Other facilities include boat sales, rentals, and charters. Sunset Point is a public park and popular fishing spot with picnic facilities, a paved path, and wheelchair-accessible restrooms.

Quivira Basin includes a large grassy picnic area with excellent bay views, and sportfishing and marina services. The Mission Bay Park Headquarters is located at Hospitality Point; call: (619) 221-8900. The Marina Village Marina within the basin has 634 boat slips, pump-out station, boat sales, restaurant, laundry, restrooms, and showers; call: (619) 224-3125. The Mission Bay Marina provides 200 slips, boat sales, shops, and deli; call: (619) 223-5191. Also located within the basin is the Islanda Marina: (619) 224-1234, and the Seaforth Marina: (619) 224-6807. There is also a path along the San Diego River and along the south jetty of the Mission Bay entrance channel. A bicycle path runs from the basin east along Sea World Dr. and Friars Rd., and connects with another bicycle path that runs along E. Mission Bay Drive.

SEA WORLD: A 189-acre aquarium and theme park located on the southern shore of Mission Bay. Attractions include marine life shows and displays, Clydesdale horses, two rides, snack bars, a beer garden, and gift shops. Wheelchair rentals available; restrooms are wheelchair accessible. Open daily; entrance fee. Call: (619) 226-3901 (recorded information) or 222-6363. The Sea World Marina at Perez Cove, located at 1660 S. Shores Rd., has 190 slips and 50 dry storage slips, restrooms, and showers; call: (619) 226-3915.

South Shores recreation area on Sea World Drive, located east of Sea World, provides parking, wheelchair-accessible restrooms, showers, sunbathing beach, 10-lane boat launching ramp and docks, and a nine-acre lagoon.

FIESTA ISLAND: The western and southern portions of this mostly undeveloped island are used for jet-skiing and fishing, and for viewing special aquatic events such as speedboat races. There are several swimming beaches around the island with fire rings; a youth campground is located on the eastern portion of the island.

EAST SHORE: Facilities include the San Diego Visitor Information Center; call: (619) 276-2071; and landscaped picnic areas, paved paths, wheelchair-accessible restrooms, playgrounds, physical fitness course, basketball court, snack bar, gift shop, and hotel; swimming and fishing at sandy beaches, which include Playa Pacifica, Leisure Lagoon, and Tecolote Shores. A barrier-free playground called Tecolote Shores Playground is located on E. Mission Bay Drive, south of the Hilton Hotel.

DE ANZA COVE: Includes a sandy beach and swimming area with lifeguard services, large grassy picnic areas, volleyball, playground, paved path, four-lane concrete boat ramp, and dock; the cove is used by swimmers and water-skiers. A trailer and R.V. park is located at the west side of De Anza Point. A private campground with both tent and trailer sites and a marina is near the cove at 2211 Pacific Beach Dr.; call: 1-800-422-9386.

Mission Bay, Ski Beach

Diving

California's lengthy shoreline, and diverse marine environments, coupled with generally favorable weather statewide, provide numerous opportunities for people to skin and scuba dive along the coast.

Skin diving is a relatively inexpensive way to view marine life and, where legal, to collect specimens. Typical skin diving equipment consists of swim fins, a mask, and a snorkel. A snorkel is a breathing tube that allows a person to swim at the water's surface and to look underwater without having to hold his or her breath. When diving below the range of the snorkel, however, one's breath must be held, and upon surfacing, the diver must first exhale to clear water out of the snorkel tube.

In some Southern California locations the water is warm enough to dive without a wetsuit; however, for deeper dives, prolonged periods in the water, or dives in Northern California, a wetsuit is a necessity. Any diver who becomes cold should get out of the water; diving while in poor physical condition can be dangerous.

The term "scuba" is an acronym for the words "self-contained underwater breathing apparatus." Scuba gear was developed in the 1940s by Emil Gagnan and Jacques Cousteau. Basic scuba gear consists of a wetsuit, fins, an air tank that is strapped to a diver's back, and an air hose and regulator system that supplies and controls the flow of air to a diver's mouth. Scuba gear allows people to dive deeper and stay submerged longer than is possible when skin diving.

Scuba diving involves certain potential hazards. For example, if a diver surfaces from a dive too quickly, air embolism (blockage of blood to the brain or heart by an air bubble) may occur, and the rapid change in water pressure may cause decompression sickness, commonly called "the bends." No one should scuba dive without first receiving expert instruction.

Diving is a safe, enjoyable experience for trained, qualified divers who abide by basic safety rules. Always dive with a companion, and be aware of your own physical limitations as well as those of your companion. Use a diver's flag (a red flag with a white diagonal stripe) when diving is in progress; the flag notifies boaters that divers are in the water and to take heed.

All divers should be familiar with decompression procedures, tables, and emergency action.

In a diving-related medical emergency, assistance by a physician trained in diving emergencies is absolutely necessary. Divers should be aware of the nearest medical facility equipped to handle such emergencies. The U.S. National Diving Accident Network (DAN), sponsored by Duke University in North Carolina, offers a 24-hour hotline service; the physicians on call will refer the caller to the nearest recompression chamber and trained staff. Call: (919) 684-8111 for emergencies only, after first calling local emergency medical services. For general information on diving, call: 1-800-446-2671, Monday–Friday.

Divers should also be aware of fish and game laws pertaining to the area where they are diving. Many coastal communities contain dive shops that provide services, supplies, and information. Skin and scuba diving lessons are offered by private diving schools, at a number of colleges, and as special programs by many community recreation departments.

San Diego County

OCEAN BEACH

NAME	LOCATION	Entrance/Parking Fee	Parking	Restrooms	Lifeguard	Campground	Showers	Firepits	Stairs to Beach	Path to Beach	Bike Path	Hiking Trail	Facilities for Disabled	Boating Facilities	Fishing	Equestrian Trail	Sandy Beach	Dunes	Rocky Shore	Upland from Beach	Stream Corridor	Bluff	Wetland
Robb Field and Playground	W. of Sunset Cliffs Blvd., along W. Point Loma Blvd., Ocean Beach		•	•									•								•	•	
Dog Beach	N. of Voltaire St., Ocean Beach		•	•													•						
Ocean Beach Park	Between the ends of Niagara Ave. and Voltaire St., Ocean Beach		•	•	•			•					•		•		•						
Ocean Beach Municipal Fishing Pier	End of Niagara Ave., Ocean Beach		•	•						•			•		•								
Ocean Beach City Beach	Between Ocean Beach Pier and the end of Pescadero Ave., Ocean Beach		•						•	•							•		•				
Sunset Cliffs Park	Along Sunset Cliffs Blvd., from Pt. Loma Ave. to Ladera St., Ocean Beach		•	•								•					•					•	
Elliott (Point Loma) International Youth Hostel	3790 Udall St., Ocean Beach	•		•			•																

ROBB FIELD AND PLAYGROUND: A large park and athletic field at the mouth of the San Diego River with picnic areas, a wheelchair-accessible path, restrooms, and facilities for skateboarding, baseball, tennis, and basketball. Information and reservations: (619) 531-1563. A bike path runs along the river's south bank.

DOG BEACH: On the north end of Ocean Beach past Voltaire St., dogs are permitted without a leash at all hours; dog owners are responsible for control and cleanup of their pets. Call: (619) 221-8901.

OCEAN BEACH PARK: Sandy beach with tidepools, a small grassy picnic area, and a paved promenade. The beach is noted for excellent surfing, but swimming and surf fishing are also popular; lifeguard service. Facilities include parking at Newport St., an outdoor shower, and wheelchair-accessible restrooms.

OCEAN BEACH MUNICIPAL FISHING PIER: The 2,100-foot-long T-shaped pier, located at the south end of Ocean Beach Park, is the longest pier on the west coast. Facilities include parking, wheelchair-accessible restrooms, bait and tackle shops, and a fish cleaning area. Stairs lead from the pier to the sandy beach. Call: (619) 221-8901.

OCEAN BEACH CITY BEACH: A series of small pocket beaches and tidepools along the coast from the Ocean Beach pier south to the end of Pescadero Ave.; sunbathing and surfing. A steep concrete ramp at Cable St. and Orchard Ave. leads to a rocky beach. Parking is available at Del Monte Avenue. Stairs at Santa Cruz, Bermuda, Orchard, and Narragansett avenues lead to a rough, uneven concrete path that follows the shoreline.

SUNSET CLIFFS PARK: A dirt path along dangerous cliffs provides a spectacular view of the coastline. There are parking areas along Sunset Cliffs Blvd.; a parking lot and pit toilets are located at the end of Cornish Dr. at the south end of the park. Several steep trails lead from the parking lot down the cliff face to pocket beaches frequently used by experienced divers and surfers; these trails are highly eroded and can be dangerous. A stairway at the end of Ladera St. leads to a rocky beach. The park also contains some upland hiking trails among eucalyptus trees, providing views of the beaches and tidepools below the cliffs.

ELLIOTT (POINT LOMA) INTERNATIONAL YOUTH HOSTEL: Centrally located between Mission Bay and San Diego Bay, about a mile east of Ocean Beach; accommodates 76 guests in 2- to 8-person rooms. Fully equipped community kitchen; open 4:30 PM-10 PM. Overnight fee. For reservations send a deposit for the first night's lodging to: Reservation Desk, Point Loma Hostel, 3790 Udall St., San Diego 92107. For information, call: (619) 223-4778.

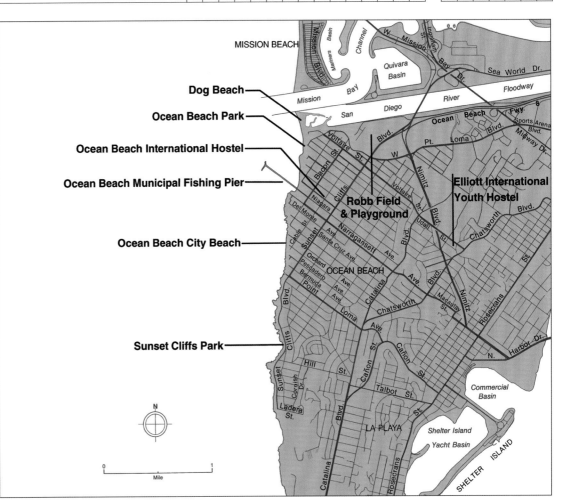

Kelp Harvesting

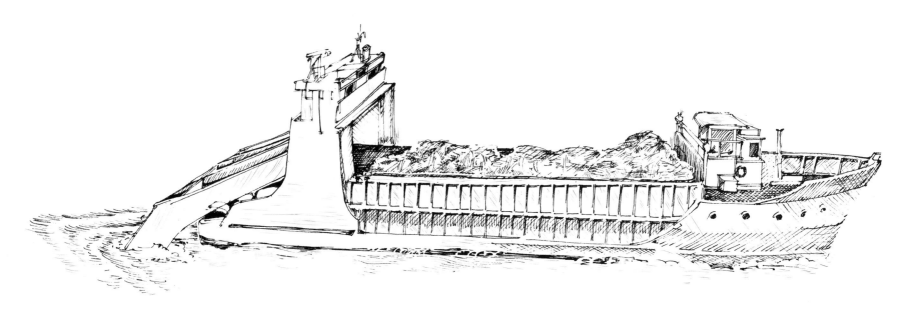

Kelp, a marine plant that provides food and habitat for diverse species of fish, marine invertebrates, and marine mammals, is also a valuable resource used in the manufacture of numerous everyday products.

Kelp is a type of alga. A number of species of kelp grow off the California coast, but giant kelp (*Macrocystis pyrifera*) is the most commercially important species. Giant kelp grows outside the surf zone to depths of 100 feet; the amount of sunlight that extends below 100 feet is not generally sufficient to sustain kelp growth. Giant kelp has no true roots, but uses root-like structures called holdfasts to anchor itself to hard rocky surfaces. Whereas terrestrial vegetation absorbs most of its nutrients through its root system, the entire kelp plant absorbs food; therefore, some of the largest and healthiest kelp beds grow in areas where strong ocean currents provide the kelp with a continuous supply of nutrients.

Giant kelp has an interesting reproductive cycle. Microscopic spores are released from the kelp beds, attach themselves to hard surfaces below the water surface, and germinate and grow into microscopic, filament-like male and female plants.

Female eggs are fertilized by sperm released by the male plants; the fertilized egg then grows rapidly, and within a year develops into a plant recognizable as giant kelp. Young kelp plants double in size every three weeks. For a period of five to ten years a kelp plant continously sends up new stalks, called stipes; giant kelp stipes live approximately six months and may grow as much as two feet in a day.

Giant kelp contains numerous useful elements and chemical compounds such as iodine, minerals, vitamins, and carbohydrates, making it a nutritious food supplement for humans and livestock. Kelp also contains very high concentrations of potassium, an important constituent in fertilizer and soaps.

Perhaps the most important substance found in kelp is a versatile compound called algin, which is used in many manufacturing processes. Since algin absorbs large quantities of water relative to its volume, it is used in preparing commercial ice cream to prevent the formation of large ice crystals. Algin is able to suspend various compounds in solutions and therefore is useful in the manufacture of antibiotics, polishes, and paint products. Rubber companies use algin as a thickening and stabilizing agent in the manufacture of synthetic rubber

products, such as latex. These are just a few examples of the more than one hundred food, industrial, and pharmaceutical products that contain algin.

In the United States the kelp harvesting industry is centered in Southern California. The waters off the San Diego coast are an attractive harvesting location because expansive giant kelp beds and favorable weather conditions allow virtual year-round harvesting. Kelp harvesting is regulated by the California Department of Fish and Game; some kelp beds are leased by the state to individual companies through a competitive bidding process, while others are open to all harvesting companies. Harvesting companies pay the state a royalty based on the amount of kelp harvested.

Today kelp is harvested by specially designed barges that cut and collect as much as 300 tons of wet kelp in one outing. By law, kelp can be cut to a maximum depth of four feet below the water surface. Harvesting kelp in this manner promotes new kelp growth, because the removal of the top layer of kelp allows more sunlight to reach the rest of the plant. After the kelp is harvested, it is transported to onshore processing plants where it is converted into its various marketable forms.

San Diego County

POINT LOMA

NAME	LOCATION	Entrance/Parking Fee	Parking	Restrooms	Lifeguard	Campground	Showers	Firepits	Stairs to Beach	Path to Beach	Bike Path	Hiking Trail	Facilities for Disabled	Boating Facilities	Fishing	Equestrian Trail	Sandy Beach	Dunes	Rocky Shore	Upland from Beach	Stream Corridor	Bluff	Wetland
Commercial Basin	E. of Rosecrans St., N. of Shelter Island Dr., San Diego		•											•	•								
Shelter Island Yacht Basin	E. of Rosecrans St., S. of Shelter Island Dr., San Diego		•											•	•								
Shelter Island	Along Shelter Island Dr., San Diego			•						•		•	•	•			•						
La Playa	Along San Antonio Ave., between Bessemer and Kellogg Streets, San Diego								•								•						
Cabrillo National Monument	S. end of Cabrillo Memorial Dr., Point Loma	•	•	•									•	•					•	•		•	
Point Loma Ecological Reserve	Offshore, W. side of Pt. Loma														•								

COMMERCIAL BASIN: Home of a large fleet of commercial fishing vessels; can accommodate more than 800 small craft. Facilities include sportfishing services, three municipal fishing piers, fuel dock, launching ramp, marine railway, marine supplies, and boat sales and service. Whale-watching trips are commercially available from Dec.-March. The basin can be viewed from a public observation deck located on the west side of Shelter Island Dr., 400 feet north of the traffic circle.

Two docks are open to public use: the gangway and boat dock adjacent to 2515 Shelter Island Dr. may be used for viewing of the basin and moored boats; the wooden dock north of the intersection of Anchorage Lane and Shelter Island Dr., and adjacent to the Kettenburg Marine facilites, may be used for berthing of dinghies and viewing of the shipyards. Parking is available at Kettenburg Marine's peripheral parking areas after business hours and at the public parking lots on Shelter Island Drive.

SHELTER ISLAND YACHT BASIN: Contains a number of marinas and yacht clubs; can accommodate almost 2,000 small craft. Boating facilities include boat launches, a fuel dock, transient berthing and mooring areas, marine supplies, and boat sales and service.

SHELTER ISLAND: The beach area of the island is popular for swimming, fishing, water-skiing, and picnicking. A landscaped, wheelchair-accessible pedestrian/bicycle path runs the length of the island, just inland from a sandy beach; other features include picnic areas, public moorings, hoists, a fishing pier, and a 24-hour boat launch with ten concrete lanes. A Coast Guard station is located at the south end of the island.

LA PLAYA: Narrow, sandy beach located on the Point Loma peninsula just west of Shelter Island; narrow trail along the shoreline provides excellent views of the bay and yacht basin. Access at the ends of Bessemer, Perry, Owens, McCall, Lawrence, and Kellogg streets. Swimming is prohibited within the marked yacht channel.

CABRILLO NATIONAL MONUMENT: Commemorates Cabrillo's discovery of San Diego Bay. The monument features a visitor center with exhibits, interpretive programs, and a gift shop; the old Point Loma Lighthouse, built in 1854; and a telescope-equipped scenic overlook, where migration of the gray whale can be viewed from Dec.-Mar. each year. Hiking trails overlooking the ocean and the bay lead to some of the finest tidepools in Southern California. Walkways and trails, including the 1.5-mile-long Bayside Trail, feature wayside interpretive displays; a hard-packed sand and gravel path on the ocean bluffs is wheelchair negotiable. The monument is open 9 AM-5:15 PM daily. For information, call: (619) 557-5450.

POINT LOMA ECOLOGICAL RESERVE: Underwater reserve along the western shore of Pt. Loma, adjacent to Cabrillo National Monument . Only fin-fish can be taken from the reserve.

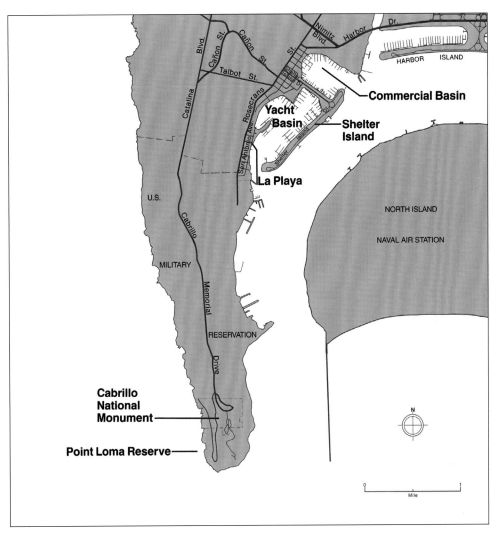

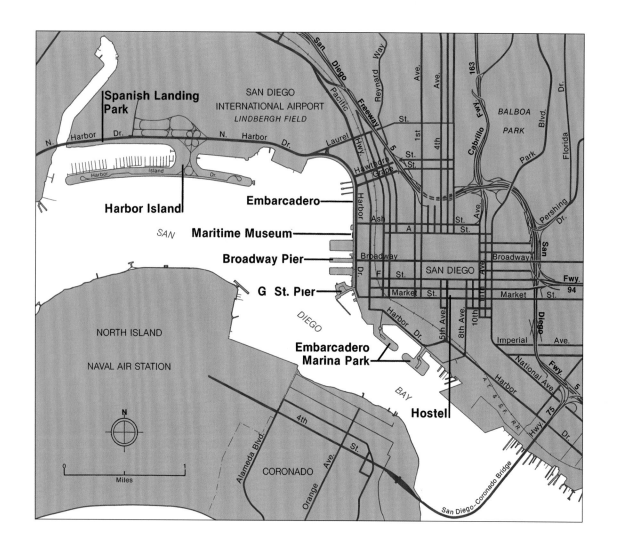

Spanish Landing Park

SAN DIEGO INTERNATIONAL AIRPORT
LINDBERGH FIELD

BALBOA PARK

Harbor Island

Embarcadero

SAN

Maritime Museum

Broadway Pier

G St. Pier

SAN DIEGO

DIEGO

NORTH ISLAND

NAVAL AIR STATION

Embarcadero Marina Park

BAY

Hostel

N

Miles

CORONADO

San Diego-Coronado Bridge

Embarcadero Marina Park

G Street Pier

San Diego County

CITY OF SAN DIEGO

NAME	LOCATION	Entrance/Parking Fee	Parking	Restrooms	Lifeguard	Campground	Showers	Firepits	Stairs to Beach	Path to Beach	Bike Path	Hiking Trail	Facilities for Disabled	Boating Facilities	Fishing	Equestrian Trail	Sandy Beach	Dunes	Rocky Shore	Upland from Beach	Stream Corridor	Bluff	Wetland
Spanish Landing Park	W. of Lindbergh Field, on N. Harbor Dr., San Diego		•	•			•						•				•						
Harbor Island	S. of Lindbergh Field, on Harbor Island Dr., San Diego		•	•									•	•	•								
Embarcadero	Along Harbor Dr., at the end of Hawthorne St., San Diego		•	•									•	•									
Maritime Museum	W. of Harbor Dr., at the end of Ash St., San Diego	•	•																				
Broadway Pier	W. end of Broadway, San Diego	•																					
G Street Pier	W. of N. Harbor Dr., at the end of G St., San Diego	•	•												•								
Embarcadero Marina Park	Harbor Dr. at the end of Kettner Blvd., San Diego		•	•									•	•	•								
San Diego Hostel International	521 Market St., San Diego		•	•			•																

SPANISH LANDING PARK: The sandy beach located just south of the San Diego International Airport is a popular swimming spot; there is also a grassy picnic area near the shore. A wheelchair-accessible path along a seawall provides views of the bay; restrooms are wheelchair accessible. Outdoor shower at the beach.

HARBOR ISLAND: Like Shelter Island to the southwest, Harbor Island serves primarily as a boating center; the various marinas can accommodate a total of about 1,000 vessels, and some marinas have guest berths or moorings. Other facilities include hotels, restaurants, shops, and boating services. A landscaped wheelchair-accessible path with benches runs the length of the island, and provides fine views of the bay; restrooms are wheelchair accessible.

EMBARCADERO: A paved pedestrian/bicycle path with benches runs along the waterfront on the east side of the bay; path and restrooms are wheelchair accessible. The Embarcadero also features restaurants, fish markets, shops, and harbor excursion tours; commercial fishing activities are concentrated in the area south of Hawthorne Street.

MARITIME MUSEUM: Three historic ships moored along the Embarcadero at the end of Ash Street. The most famous of the ships is the 1863 *Star of India*, a three-masted bark that once transported emigrants from London to New Zealand. Ships can be toured from 9 AM-8 PM every day of the year, open until 9 PM in summer; admission fee. For information, call: (619) 234-9153.

BROADWAY PIER: The walkway along the pier offers fine views of the bay. Ferry service to Coronado operates daily; call: (619) 234-4111. The U.S. Customs office is located at the end of the pier. Public tours of several U.S. Navy ships are available on most weekend afternoons, call: (619) 532-1431. For information on aircraft carrier tours on North Island, call: (619) 545-1138. Tours of ships at the 32nd St. Naval Station are available by reservation for groups of ten or more; call: (619) 437-2735. A water taxi service offers on-call transportation along San Diego Bay; call: (619) 235-8294.

G STREET PIER: Popular fishing spot; walkway with bay views.

EMBARCADERO MARINA PARK: The park is divided into two sections, but is connected by a walkway along a seawall on the bay. The northern section has a lawn, picnic tables, paved path, and drinking fountains; popular activities include fishing and jogging. Enter from the Seaport Village Shopping Center on Harbor Dr., where there is a metered parking lot. The entrance to the southern section is at the intersection of Harbor Dr. and Harbor St.; facilities include a grassy picnic area, paths, fishing pier, basketball courts, and par course.

SAN DIEGO HOSTEL INTERNATIONAL: An AYH-operated dormitory hostel provides lodging for both members and non-members; linens provided. Storage of backpacks and other items available for a fee. Facilities include showers, full kitchen, and coin-operated laundry. Hostel open daily 7 AM–midnight year-round. Also in the building is a fully-equipped athletic facility, available on a membership or daily fee basis. For further information, call: (619) 525-1531.

Spanish Landing

Shellfish

Shellfish comprise a significant and highly desirable resource off the California coast. Included in this category are crustaceans such as shrimp, prawns, crabs, and lobsters; and mollusks such as clams, oysters, abalone, mussels, and squid.

One of the most popular shellfish is abalone, many species of which are found along the state's entire coast. Adult abalone feed on algae such as kelp, and usually remain on a single rock for the duration of their lives. The shells are noted for their beauty and brilliance; archaeological evidence indicates that Native Americans used them for barter. Today, commercial abalone fishing is one of the largest shellfish industries in Southern California; in Northern California abalone are taken in large numbers by sport divers.

Typical predators of abalone, other than people and sea otters, are rock crabs, who frequently pick the thick shells apart to get at the meat inside. Inhabitants of inshore rocky areas, rock crabs are common catches of pier and shore net fishermen. The larger and more commercially popular crab sold in restaurants and stores is the Dungeness crab, found primarily in the San Francisco and Humboldt Bay areas.

California spiny lobster populations are located off the San Diego and Southern California coasts, providing an important commercial shellfish. Spiny lobsters, in large groups of up to several hundred, typically inhabit rocky crevices and dens during the daytime; at night they forage singly, usually feeding on other crustaceans and mollusks.

The most commonly found mollusk in rocky areas is the mussel, often used for bait by fishermen, but also popular for eating. Mussels are filter feeders, and subsist on phytoplankton. During summer months, mussels and other bivalves may ingest and accumulate large amounts of certain toxic plankton, causing them to become poisonous to humans. Shellfish poisoning is not limited to mussels; sand crabs and razor, Washington, and gaper clams may also accumulate toxic plankton if they are in or near the ocean. Shellfish taken in inland bays are generally safe. However, always check with the environmental health sections of local or state health services agencies concerning quarantines before removing any of these

shellfish. Mussels are usually quarantined between May 1 and October 31.

Although only one oyster species is native to California, several introduced species are grown in California using mariculture techniques. Natural reproduction of these non-native oysters is inhibited in California because of low water temperatures, requiring commercial oyster farms to artificially plant juvenile oysters. The Pacific oyster is the most successful in California, introduced originally from Japan; commercial Pacific oyster beds are grown in Encina Lagoon, Drakes Estero, and Humboldt, Tomales, and Morro bays.

Squid, considered a shellfish because of its small internal shell, is one of the most important commercial catches off the Monterey coast. Schools of squid are typically caught at night, using floodlights or torches held above the water to attract them into submerged nets. California squid grow to as much as 12 inches in length, and feed on small fishes, shrimp, and other squid.

Bag limits, equipment, and fishing seasons for shellfish are regulated by the California Department of Fish and Game. For information, see the annual California Marine Sport Fishing Regulations available from the Department of Fish and Game or at sporting goods stores.

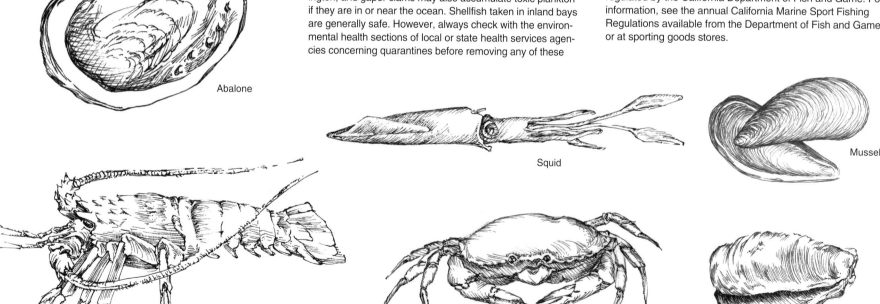

Abalone

Squid

Mussel

Spiny lobster

Rock crab

Pacific oyster

San Diego County

CORONADO

NAME	LOCATION	Entrance/Parking Fee	Parking	Restrooms	Lifeguard	Campground	Showers	Firepits	Stairs to Beach	Path to Beach	Bike Path	Hiking Trail	Facilities for Disabled	Boating Facilities	Fishing	Equestrian Trail	Sandy Beach	Dunes	Rocky Shore	Upland from Beach	Stream Corridor	Bluff	Wetland
Bay View Park	"I" and 1st streets, Coronado												•									•	
Harbor View Park	1st and E streets, Coronado				•																	•	
Centennial Park	1st and Orange streets, Coronado	•								•	•		•				•					•	
Coronado Tidelands Regional Park	Foot of 3rd St., along Mullinix Dr., Coronado	•	•							•	•		•				•					•	
Coronado Municipal Golf Course	Visalia Row and Glorietta Blvd., Coronado	•						•					•				•						
Glorietta Bay Marina	1715 Strand Way, Coronado	•												•	•								
Glorietta Bay Park	1975 Strand Way, Coronado	•	•				•						•	•			•					•	
Coronado City Beach	W. of Ocean Blvd., Coronado		•	•			•								•		•						
Coronado Shores Beach	Seaward of Coronado Shores Condominiums on Hwy. 75, Coronado	•								•	•		•				•						

BAY VIEW PARK: Small, beautifully landscaped park with benches and a paved, wheelchair-accessible viewpoint; provides a good view of the bay and the San Diego skyline.

HARBOR VIEW PARK: Small, grassy park with benches; view of the downtown waterfront. Near the park, on 1st St. between E and F streets, is a narrow pathway with stairs leading to the bay. Private property adjoins the path; do not trespass. On-street parking only.

CENTENNIAL PARK: Bayfront park with lawns, benches, paved paths (wheelchair accessible), and views of the bay and downtown San Diego. Bicycle rentals are available. Small sandy beach; fishing pier at landing for pedestrian/bicycle ferry to San Diego. Recreational boat pier at the Old Ferry Landing complex of retail shops.

CORONADO TIDELANDS REGIONAL PARK: Located north of the toll plaza off Glorietta Boulevard. Large park with baseball diamond, play fields, par course, pedestrian and bicycle paths, lawns, benches, picnic tables, wheelchair-accessible restrooms, and paved paths along the water. Small sandy beach used for swimming; no lifeguard.

CORONADO MUNICIPAL GOLF COURSE: A hard-packed dirt road adjacent to the golf course office provides pedestrian access to the narrow sandy beach along the bay.

GLORIETTA BAY MARINA: Offers slips, power and sailboat rentals, and charter fishing boats; home of the Coronado Yacht Club. Call: (619) 435-5203.

GLORIETTA BAY PARK: Playground, grassy picnic area, small sandy beach, and boat launch; restrooms and some picnic tables are wheelchair accessible. The adjacent Coronado Municipal Swimming Pool offers classes and programs; call: (619) 435-4179.

CORONADO CITY BEACH: Wide, sandy beach used for swimming, surfing, and surf fishing. Fire rings and a dog run at the north end of beach; restrooms and main lifeguard tower near the foot of F Street. A large grassy picnic area is located at Sunset Park, at the north end of Ocean Boulevard. On-street parking only. Call: (619) 522-7342.

CORONADO SHORES BEACH: The entrance and a public parking area are located at the Coronado Shores condominium development. Wide, sandy beach heavily used for surfing, swimming, and surf fishing. A concrete promenade that runs along the top of a seawall adjacent to the shore affords excellent views. The promenade is accessible from Avenida del Sol, Avenida de las Arenas, and Avenida Lunar; a parking lot and wheelchair ramp are at the end of Avenida de las Arenas.

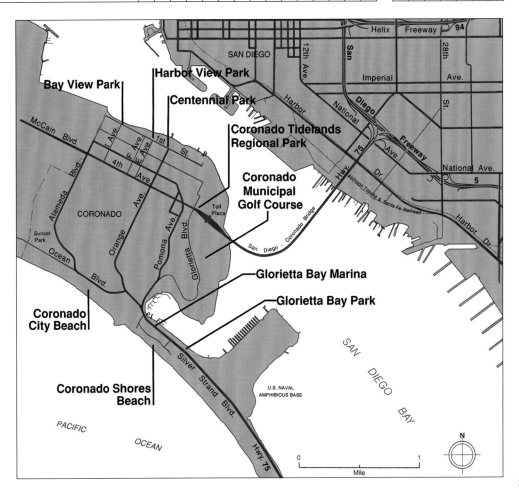

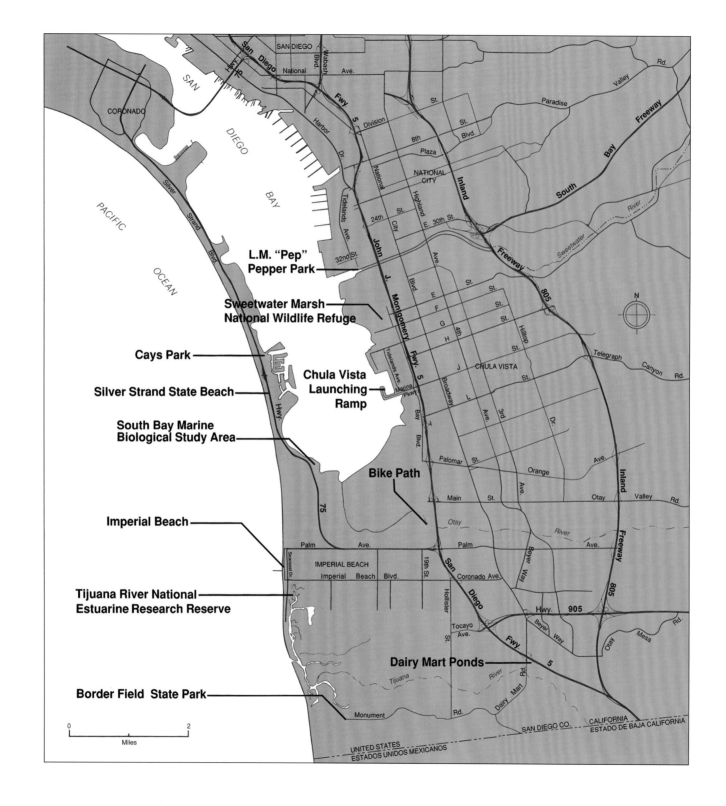

San Diego County

SOUTH BAY / IMPERIAL BEACH

FACILITIES   **ENVIRONMENT**

NAME	LOCATION	Entrance/Parking Fee	Parking	Restrooms	Lifeguard	Campground	Showers	Firepits	Stairs to Beach	Path to Beach	Bike Path	Hiking Trail	Facilities for Disabled	Boating Facilities	Fishing	Equestrian Trail	Sandy Beach	Dunes	Rocky Shore	Upland from Beach	Stream Corridor	Bluff	Wetland
L.M. "Pep" Pepper Park	S. end of Tidelands Ave., National City		•	•				•					•	•	•								•
Sweetwater Marsh National Wildlife Refuge	E. shore of San Diego Bay, National City and Chula Vista	•		•									•										•
Chula Vista Launching Ramp	W. end of Marina Way, off Marina Parkway, Chula Vista		•	•				•					•	•									•
Cays Parks	Off Hwy. 75 (Silver Strand Blvd.), along Coronado Cays Blvd., Coronado		•	•																			
Silver Strand State Beach	5000 Hwy. 75, Coronado	•	•	•	•	•	•	•					•		•		•	•					•
South Bay Marine Biological Study Area	E. of Hwy. 75, 1 mi. N. of Imperial Beach city limits, Coronado		•																			•	•
Bike Path	E. of Imperial Beach, from W. end of Main St. to N. end of 19th St., San Diego										•											•	•
Imperial Beach	W. of Ocean Lane, from Carnation to Encanto avenues, Imperial Beach		•	•	•		•						•		•		•						
Tijuana River National Estuarine Research Reserve	Off Hwy. 5 at Coronado Ave., Imperial Beach		•	•								•			•	•	•	•		•	•		•
Dairy Mart Ponds	S. of I-5, both sides of Dairy Mart Rd., San Diego											•				•							•
Border Field State Park	W. end of Monument Rd., Imperial Beach		•	•				•				•	•		•		•	•		•		•	•

L. M. "PEP" PEPPER PARK: Public ramp operated by the Port of San Diego. Facilities include a ten-lane concrete boat ramp, a small dock, trailer space, a landscaped picnic area with barbecue grills, a fishing platform, and wheelchair-accessible restrooms. Open 6:30 AM–10:30 PM. From I-5 take the 24th St. off-ramp west to Tidelands Ave., and south on Tidelands to end. Call: (619) 686-6200.

SWEETWATER MARSH NATIONAL WILDLIFE REFUGE: Established in 1988, the 316-acre refuge protects the largest salt marsh in San Diego Bay. The wetlands are frequented by about 224 species of birds, including the endangered California least tern, California light-footed clapper rail, and Belding's savannah sparrow. The Chula Vista Nature Center, located at Gunpowder Point, offers guided tours of Sweetwater Marsh; there are also shorebird and raptor aviaries, butterfly and hummingbird gardens, natural history exhibits, aquaria, terraria, and observation platforms; special programs are available to groups and classes. Hours are 10 AM–5 PM Tuesday-Sunday. Entrance fee. The Center is not accessible to pedestrians or private cars; catch the free "Nature Center" shuttle at the parking lot at the intersection of E St. and Bay Blvd. or at the Chula Vista Visitor's Center at E St. and I-5, or take the San Diego Trolley to the Bayfront/E Street Station. The shuttle and the Center are wheelchair accessible. Call: (619) 409-5900.

CHULA VISTA LAUNCHING RAMP: Concrete boat launch with ten lanes; open 6:30 AM–10:30 PM. Facilities include docks, a hoist, picnic areas with barbecue grills, a playground, and wheelchair-accessible restrooms. Noted for swimming, fishing, and water-skiing; however, swimming and skiing are prohibited within the launch basin. Bayside Park, on the north edge of the boat basin, also has picnic facilities; Marina View Park, along Marina Parkway, has a playground. A paved pedestrian and bicycle path runs along the bay shore. The ends of G and F streets provide pedestrian access to South Bay tidal areas. For information, call: (619) 686-6200.

CAYS PARKS: There are three segments: North Cays, Central Cays, and South Cays. Central Cays, located south of Cays entrance between Hwy. 75, Coronado Cays Blvd., and Grand Caribe causeway, offers the most facilities, with tennis courts, volleyball, a baseball diamond, a multi-purpose field, benches, and restrooms.

SILVER STRAND STATE BEACH: Sandy beach noted for swimming, surfing, clamming, surf fishing, and catching grunion; beachcombing is also popular because of the large quantity of shells found on the beach. Pedestrian tunnels connect the ocean beach with the bay shore, where there is a calm-water swimming basin. The ranger station contains exhibits of local plant and animal life. The state beach also provides first-aid stations, lifeguard service, covered picnic areas, firepits, and 125 enroute campsites for self-contained R.V.'s only; restrooms are wheelchair accessible. Fees for day and overnight use. This beach supports snowy plovers, a threatened species, and there may be some restrictions on public access in this area to protect the habitat. For information, call: (619) 435-5184.

SOUTH BAY MARINE BIOLOGICAL STUDY AREA: This nature refuge provides an opportunity to observe the plant and animal life of a wetland environment, and is a popular bird-watching area. Located adjacent to a nesting area for endangered California least terns. Provides paved parking lot, and walkway into study area. A paved bike path that runs along Hwy. 75 (Silver Strand Blvd.) from Coronado to Imperial Beach passes through the study area. Stay on established trail and walkway. Affords panoramic views of southern San Diego Bay.

BIKE PATH: The path follows the perimeter of the Otay River Marsh and runs through the salt flats along the edge of the bay.

IMPERIAL BEACH: Wide, sandy beach popular for swimming, body surfing, and surf fishing. There is a metered parking lot at the intersection of Evergreen and 1st streets; street ends from Palm Ave. to Encanto Ave. also provide access and some on-street parking. The Imperial Beach Municipal Pier, at the foot of Evergreen Ave., has a restaurant, fish-cleaning area, and wheelchair-accessible restrooms. The beach also has restrooms and outdoor showers. Portwood Pier Plaza located at Seacoast Dr. and Elder Ave. features Surfhenge, a giant colored acrylic sculpture, and ten surfboard benches, playground, picnic tables, and an outdoor performance area. Dunes Park has a sandy beach, volleyball, picnic areas, a playground and a Wyland sculpture of three dolphins riding a wave. A motorized beach wheelchair is available at the main lifeguard tower.

TIJUANA RIVER NATIONAL ESTUARINE RESEARCH RESERVE: The reserve comprises 2,531 acres of salt marsh, riparian habitat, sandy beach, dunes, and upland areas in the floodplain of the Tijuana River; includes the Tijuana Slough National Wildlife Refuge and Border Field State Park. The reserve is a good place to view wetland wildlife, most notably migratory and resident water birds. Public access to the reserve is allowed only on established trails to protect sensitive habitat. Sand dunes are off-limits year-round; other areas are closed seasonally for nesting birds such as the threatened snowy plover. Collection for scientific puposes is permitted in the national wildlife refuge by permit only. Fishing is allowed only along the shoreline in open ocean surf.

The Visitor's Center, located at 301 Caspian Way off 3rd St. in Imperial Beach, has trails, restrooms, exhibits, a bookstore, and a small library. The facility and some trails are wheelchair accessible. Guided tours and educational programs, including an active school program with workshops, are available. The center is open 10 AM–5 PM daily; for information, call: (619) 575-3613.

DAIRY MART PONDS: Hiking and equestrian trails around the freshwater ponds provide opportunities to view wildlife, particularly migratory birds. The ponds will be included in a future county park that will adjoin Border Field State Park.

BORDER FIELD STATE PARK: Located on the U.S. border with Mexico, the park features a two-mile-long stretch of sandy beach used for swimming, clamming, and surf fishing. Provides picnic areas, barbecue pits, wheelchair-accessible restrooms, interpretive signs, and hiking and equestrian trails; corrals are located outside the park gate, and several stables in the area rent horses. This park supports snowy plovers, a threatened species, and there may be some restrictions on public access in this area to protect the habitat. The Tijuana River Estuary within the park is an important habitat for rare and endangered species of plants and animals. Border Field State Park is open 9:30 AM–7 PM Thursday–Sunday. For information, call: (619) 575-3613.

Afterword

This revised, expanded Guide is accurate to the best of our knowledge. All accessways were visited, and all information was verified using sources such as Local Coastal Programs, which were prepared by local governments in compliance with the Coastal Act of 1976. We also incorporated into this updated edition the comments and corrections supplied by various agencies and members of the public who wrote to us regarding the information in earlier editions. Nevertheless, we are aware that conditions on the coast are constantly changing and that inaccuracies may still exist in the text.

If you think something is incorrect or has been omitted, please let us know. The Commission intends to continue publishing revised and expanded guides in the future and would therefore appreciate any additional information you can provide. Please remember, however, that the Guide includes only those beaches that are managed for public use.

Address all comments to:

California Coastal Commission
45 Fremont Street, Suite 2000
San Francisco, CA 94105

or e-mail to: coast4U@coastal.ca.gov

Acknowledgements

Louise McCorkle Adams	Allyson Hitt	Charles Posner
Rich Allen	Mary Hudson	James Raives
Christine Baird	Tom Jackson	Darryl Rance
Deborah S. Benrubi	James Johnson	Victoria Randlett
Gina M. Bentzley	Michael Julian	Rick Rayburn
Michael Buck	Chris Kern	Mindy S. Richter
Sarah Christie	Anna Kondolf	Mishell Rose
Karen Clement	Evelyn Lee	Sherilyn Sarb
Elinor Craig	John Lentz	Sylvia Sherman
Trevor Cralle	Jack I. Liebster	Richard L. Snyder
Pam Emerson	Cynthia K. Long	Paul Thayer
Lesley Ewing	Daniel S. Miller	Mary Travis
Judy Feins	Gianmaria Mussio	Martha R. Weiss
Lillian Ford	Carol Opotow	Jayson Yap
Jack Gregg	Lee Otter	Cy Yee
Miriam Gordon	Christiane Parry	Cindy Young
Melanie Hale	Dan Porter	Erica Young

The editors wish to thank the staff of the University of California Press for assistance in preparation and revision of the guide, and to thank the staff of the many state, federal, local, and private agencies who helped us bring the guide up-to-date.

Special thanks to Don Neuwirth, manager of the Coastal Access Program from 1979 to 1983, for his great imagination, creativity, perseverance, and enthusiastic leadership.

With grateful acknowledgement to Pat Stebbins, manager of the Coastal Access Program from 1983 to 1990, for her guidance and inspiration.

Selected References

Allen, Richard K. *Common Intertidal Invertebrates of Southern California.* Palo Alto, Ca.: Peek Publications, 1969.

American Youth Hostels Handbook 1990. Washington, D.C.: American Youth Hostels, Inc., 1983.

Anthrop, Donald F. *Redwood National and State Parks.* Happy Camp, Ca.: Naturegraph, 1977.

Ashley, Frank and Mary Ashley. ìExploring Coronadoó San Diegoís ìIsolatedì Peninsula.ì *California Traveler* (July 1982).

Bakker, Elna. *An Island Called California.* 2d ed. Berkeley: University of California Press, 1984.

Bancroft, Hubert Howe. *The History of California.* 7 Vol. San Rafael, Ca.: Bancroft Press, 1888.

Bascom, Willard. *Waves and Beaches: The Dynamics of the Ocean Surface.* rev. ed. Garden City, N.Y.: Anchor Books, Anchor Press/Doubleday, 1980.

Beachcombers Guide to the Pacific Coast. Menlo Park, Ca.: Sunset Books, Lane Publishing Co., 1966.

Benet, James. *A Guide to San Francisco and the Bay Region.* New York: Random House, 1963.

Berssen, William, ed. *Pacific Boating Almanac.* Ventura, Ca.: Western Marine Enterprises, Inc., 1981.

Bies, Frank. ìSurf Wars.ì *New West* (May 1981).

Brusca, Gary J. and Richard C. Brusca. *A Naturalistís Seashore Guide: Common Marine Life of the Northern California Coast and Adjacent Shores.* Eureka, Ca.: Mad River Press, 1978.

California Coastal Commission. *California Coastal Resource Guide.* Berkeley: University of California Press, 1987.

California, A Guide to the Golden State. Federal Writers Project, State of California, American Guide Series. New York: Hastings House, 1939.

California Sport Fishing Regulations, 1990. Sacramento: California Department of Fish and Game, 1990.

California State Parks. Menlo Park, Ca.: Sunset Books, Lane Publishing Co., 1972.

Californiaís Living Marine Resources and Their Utilization. Sacramento: California Department of Fish and Game, 1971.

Campground and Trailer Park Guide to the United States, Canada, and Mexico. Skokie, Ill.: Rand McNally, 1980.

Catalog of California Seabird Colonies. Washington, D.C.: U.S. Department of the Interior, Fish and Wildlife Service, 1980.

Clark, Eugenia. ìSharks: Magnificent and Misunderstood.ì *National Geographic* (August 1981).

Clausen, Lucy W. *Insect Fact and Folklore.* New York: Collier Books, 1954.

Coastal Wetlands of San Diego County. California State Coastal Conservancy, 1989.

Cogswell, Howard L. *Water Birds of California.* Berkeley: University of California Press, 1977.

Concept Plan for Waterfowl Wintering Habitat Preservation: California Coast. Portland, Or.: U.S. Department of the Interior, Fish and Wildlife Service, 1979.

Cralle, Trevor. *The Surfiníary: A Dictionary of Surfing Terms and Surfspeak.* Berkeley: Ten Speed Press, 1991.

Cuanang, Abe. ìShark! Big Game in the Bay.ì *Western Saltwater Fisherman* (December 1981).

Culliney, John L. and Edward S. Crockett. *Exploring Underwater, The Sierra Club Guide to Scuba and Snorkeling.* San Francisco: Sierra Club Books, 1980.

Dawson, E. Yale. *Seashore Plants of Northern California.* Berkeley: University of California Press, 1979.

Dawson, E. Yale. *Seashore Plants of Southern California.* Berkeley: University of California Press, 1962.

Dixon, Sarah and Peter Dixon. *West Coast Beaches: A Complete Guide.* New York: E.P. Dutton, 1979.

Doss, Margot Patterson. *San Francisco At Your Feet.* New York: Grove Press, 1964.

Doss, Margot Patterson. *The Bay Area At Your Feet.* San Rafael, Ca.: Presidio Press, 1981.

Dougherty, Anita E. *Marine Mammals of California.* Sacramento: California Department of Fish and Game, 1972.

Durrenberger, Robert W. and Robert B. Johnson. *California: Patterns on the Land.* 5th ed. Palo Alto, Ca.: Mayfield Publishing Co., 1976.

Easterbrook, Don J. *Principles of Geomorphology.* New York: McGraw-Hill, 1969.

Engbeck, Joseph Jr., ed. *A Visitorís Guide to California State Parks.* Santa Barbara, Ca.: Sequoia Publications, 1990.

Explore, a series of pamphlets for the Year of the Coast. San Francisco: U.S. Army Corps of Engineers, 1980.

Fagan, Brian M. and Graham Pomeroy. *Cruising Guide to the Channel Islands.* Santa Barbara and Van Nuys, Ca.: Capra Press and Western Marine Enterprises, 1980.

Fire Weather. Washington, D.C.: U.S. Department of Agriculture, Forest Service, 1970.

Fitch, John E. *Common Marine Bivalves of California.* Fish Bulletin 90. California Department of Fish and Game, 1953.

Fitch, John E. and R. Lavenberg. *Tidepool and Nearshore Fishes of California.* Berkeley: University of California Press, 1975.

Gordon, Burton L. *Monterey Bay Area: Natural History and Cultural Imprints.* Pacific Grove, Ca.: Boxwood Press, 1974.

Griggs, Gary B. and John A. Gilchrist. *The Earth and Land Use Planning.* Belmont, Ca.: Duxbury Press, 1977.

Gudde, Erwin G. *California Place Names.* Berkeley: University of California Press, 1974.

Guide to California Boating Facilities. Sacramento: California Department of Boating and Waterways, 1976.

Harrison, Richard J. and Judith E. King. *Marine Mammals.* London: Hutchinson and Co., 1965.

Hayden, Mike. *Exploring the North Coast from the Golden Gate to the Oregon Border.* San Francisco: Chronicle Books, 1976, 1982.

Hayden, Mike. *Guidebook to the Northern California Coast.* Los Angeles: Ward Ritchie Press, 1970.

Hedgpeth, Joel W. and Sam Hinton. *Common Seashore Life of Southern California.* Happy Camp, Ca.: Naturegraph Publishers, Inc., 1961.

Hedgpeth, Joel W. *Introduction to Seashore Life of the San Francisco Bay Region and the Coast of Northern California.* Berkeley: University of California Press, 1962.

Hinton, Sam D. *Seashore Life of Southern California.* Berkeley: University of California Press, 1969.

Hittell, Theodore H. *History of California.* 4 Vol. San Francisco: N.J. Stone & Company, 1897.

Hoover, Mildred Brooke, Hero Eugene Rensch, and Ethel Grace Rensch. *Historic Spots in California.* 3d ed., revised by William N. Abeloe. Stanford, Ca.: Stanford University Press, 1966.

Hutchinson, W.H. *California: The Golden Shore by the Sundown Sea.* Palo Alto, Ca.: Star Publishing Company, 1980.

Iacopi, Robert, ed. *Redwood Country and the Big Trees of the Sierra.* Menlo Park, Ca.: Sunset Books, Lane Publishing Co., 1969.

Inventory of California Boating Facilities. Management Consulting Corporation for California Department of Navigation and Ocean Development, Sacramento, 1977.

Jackson, Ruth A. *Combing the Coast: San Francisco through Big Sur.* San Rafael, Ca.: Presidio Press, 1977.

Knox, Maxine and Mary Rodriguez. *Making the Most of the Monterey Peninsula and Big Sur.* San Rafael, Ca.: Presidio Press, 1978.

Le Boeuf, Burney J. and Stephanie Kaza, eds. *The Natural History of A o Nuevo.* Pacific Grove, Ca.: Boxwood Press, 1981.

Lee, Georgia et al. *An Uncommon Guide to San Luis Obispo County, California.* San Luis Obispo, Ca.: Padre Publications, 1977.

Los Angeles: A Guide to the City and its Environment. Federal Writers Project, State of California. New York: Hastings House, 1951.

Los Angeles & Orange County Guide. Rand McNally & Company, 1980.

Lussier, Tomi Kay. *Big Sur: A Complete History and Guide.* Monterey, Ca.: Big Sur Publications, 1979.

Magary, Alan and Kerstin Fraser Magary. *Across the Golden Gate.* New York: Harper and Row, 1980.

Mallan, Chicki. *Guide to Catalina and Californiaís Channel Islands.* 3d ed. Chico, Ca.: Moon Publications, 1990.

Marcus, Laurel and Anna Kondolf. *The Coastal Wetlands of San Diego County.* Oakland, Ca.: California State Coastal Conservancy, 1989.

Marine Mammal and Seabird Survey of the Southern California Bight. University of California at Santa Cruz for U.S. Department of the Interior, Bureau of Land Management, Washington, D.C., 1978.

Mason, Jack. *Point Reyes, the Solemn Land.* Inverness, Ca.: North Shore Books, 1970.

McWilliams, Carey. *Southern California: An Island on the Land.* Santa Barbara, Ca. and Salt Lake City: Peregrine Smith, Inc., 1973.

Meyers, Carole Terwilliger. *Weekend Adventures for City-weary People: Overnight trips in Northern California.* Albany, Ca.: Carousel Press, 1988.

Mikiten, Erick. *A Wheelchair Riderís Guide to San Francisco Bay and Nearby Shorelines.* Oakland, Ca.: California State Coastal Conservancy, 1990.

Miller, Daniel and Robert N. Lea. *A Guide to the Coastal Marine Fishes of California.* Fish Bulletin 157. California Department of Fish and Game, 1972.

Miller, Daniel J. ìShark Attacks.î *Outdoor California* (November-December 1981).

Muller, Barbara D. *The Mendocino Coast.* Mendocino, Ca.: Mendocino Community Land Trust, Inc., 1981.

Munz, Phillip A. *Shore Wildflowers of California, Oregon, and Washington.* Berkeley: University of California Press, 1973.

Nathenson, Si. ìSharks as Food.î *Outdoor California* (November-December 1981).

Norris, Robert and Robert Webb. *Geology of California.* New York: John Wiley & Sons, Inc., 1976.

North, Wheeler J. *Underwater California.* Berkeley: University of California Press, 1976.

Oakeshott, Gordon. *Californiaís Changing Landscapes: A Guide to the Geology of the State.* New York: McGraw-Hill Book Co., 1971.

Orr, Robert T. *Marine Mammals of California.* Berkeley: University of California Press, 1976.

Pacific Coast Bicentennial Bike Route Guide. Sacramento: California Department of Transportation, 1976.

Power, Dennis, ed. *The California Islands: Proceedings of a Multi-disciplinary Symposium.* Santa Barbara, Ca.: Santa Barbara Museum of Natural History, 1980.

A Primer of Offshore Operations. Austin, Texas: Petroleum Extension Service, University of Texas at Austin, 1976.

Redwood Empire Visitorís Guide. San Francisco: Redwood Empire Association, 1980, 1981.

Ricciuti, Edward R. *Dancers on the Beach, The Story of Grunion.* New York: Thomas Y. Crowell Co., 1973.

Ricketts, Edward F., Jack Calvin, and Joel W. Hedgpeth. *Between Pacific Tides.* 4th ed., Stanford, Ca.: Stanford University Press, 1968.

Salitore, Edward, ed. *California Information Almanac: Past, Present, Future.* Lakewood, Ca.: California Almanac Company, 1973.

Smith, Clifton F. *A Flora of the Santa Barbara Region, California.* Santa Barbara, Ca.: Santa Barbara Museum of Natural History, 1976.

Smith, Emil J. Jr. et al. *The Marine Life Refuges and Reserves of California.* Marine Resources Information Bulletin No. 1. Sacramento: California Department of Fish and Game, 1979.

Smith, Ralph I. and James T. Carlton, eds. *Lightís Manual: Intertidal Invertebrates of the Central California Coast.* Berkeley: University of California Press, 1975.

Sowls, Arthur L. et al. *Catalog of California Seabird Colonies.* U.S Fish and Wildlife Service, Office of Biological Services, Coastal Ecosystems Project (FWS/OBS-80/37), 1980.

A Summary of Knowledge of the Central and Northern California Coastal Zone and Offshore Areas. Winzler and Kelly for U.S. Department of the Interior, Bureau of Land Management, Washington, D.C., 1977.

A Summary of Knowledge of the Southern California Coastal Zone and Offshore Areas Vol. 1. Southern California Ocean Studies Consortium of the California State University and Colleges for U.S. Department of the Interior, Bureau of Land Management, Washington, D.C., 1974.

Sunset Travel Guide to Northern California. Menlo Park, Ca.: Sunset Books, Lane Publishing Co., 1975.

Sunset Travel Guide to Southern California. Menlo Park, Ca.: Sunset Books, Lane Publishing Co., 1979.

A Survey of the Marine Environment from Fort Ross, Sonoma County, to Point Lobos, Monterey County. Sacramento: California Department of Fish and Game, 1968.

Tasto, Robert N. *Marine Bivalves of the California Coast.* Marine Resources Leaflet No. 6. Sacramento: California Department of Fish and Game, 1974.

ìVanishing Giants: a Biography [of the California Gray Whale].î *Audubon* (January 1975).

Weber, Michael and Richard Tinney. *A Nation of Oceans.* Washington, D.C.: Center for Environmental Education, Inc., 1986.

Welles, Annette. *The Los Angeles Guide Book.* Los Angeles: Sherbourne Press, 1972

Whitaker, Thomas W., ed. *Torrey Pines State Reserve.* La Jolla, Ca.: The Torrey Pines Association, 1964.

The Why and How of Undersea Drilling (pamphlet). Washington, D.C.: American Petroleum Institute, 1973.

Wieman, Harold. *Nature Walks on the San Luis Coast.* San Luis Obispo, Ca.: Padre Publications, 1980.

Winlund, Edmond. *ChartGuide for Southern California.* Anaheim, Ca.: ChartGuide, 1978.

Wood, Basil C. *The What, When and Where Guide to Southern California.* rev. ed. Garden City, N.Y.: Doubleday, 1979.

Wurman, Richard Saul. *LA/Access, Official Publication of the Los Angeles 200 Committee.* Los Angeles: Access Press, Inc., 1981.

Index

C

D